International Conference on Environment and Technology (IC-Tech 2017)

Advancing Technology for Sustainable Environment

IOP Conference Series: Earth and Environmental Science
Volume 97

Pekanbaru, Indonesia
26 – 27 July 2017

ISBN: 978-1-5108-5489-5
ISSN: 1755-1307

Printed from e-media with permission by:

Curran Associates, Inc.
57 Morehouse Lane
Red Hook, NY 12571

Some format issues inherent in the e-media version may also appear in this print version.

Printed by Curran Associates, Inc. (2018)

For permission requests, please contact the Institute of Physics
at the address below.

Institute of Physics
Dirac House, Temple Back
Bristol BS1 6BE UK

Phone: 44 1 17 929 7481
Fax: 44 1 17 920 0979

techtracking@iop.org

Additional copies of this publication are available from:

Curran Associates, Inc.
57 Morehouse Lane
Red Hook, NY 12571 USA
Phone: 845-758-0400
Fax: 845-758-2633
Email: curran@proceedings.com
Web: www.proceedings.com

Table of contents

Volume 97

International Conference on Environment and Technology (IC-Tech) 2017
26–27 July 2017, Pekanbaru, Indonesia

Accepted papers received: 23 November 2017
Published online: 14 December 2017

Preface

Papers

IOP Conf. Series: Earth and Environmental Science **97** (2017) 011001 doi:10.1088/1755-1315/97/1/011001

ICE-Tech 2017
http://icetech.unilak.ac.id/

A. Background

In response to the increasing research and issues on the field of environment and technology in Southeast Asia, Universitas Lancang Kuning, Indonesia in collaboration with Universiti Teknologi Malaysia and Universitas Kebangsaan Malaysia will conducting International Conference on Environment and Technologies (ICE-Tech 2017) focusing on theme "***Advancing Technology for Sustainable Environment***". This Conference provides a chance for academic and industry professionals to discuss recent progress in the area of Environment and Technology. We expect that the conference and its publications will be a trigger for further related research and technology improvements in this important subject.

ICE-Tech 2017 will be the most comprehensive conference focused on the various aspects of advances in Interdisciplinary of Environment and Technologies. The conference will take place in Pekanbaru, Indonesia, on 25-26 July 2017.

B. Scope and Themes

ICE-Tech 2017 will be the most comprehensive conference focused on the various aspects of advances Interdisciplinary on Environment and Technology. It will cover vital issues in Environment and Technology under multiple sub-themes such as Agriculture, Forestry, Computer Sciences and, Engineering

AGRICULTURE: Agribusiness, (off-on farm) , Agricultural Institutional, Agricultural Policy, Sustainable Agriculture, Organic Agricultural, Agricultural Technology, Marketing of Agriculture, Agricultural Extension and Communication, Agro industry

FORESTRY: Forest Management, Forest and Climate Change, Forest Bio-diversity, Ecosystem Service and Conservation, Silviculture and Agroforestry, Forestry and Environmental technology, Forest biotechnology and tree improvement

COMPUTER SCIENCES: Artificial intelligence, Digital Humanities, Computer Vision, Robotic, Mobile Application, Network Security, E-Government, E-Commerce, Data-mining, Geography Information system, Cloud and Grid Computing, Bio Informatika, Decision, Support System

ENGINEERING: Architectural Studies, Sustainable Architecture, Architecture Urban Conservation, Environment-Friendly Construction and Development, Safety Management, Heritage Culture Tourism, Urban Planning, Building and Construction Technologies, Mechanical, Energy and Machine Engineering, Electronics, System and Statistics Engineering, Renewable Energy Technologies, Electric Drive and Energy Conversion, Information System and Management, Mechatronic, Robotic and Automation

More information : http://icetech.unilak.ac.id/Tracks.htm

IOP Conf. Series: Earth and Environmental Science **97** (2017) 011001 doi:10.1088/1755-1315/97/1/011001

C. Important Dates

DESCRIPTIONS	DEADLINES
Abstract Submission Duedate:	25 April 2017
Notification of Acceptance:	10 Mei 2017
Full Paper Submission:	10 June 2017
Conference Day:	25-26 July 2017

D. KEYNOTE SPEAKER

Prof. Dr. Imam Robandi (Institut Teknologi Sepuluh November, Indonesia)
Prof. Dr. Ekhwan Toriman (Universiti Kebangsaan Malaysia)
Prof. Dr. Mohz Wazir Muztafa (Universiti Teknologi Malaysia)

International Advisory Board

Prof. Dr. Joran (Linkoping University, Swedia)
Prof. Dr. Amran bin Hamzah (Universitas Teknologi Malaysia, Malaysia)
Prof. Dr. Khaerurijal (Institut Teknologi Bandung, Indonesia)
Prof. Dr. Asep Saefuddin (Institut Pertanian Bogor, Indonesia)
Prof. Dr. Mauridi (Institut Teknologi Sepuluh November, Indonesia)
Prof. Dr. Adi Soeprijanto (Institut Teknologi Sepuluh November, Indonesia)
Prof. Dr. Syafrani (Universitas Lancang Kuning, Indonesia)
Prof. Dr. Hariadi Kartodiharjo (Institut Pertanian Bogor, Indonesia)
Prof. Dr. Sumbangan Baja (Universitas Hasanuddin, Indonesia)
Prof. Wiendu Nuryanti Ph.D (Universitas Gajah Mada, Indonesia)
Prof. Dr. Chairul Saleh (Universitas Islam Indonesia, Indonesia)
Prof. Dr. Richardus Eko Indrajit (STMIK Perbanas, Indonesia)

-Scientific Committee:

- Assoc. Prof. Dr. Rila Mandala (Institut Teknologi Bogor, Indonesia)
- Assoc. Prof. Dr. Romi Satria Wahono (Universitas Dian Nuswantoro, Indonesia)
- Assoc. Prof. Dr. Phil. Arinafril (Universitas Sriwijaya, Indonesia)
- Assoc. Prof. Dr. Arachchilage Aruna Shanta (Sabaragamuwa University of Sri Lanka)
- Assoc. Prof. Dr. Abdurrahman Radin Haggi (Sultan Sharif Ali Islamic University, Brunei)
- Assoc. Prof. Dr. Dedi Zargustin (Universitas Lancang Kuning, Indonesia)
- Assoc. Prof. Dr. Hamzah (Universitas Lancang Kuning, Indonesia)
- Assoc. Prof. Dr. Sitti Latifah (Universitas Mataram, Indonesia)
- Assoc. Prof. Dr. Marhamah Nadir (Universitas Hasanuddin, Indonesia)
- Assoc. Prof. Dr. Indah Raya (Universitas Hasanuddin, Indonesia)
- Assoc. Prof. Dr. Wasilah Sahabuddin (UIN Alaudin Makassar, Indonesia)
- Assoc. Prof. Dr. Badrussaman (Institut Teknologi Sepuluh November, Indonesia)
- Sofiane Hamrioui, Ph.D(Universite de Haute Alsace, France)
- Amine Brahmia, Ph.D (eXia. CEsi School of Computer and IT, Strasbourg, France)
- Onrizal, S.Hut, M.Si, PhD (Universitas Sumatera Utara, Indonesia)

Organizing Committee :

- Advisor :
  Assoc. Prof. Dr. Junaidi, M.Hum (Universitas Lancang Kuning, Indonesia)
- General Chair :
  Assoc. Prof. Ismail Suardi Wekke, Ph.D (STAIN Sorong, Indonesia)
- Program Chairman
  DR. Eno Suwarno, M.Si
- General Co-Chair :
  Budianto Hamuddin, M.Esl (Universitas Lancang Kuning, Indonesia)
- Secretary :
  Aziz Akbar, M.Si (Universitas Gajah Mada, Indonesia)

Documentations

IOP Conf. Series: Earth and Environmental Science **97** (2017) 011001 doi:10.1088/1755-1315/97/1/011001

IOP Conf. Series: Earth and Environmental Science **97** (2017) 011002 doi:10.1088/1755-1315/97/1/011002

Rector Welcoming Remarks

It is our great pleasure to welcome you to the International Conference on Environment and Technologies (ICE-Tech 2017) which will be held during **July 25 to 27, 2016 in Pekanbaru, Indonesia** with the theme "***Advancing Technology for Sustainable Environment***". This conference is jointly organized by Universitas Lancang Kuning, Indonesia, Universiti Kebangsaan Malaysia and Universiti Teknologi Malaysia

the International Conference ICE-Tech 2017 focusing on special topic in Environment and Technology with aims to bring together researchers, scientist, practitioners and scholar student to exchange and share their experiences, new ideas, and research result. Attending ICE-Tech 2017 means you also will experience Pekanbaru, a beautiful city renowned for its international flavor and inspiring diversity.

ICE-Tech 2017 will be an exceptional conference for sharing the latest academic insight as well as experiencing the unique culture of Malay Hospitality in Pekanbaru City. We hope you will join us for a symphony of outstanding science, and take a little extra time to enjoy the spectacular and unique beauty of this region. We thank you for your participation and look forward to seeing you in Pekanbaru, Indonesia.

With best wishes

Assoc. Prof. Dr. Hj. Hasnati, MH,

IOP Conf. Series: Earth and Environmental Science **97** (2017) 011003 doi:10.1088/1755-1315/97/1/011003

Peer review statement

All papers published in this volume of *IOP Conference Series: Earth and Environmental Science* have been peer reviewed through processes administered by the proceedings Editors. Reviews were conducted by expert referees to the professional and scientific standards expected of a proceedings journal published by IOP Publishing.

Species Diversity and Bird Feed in Residential Complex

Hadinoto[1] and Eni Suhesti[2]
[1]E-mail: hadinoto@unilak.ac.id
Tel/Fax: +62761 54092
Universitas Lancang Kuning, Pekanbaru, 28265, Indonesia,
[2]E-mail: hesti1170@yahoo.co.id
Universitas Lancang Kuning, Pekanbaru, 28265, Indonesia

Abstract. Bird is one component of the ecosystem which has an important role in supporting the occurrence of an organism's life cycle. Therefore, the presence of birds in an area is important, because it can affect the existence and distribution of plant species. The purpose of this study is to calculate the diversity of bird species and identify the source of bird feed in the compound. This study was conducted by field surveys in the residential complex. In addition to the birds as a research object vegetation as habitat / foraging birds were also observed. Data were analyzed by using the bird diversity index, richenes index, bundance index, dominance analysis, analysis of bird distribution and analysis of the level of meeting types, while vegetation will be analyzed based on the type and part of what is eaten by birds. In Pandau Jaya housing complex, found as many as 12 species of birds which consists of seven families. Bird species often present is Cucak Kutilang (Pycnonotus aurigaster) of 20 individuals, Bondol Peking (Lonchura punctulata) 14 individuals and Perkutut Jawa (Geopelia striata) 10 individuals. Bird species diversity (H ') in Pandau Jaya housing complex is still relatively moderate with a value of 2.27, while the Evenness Index (E) of 0.91 and Richenes Index (R) of 2.45. Types of vegetation as a food source, among others: mango, guava, cherry, jackfruit, ketapang, coconut, areca, palm, banana, papaya, flowers and grasses.

Keywords: *bird, feed, housing complexes*

1. Introduction

Birds are wildlife that is easily found in almost every vegetated environment. Habitats can include different types of ecosystems, from natural ecosystems to artificial ecosystems. This wide spread makes birds as one of the potential sources of Indonesia's biological wealth. In addition to playing a role in the balance of bird ecosystems can be an indicator of environmental change.

Pekanbaru is one of the city that is currently growing rapidly both in the development of facilities and infrastructure and the increase of population. The development of Siak Hulu sub-district has an influence on the development of the surrounding area as a development area. It will certainly have an impact on the availability of green open space or land that can be used as bird habitat becomes less.
Indonesia has a fairly high diversity of birds. Based on this function, the diversity of bird species is also closely related to the diversity of habitat types as well as the diverse functions of each type of habitat present in the housing complex. The sustainability of birds can be maintained by conserving the species preceded by various studies or studies of these animals, including the population, habitat and the environment that affect it. Residential complexes not only serve as human habitation but also

as a place for birds to find food / habiat on the existing vegetation. With the diversity of existing plant species, housing complexes are expected to support the lives of various species of birds.
Bird is one component of the ecosystem that has an important role in supporting the life cycle of an organism. This situation can be seen from food chains and life webs that make up their living systems with other ecosystem components such as plants and insects. Therefore, the existence of birds in a region is very important, because it can affect the existence and spread of plant species.

1.1 Research purposes
The purposes of this study are: (1). Identify the diversity of bird species in residential complexes; (2). Identify the source of bird feed that is in the housing complex

1.2. Benefits of research
This study is expected to provide benefits as a source of information supporting the conservation of bird species that contain residential complexes and considerations for the development of regional development with due attention to ecological functions.

2. Research Methods
2.1. Place and time
This research was conducted in residential complex of Pandau Jaya Village, Siak Hulu Sub-district, Kampar District (Pandau Permai Housing and Mahang Raya Housing). Study time October 2016 - January 2017.
Objects studied in this study are the types of birds, vegetation as a habitat / bird feeding place in the housing complex. Thally sheet as the record material data. The equipment used is: Binoculars (binoculars) with size 30 x 60; Digital SLR camera with lens size 75 - 300 mm; Stationery writing; Recording device for recording sound / bird chirping; Timepiece; Field Guide Introduction Birds: Birds In Java and Bali (MacKinnon, 1991); Birds in Sumatra and Kalimantan (Holmes, 1999); List of Birds Indonesian No.2 (Sukmantoro, 2007); Birds in Sumatra, Kalimantan, Java and Bali (MacKinnon, 2010); Additional Information Birds in Sumatra, Java, Bali and Kalimantan (Balen, 2010); Nurturing the Popular Birds of Speaks (Date, 2010).

2.2. Bird Data Collection Method
All bird data is taken using point count method (Bibby, 2000) with circular plot shape with plot diameter □ 40 meters at the point that has been determined with plot number of 6 plots for each location. The data were collected by observing the birds at the study site by taking all the data of bird species, the number of individuals, the time of encounter, the behavior and activity of birds, and the type of food.

Observations are made in the morning and afternoon between 06.00 - 08.00 WIB and at 16.00.00 - 18.00 WIB in good weather, in the observation track that has been determined with the observation time every 20 minutes. Special observations were made for nocturnal birds. Especially for night birds data taken is the type and amount, for one region. To ensure visibility and recognition of bird species, the observation is done at night at 19.00 - 21.00 WIB by recognizing his voice or seeing the bird species directly.

Recording of bird species is done by direct and indirect combination method. The method of recording is directly done by looking at birds (both visible and using binoculars) with the help of the Bird Recognition Field Manual (Holmes, 1999; Sukmantoro 2007: MacKinnon 1991, 2010) and indirectly based on the sound of birds and nests.

Observation Stage (Bibby 2000): Observers observe birds on each plot; The observer records the name and number of species present in each plot; Recording done during 20 minutes interval; Observer moves on another plot; Every bird observation plot is observed twice a day in the morning and afternoon.

IOP Conf. Series: Earth and Environmental Science **97** (2017) 012001 doi:10.1088/1755-1315/97/1/012001

2.3. Data analysis

Diversity Index

To determine the diversity of species Shannon-Wiener diversity index used with the formula: $H' = -\sum p_i \ln p_i$ *(H '= Shannon diversity index; Pi = (ni / N); Ni = number of individual i; N = Total number of individuals; Ln = natural logarithm).* According to Maguran (1998) in Rahmawaty (2006), the index value of species diversity ranges from 1.5 to 3.5. A value of <1.5 indicates a low species diversity, then a value of 1.5 to 3.5 indicates a moderate diversity of>> 3.5 values indicates high biodiversity.

Evenness Index

To determine the proportion of species abundance present in each type of urban forest, the Index of Eveness is used, ie the number of individuals of a species or abundance of each species in a community (Setiawan 2006; Dewi 2007) : $E = H'/\ln S$ *(E = fairness index (range 0 - 1); H '= Shannon diversity index; S = number of types; Ln = natural logarithm)*

Richness Index

To calculate species richness at each location using Margalef Index (Setiawan 2006) as follows: $R = \frac{(S-1)}{\ln N}$ *(R = Margalef Index; S = Number of Types; N = Number of Individuals; Ln = Natural logarithm)*

Domain Type Domination Analysis

The dominant analysis of bird species was used to see the dominant, sub dominant and rare bird species composition in observed bird communities. The analysis used relative density parameters according to the category proposed by Jorgensen (1974), the dominant bird category when relative density> 5%, sub dominant if relative density between 2% - 5% and rarely if relative density <2 %. The formula used to analyze was dominance: Density Type (K) = The number of bird species/ Sample plot example; Relative Density (KR) = (Density of a type/ Density of all types) x 100%.

Bird Spread Analysis

Spread analysis is used to view the horizontal distribution of each observation habitat by using the frequency value of the bird species found in the sample plots. The formula used is: Frequency Type (Fj) = The number of plots found in bird species/ number of whole sample plots; Relative Frequency (FR) = (Frequency of a type/ Frequency of all types) x 100%

Type of Meeting Rate Analysis

This analysis was used to calculate the level of encounter for each species of bird present in the study site, it was by dividing the total number of each bird species recorded by the total observation time multiplied by ten (10), resulting in a meeting rate of each bird species per ten hours of observation. Then the results of the calculation are categorized, whether including "rare", "not common", "often", "general" or "overflow". The division of the categories uses the table of bird type encounter rates by Lowen et al., (1996) in Bibby (2000) as in Table 1.

Table 1. Use of meeting level to show the scale of the abundance sequence (Lowen et al. 1996 in Bibby 2000)

Abundance category (individual number per 10 hours observation	Abundance Value	Order scale
< 0,1	1	Rare
0,1 – 2,0	2	General Not
2,1 – 10,0	3	Often
10,1 – 40,0	4	General
> 40,0	5	Overflow

IOP Conf. Series: Earth and Environmental Science **97** (2017) 012001 doi:10.1088/1755-1315/97/1/012001

3. Results and Discussion

3.1. Location observation Housing Pandau Permai

At this location found 12 species of birds belonging to the 7 families. Types of birds that are often present are Cucak Kutilang *(Pycnonotus aurigaster)* as many as 12 individuals, Bondol Peking *(Lonchura punctulata)* with the number of individuals found each of 9 individuals and Merbah Cerukcuk (*Pycnonotus flavescens)* as many as 8 individuals. The bird species at Panda Permai Housing observation location are listed in Table 2.

Table 2. Bird species at Pandau Permai Housing observation location

No.	Local Name	Scientific Name	Family	Numbers of Individuals
1	Perkutut Jawa	*Geopelia striata*	Columbidae	6
2	Bondol Peking	*Lonchura punctulata*	Estrildidae	9
3	Burung Madu Sriganti	*Nectarinia jagularis*	Nectariniidae	2
4	Burung Madu Kelapa	*Antrheptes malacensis*	Nectariniidae	4
5	Burung Madu Polos	*Antrheptes simplex*	Nectariniidae	3
6	Pijantung Kecil	*Arachnothera longirostra*	Nectariniidae	2
7	Cinenen Kelabu	*Orthotomus ruficeps*	Silviidae	4
8	Cucak Kutilang	*Pycnonotus aurigaster*	Pycnonotidae	12
9	Merbah Cerukcuk	*Pycnonotus flavescens*	Pycnonotidae	8
10	Prenjak Jawa	*Prinia familiaris*	Sylvidae	5
11	Cabak Kota	*Caprimulgus affinis*	Caprimulgidae	2
12	Cabai Jawa	*Dicacum trchileum*	Dicaeidae	4
		Total Individu		**61**

Source: Processed Data, 2017.

3.2. Location observation Mahang Raya Housing

At this location found as many as 8 species of birds belonging to 6 families. Types of birds that are often present are Cucak Kutilang *(Pycnonotus aurigaster*) of 8 individuals, Bondol Peking (*Lonchura punctulata*) as many as 5 individuals and Perkutut Jawa (*Geopelia striata*) as many as 4 individuals as listed in Table 3.

Table 3. Types of birds in the observation location of Mahang Raya Housing

No.	Local Name	Scientific Name	Family	Number of Individu
1	Perkutut Jawa	*Geopelia striata*	Columbidae	4
2	Bondol Peking	*Lonchura punctulata*	Estrildidae	5
3	Burung Madu Kelapa	*Antrheptes malacensis*	Nectariniidae	2
4	Burung Madu Polos	*Antrheptes simplex*	Nectariniidae	3
5	Cinenen Kelabu	*Orthotomus ruficeps*	Silviidae	2
6	Cucak Kutilang	*Pycnonotus aurigaster*	Pycnonotidae	8
7	Prenjak Jawa	*Prinia familiaris*	Sylvidae	2
8	Cabai Jawa	*Dicacum trchileum*	Dicaeidae	2
		Total Indivdiu		**28**

Source: Processed Data, 2017.

3.3. Types of birds at all observation sites

From the results of collecting data on bird species observation in 2 locations found as many as 12 species of birds consisting of 7 families. The types of birds that are often present are Cucak Kutilang (*Pycnonotus aurigaster*) of 20 individuals, Bondol Peking (*Lonchura punctulata*) as many as 14 individuals and Perkutut Jawa (*Geopelia striata*) as many as 10 individuals as listed in Table 4.

Table 4. Bird species at all observation sites

No.	Local Name	Scientific Name	Family	Number of Individu
1	Perkutut Jawa	*Geopelia striata*	Columbidae	10
2	Bondol Peking	*Lonchura punctulata*	Estrildidae	14
3	Burung Madu Sriganti	*Nectarinia jagularis*	Nectariniidae	2
4	Burung Madu Kelapa	*Antrheptes malacensis*	Nectariniidae	6
5	Burung Madu Polos	*Antrheptes simplex*	Nectariniidae	6
6	Pijantung Kecil	*Arachnothera longirostra*	Nectariniidae	2
7	Cinenen Kelabu	*Orthotomus ruficeps*	Silviidae	6
8	Cucak Kutilang	*Pycnonotus aurigaster*	Pycnonotidae	20
9	Merbah Cerukcuk	*Pycnonotus flavescens*	Pycnonotidae	8
10	Prenjak Jawa	*Prinia familiaris*	Sylvidae	7
11	Cabak Kota	*Caprimulgus affinis*	Caprimulgidae	2
12	Cabai Jawa	*Dicacum trchileum*	Dicaeidae	6
	Total Individu			**89**

Source: Processed Data, 2017.

Causes of large numbers of individuals Cucak Kutilang (*Pycnonotus aurigaster*), Bondol Peking (*Lonchura punctulata*), Perkutut Jawa (*Geopelia striata*) and Merbah Cerukcuk (*Pycnonotus flavescens*) are the availability of adequate food, habitat fulfillment by vegetation that allows this species of birds to live and flourish. The species of birds above are the type of fruit-eating birds, insects and grains as the main feed. In some observations, the birds were also seen eating a caterpillar, a flower honey essence or grabbing a small flying insect. This type of bird is also in its movement seen active in all observation points that exist in 2 locations of observation.

The number of individuals, types and families of each location of observation for the whole is dominated by Pandau Permai Housing with the total of 68 individuals, 12 species and 7 families while for Mahang Raya Housing with the number of 28 individuals, 8 species and 6 families. This indicates that the location of observations in Pandau Permai Housing the number of individuals and species of birds is higher than in the Mahang Raya housing location but for the number of families is almost the same. This is because in the location of Housing Pandau Permai vegetation quite a lot and varied when compared with Housing Mahang Raya

In addition to more vegetation conditions in Pandau Permai Housing and become a place of living necessities for birds is available. This happens because Housing Pandau Permai has been built long enough, so the vegetation, especially the trees are quite old. In Mahang Raya housing location the number of vegetation is relatively young and small because the housing is still not long built so the lack of shelter space for birds and coupled with the existence of vehicles that often pass that can disturb the existence of birds.

3.4. The diversity of bird species in the Housing Complex

After analyzing the data on bird species, the highest index of bird species diversity was found on the location of Pandau Permai Housing (2.32) and in Mahang Raya Estate (1.94). However, the overall index of species diversity reached 2.27. The index value of species diversity at all observation sites showed moderate species diversity level (H value 1.5 - 3.5), quite high index of species diversity at Pandau location because the location is a relatively long residential area built. In addition, Pandau Permai Housing area also has a more diverse vegetation compared with other locations that are still considered low and still small in number. In addition to the vegetation conditions that still have little influence in the fulfillment of feeding needs for birds, diverse bird species implies that the existing vegetation is able to provide adequate food with different classes in producing fruit, young leaves for insect food and so on. This fact leads to competition in the seizure of food, space and security is not met, causing the number and types of birds are not much and the index of diversity to be low. This is also in line with the results of Hadinoto, et.al (2012) research in Pekanbaru Urban Forest that species diversity is positively correlated with tree species diversity. The higher the diversity of tree species, the diversity of bird species encountered the higher.

Trees as a component of bird habitat, can serve as cover (shelter from weather and predators, nesting, resting play). In addition to providing tree parts (leaves, flowers, and fruit) a tree can serve as a habitat for various other types of organisms that are food available to birds such as insects. The index of the community structure of birds in each location is relatively the same, except in Tahura Minas which is higher as listed in Table 5

Table 5. Index of the bird community structure in each location

Location	Species	Individu	Family	H'	E	R
Pandau Permai Housing Complex	12	61	7	2,32	0,93	2,68
Mahang Raya Housing Complex	8	28	6	1,94	0,93	2,10

Data Table 5 shows that the wealth index for Pandau Permai Housing with Mahang Raya Housing is relatively the same. The diversity index of Pandau Permai Housing is higher than Mahang Raya Housing. The Evennes Index (E) is used as an indicator of the symptom of dominance among each species within a community. If each type has the same number of individuals, then the community has a maximum "EVENNESS" value. Conversely, when the equity value is small, then in those communities are the dominant species, sub-dominant and the dominated species, the community has a "Evenness" minimum. From the data analysis known bird species evenness highest in observation locations Pandau Housing and Housing Permai Raya Mahang same, namely 0.93. Richness (R) contained in the highest bird observation sites Pandau Permai Housing (2.68) and Housing Mahang Kingdom (2.10), when calculated as a whole, the species richness (R) is a bird of 2.45.

3.5. Density, Dominance and Scale of Bird Sequence in Housing Complex

Based on the observation that there are two species of birds have Relative density value (KR) highest Cucak Kutilang (*Pycnonotus aurigaster*) is 22.47% and the Bondol Peking (*Lonchura punctulata*) is 15.73% and the value of the smallest KR is Burung Madu Seriganti (*Nectarinia jagularis*), Pijantung Kecil (*Arachnothera longirostra*) and Cabak Kota (*Caprimulgus affinis*) is 2.25%. This is influenced by habitat conditions were pretty good and the proliferation of this type is quite good as well as the lack of a bird of prey which allows both types are high enough value KR. Some bird species have high dominance values and have common sequence value scores, as listed in Table 6 below.

Table 6. Density, dominance and sakala sequence of birds in residential complexes

No	Local Name	Scientific Name	K	KR (%)	Dominansi	Abudance Category	Abudance Value	Order Scale
1	Perkutut Jawa	*Geopelia striata*	6.63	11.24	Dominan	8.33	3	Often
2	Bondol Peking	*Lonchura punctulata*	9.28	15.73	Dominan	11.67	4	General
3	Burung Madu Sriganti	*Nectarinia jagularis*	1.33	2.25	Sub Dominan	1.67	2	General Not
4	Burung Madu Kelapa	*Antrheptes malacensis*	3.98	6.74	Dominan	5.00	3	Often
5	Burung Madu Polos	*Antrheptes simplex*	3.98	6.74	Dominan	5.00	3	Often
6	Pijantung Kecil	*Arachnother a longirostra*	1.33	2.25	Sub Dominan	1.67	2	General not
7	Cinenen Kelabu	*Orthotomus ruficeps*	3.98	6.74	Dominan	5.00	3	Often
8	Cucak Kutilang	*Pycnonotus aurigaster*	13.2 6	22.47	Dominan	16.67	4	General
9	Merbah Cerukcuk	*Pycnonotus flavescens*	5.31	8.99	Dominan	6.67	3	Often
10	Prenjak Jawa	*Prinia familiaris*	4.64	7.87	Dominan	5.83	3	Often
11	Cabak Kota	*Caprimulgus affinis*	1.33	2.25	Sub Dominan	1.67	2	General Not
12	Cabai Jawa	*Dicacum trchileum*	3.98	6.74	Dominan	5.00	3	Often

Based on Table 6 above there are 9 species of birds that have high dominance value and there are 2 species of birds that have high order scale value

3.6. Vegetation as a source of bird feed
Vegetation types in bird observation sites vary between trees, herbs, flowers, horticultural crops and grasses as listed in Table 7.

Table 7. Type of vegetation as a source of bird feed

No	Species Name	Types of vegetation	The part that birds eat	Remark
1	Mango	Tree	Fruit, insects, flower honey	In addition to foraging, vegetation is also used as a shelter and breed
2	Jambu Air	Tree	Fruit, insects, flower honey	
3	Guava	Tree	Fruit, insects, flower honey	
4	Belimbing	Tree	Fruit, insects, flower honey	
5	Klengkeng	Tree	Fruit, insects, flower honey	
6	Chery	Tree	Fruit, insects, flower honey	

7	Soursop	Tree	Fruit, insects, flower honey
8	Jackfruit	Tree	Fruit, insects, flower honey
9	Ketapang	Tree	insects, flower honey
10	Coconut	Palm	insects, flower honey
11	Pinang	Palm	insects, flower honey
12	Palms	Palm	insects, flower honey
13	Banana	Horticulture	Fruit, insects, flower honey
14	Papaya	Horticulture	Fruit, insects, flower honey
15	Flowers	Herbs	Fruit, insects, flower honey
16	Grasses	Grass	Seeds, insects, honey extract

4. Conclusions and Suggestion

After the research on Bird Species Diversity in Pandau Jaya Village Housing Complex can be summarized as follows: (1) In the Village Housing Complex Pandau Jaya, found 12 species of birds consisting of 7 families. The types of birds that are often present are Cucak Kutilang *(Pycnonotus aurigaster)* as many as 20 individuals, Bondol Peking (*Lonchura punctulata*) 14 individuals and Perkutut Jawa (*Geopelia striata*) 10 individuals. The diversity of bird species (H ') in the Pandau Jaya Village Housing Complex is still relatively moderate with a value of 2.27, while the Evenness Index (E) is 0.91 and the Wealth Index (R) is 2.45; (2) Types of vegetation as a source of feed include: mango, guava, belimbing, klengkeng, kersen, soursop, jackfruit, coconut, palm, banana, papaya, flowers and grasses

This study would like to propose some as follows: (1) To be monitored regularly to avoid the extinction of birds on various types of land use in the Village Pandau Jaya; (2) A study of bird life, both biological and ecological, includes the available population, habitat and environment.

5. References

[1] Alikodra, HS. 2002. *Wildlife Management* Volume 1. Yayasan Penerbit Fakultas Kehutanan IPB. Bogor.

[2] Bibby, C., M. Jones, and S. Marsden. 2000. *Field Expedition Techniques Bird Survey*. SMKG Mardi Yuana. Bogor

[3] Holmes, D., and S. Nash. 1999. *Birds in Sumatra and Kalimantan*. Puslitbang Biologi - LIPI. Bogor

[4] Hadinoto, A. Mulyadi & YI. Siregar. 2012. *Keanekaragaman Spesies Burung di Hutan Kota Pekanbaru*. Jurnal Ilmu Lingkungan 6 (1): 25-42.

[5] Holmes, H. 2010. *The Beautiful Blue Shrimp Blue Bird*. National Geographic Indonesia, edition February 2010. Kompas Gramedia, Jakarta

[6] MacKinnon, J. 1991. *Series of Bird and Birds Field Guide Java and Bali*. Gajah Mada University Prees Yogyakarta

[7] MacKinnon, J., K. Philip and V. Balen. 2010. *Series of Field Guide Birds of Sumatra, Borneo, Java and Bali*. Puslitbang Biologi - LIPI. Bogor

[8] Sukmantoro, W., M. Irham, W. Novarino, F. Hasudungan, N. Kemp and M. Muhtar. 2007. *Bird List Indonesian* No.2. Ornithologists Union. Bogor

An Analysis of the Expediency Social Media for Culinary Products Marketing on Micro and Middle Enterprise in Pekanbaru City

Diki Arisandi[1] and Sukri[2]
[1,2] Universitas Abdurrab, Pekanbaru, 28282, Indonesia
Telp: +62761 38762 Fax: +62761 38762
E-mail: [1]diki@univrab.ac.id, [2]sukri@univrab.ac.id

Abstract: *Today, there are many culinary business can be found Pekanbaru city. To increase the benefit, the culinary entrepreneurs are using social media such as Facebook, twitter, instagram, whatsapp and others. This research discussed about the expediency of social media analysis for culinary products marketing on micro and middle enterprise in Pekanbaru by using structural equation modeling method. The Variables were used in this research are the selection of social media, updating information, response to costumers, quality and price as variable x to culinary business marketing as variable y. Data retrieval comes from a questionnaire filled by culinary entrepreneurs who are using social media to sell their product to customer in Pekanbaru. The result of this research displayed that the value of validity to the questionnaire was minimum = 0.083 and maximum = 0.547, RMSEA = 0.08, AGFI = 0.705, CMIN / DF = 1.834, TLI = 0.813 and CFI = 0.843. From the results of these measurements can be concluded that six of seven components of the measurement value can be avowed as acceptable, so the final result in this research is all the x variables have a positive influence on y variable.*

Keywords: *Social Media, Culinary Product, SEM, Micro and middle enterprise, Marketing*

1. Introduction

Micro enterprising and middle enterprising in Indonesia is one of the priorities in growing the National economic. Moreover the enterprise is become basic economic system which is purposed to reduce the discrepancy problem on the low income class, enterprisers, the poverty alleviation and employment. And then, the growing of this enterprenur was able to enlarge the economic basic and give the significant contribution in the structural change, such as increasing the region economic and national economic [4].

Almost 99 percent, the micro enterprises and the middle enterprises generally use the local ingredient and market. That's why, they were not got the global crisis's impact. The forum economic world report 2010 stated the Indonesia's market on the 15th rank. It showed that Indonesia as the potential market for the other country. This potential was not maximally utilized by the micro and middle enterprising [6].

Micro and middle enterprising was owned by privacy and/or individual business entities which comply the micro enterprising stated in the constitution number 20 in 2008 about the middle, and micro enterprising. It also explained about the asset and turnover which is had by Micro and middle enterprenuer.

The food need is one of human's basic need. In Pekanbaru city for this present grows rapidly and high competitor. In Pekanbaru city, we can found some restaurant like Padang, Malay, Sunda, Java restaurant, fast food, warung sate, nasi uduk, mie aceh, etc. there are no limitation of pace in the culinary business because every areas have the big potency.

In this present, internet infrastructure through the social media is not only used for the interaction infrastructure for every personal. In this present the micro and middle enterpriser, the social media is also used as the infrastructure to increase the goods and services market which are ordered for all other people. The benefit of the social media can minimazed promotion cost, and enlarging the product promotion reach without knowing the place and time limitation. From the found benefit, so the culinary enterpriser in Pekanbaru also used the social media as the infrastructure to sell their product.

In this research will be discussed about how far the social media benefit by the culinary enterpriser in Pekanbaru in selling their product. The utilization analysis will be managed by using the SEM (Structural equation modeling) method. Structural Equation Modeling, or SEM, is a very general statistical modeling technique, which is widely used in the behavioral sciences. It can be viewed as a combination of factor analysis and regression or path analysis [1]. With existence of the research, hopeful giving the contribution to support the media social technology utilization effort by the micro and middle enterprising owner in selling and promoting the each culinary in Pekanbaru city. However the each culinary which is sold and promoted, it can be known by the people in or outside Pekanbaru and increasing the selling turnover for the micro and middle enterpriser.

2. Method

The research will be operated in Pekanbaru city by object is Micro middle enterprising for the culinary sector. In doing this research, the researcher guides to the research framework which can be seen on:

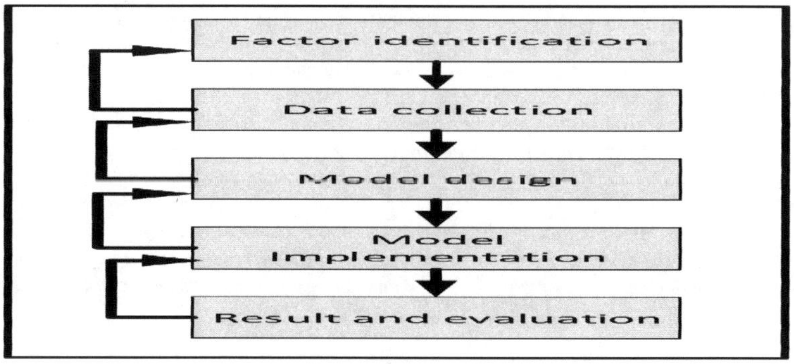

Picture 1. Research framework

2.1 Factor Identification

In this research, the used factors or variables for taking the data is used social media, culinary kinds which are ordered, costumers/ market target, turnover, and own enterprise asset.

2.2 Data collection

The data collection method in the research is questioner method which is given to the responders, they are the micro middle enterprising owner. The factor measurements are operated by using Likert scale. It is the scale to measure the somebody's or group's perception, attitude or opinion about an event or condition, based on the factor definition which have been stated by a researcher. The likert scale means it can be seen in this table:

Table 1. Likert Scale

5	4	3	2	1
SS	S	N	TS	STS

Information:
SS = Very agree
S = Agree
N = Enough
TS = Disagree
STS = Very disagree

2.3 Model Design

In doing this research, method or maintain which is done in structural equation model as this following:

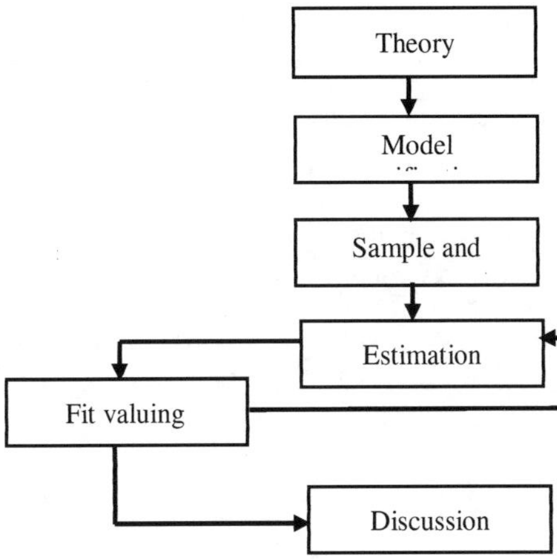

Picture 2. Model design by the structural equation modeling

Explanation of the picture is explained as this following:

- Structural equation which is decrypted by the path diagram viewed as the representation of the theory. So the correlation on the latent variable which is formed in the path diagram is forming of the theory.
- Then finding the sample and measurement, after that estimating through the model parameter.
- On the model measurement can be estimated early and followed by the structural and full model.
- Then operating the goodness fit valuing of the model and whether it must be followed by the model modification.
- If the model had been fulfill the fit criteria, so operated the discussion or result discussion.

2.4 Model implementation

The way to implement in managing the data with SEM using two techniques, they are:

- Confirmatory Factor Analysis on SEM which is used to confirm the most dominant factors in a variable group.
- Regression Weight on SEM which is used to research how big the correlation of the variables.

2.5 Evaluation

In this step was operated the examination through the matched model through each goodness of fit criteria. This is presented some accorded and cut off index value to examine if a model could be accepted or rejected. This is some accorded and cut off index value to examine if a model could be accepted or rejected:

- Statistics X2 chi square, where model viewed well or satisfied if chi-square value is low.
- RMSEA (The Root Mean Square Error of Approximation), which showed the goodness of fit hopefully if model is estimated in the population.
- GFI (Goodness of Fit Index) is non-statically measurement which has the value average among 0 (poor fit) until 1.0 (perfect fit).
- AGFI (Adjusted Goodness of Fit Index) where the receiving level which is recommended is if AGFI has the same or bigger value than 0.90.
- CMIN/DF is the sample minimum of discrepancy function which is divided by degree of freedom.
- TLI (Tucker Lewis Index) is incremental fix index which compares an examined model to a model baseline, where the value which is recommended as a reference to be accepted by model ≥ 0.
- CFI (Comparative Fit Index) / NFI (Normed Fit Index), if reaching 1, identifying the fit level which highest recommended is CFI ≥ 0.90.

3. Result and discussion

In this research which become the variable X was the media social choosing, information renewal, respond to the costumer and quality and price, whereas variable y was culinary business marketing. The data collection was done by spreading some questioners to the micro middle enterprise owner which sold the culinary product in Pekanbaru by using the social media in selling the product.

After collecting the data, operated the validity and reliability examination to determine the validity and believable level through the found data. Based on the data which was found, existence of the validity test as:

Table 2. Validity Test

	Scale Mean if Item Deleted	Scale Variance if Item Deleted	Corrected Item-Total Correlation	Squared Multiple Correlation	Cronbach's Alpha if Item Deleted
X1	70.870	60.357	.236	.337	.782
X2	71.970	55.726	.432	.669	.770
X3	70.740	58.154	.376	.316	.774
X4	71.890	55.048	.502	.746	.765
X5	71.840	54.984	.539	.608	.762
X6	71.840	57.045	.388	.521	.773
X7	70.960	57.716	.471	.450	.769
X8	71.430	55.803	.594	.613	.761
X9	71.960	60.867	.208	.381	.784
X10	71.090	56.830	.429	.436	.770
X11	71.590	56.042	.547	.580	.763
X12	72.010	58.172	.389	.515	.774
X13	70.410	62.204	.171	.733	.784
X14	70.470	61.403	.224	.716	.782
X15	70.510	61.869	.198	.730	.783
X16	70.490	62.374	.119	.500	.787
X17	71.810	60.923	.083	.307	.799
X18	70.960	58.483	.332	.612	.777
X19	71.010	58.151	.313	.728	.779
X20	71.150	56.129	.380	.710	.774

From data on the table 2, taken the guidelines that the lowest corrected value of the item-correlation (red sign) is 0.05. after data managed, the result which is found to show that the spread questioner is stated valid for all, because there is not the value under 0.05. For the reliability test, the collected data was managed and got the result as:

Table 3 Reliability test

Cronbach's Alpha	Cronbach's Alpha Based on Standardized Items	N of Items
.785	.786	20

From that table, the lowest value to stated that the managed data was stated reliable or believable was with comparing to the minimum value, 0,05. On the table 3 is seen on the cronbach's alpha based on standardized items with value 0,786 or more than minimum value, however it can be stated that the data management result can be stated reliable and believable. In addition, the data in table 2 and table 3 was from 200 respondents spread in Pekanbaru. From the data collected, the ratio to the data was 100% or in other words the respondents filled in all the questions on the questionnaire. Moreover, in measuring the validity of minimum error (alpha) or low trust level to the data used was 0,05.

After correlation model between variables has been established, validity and reliability test have been done, then the next step is to test the suitability of model based on some component / parameter according to SEM method [7]. The components in question are RMSEA, GFI, AGFI, CMIN / DF, TLI and CFI. The next test after validity and reliability test was with viewing the RMSEA value, the minimum value to examine it is 0, 08. From the managed data, it can be descripted value RMSEA is 0,092 such as in the table 4.

Table 4. RMSEA value

Model	RMSEA	LO 90	HI 90	PCLOSE
Default model	.092	.075	.108	.000
Independence model	.212	.200	.225	.000

To descript the fit value on the GFI value which is resulted from the data management. If fit reached 1 so it could include to the perfect fit category, on this research, GFI value in the table 5 is as big as 0,775 with meaning, reaching perfect fit.

Table 4. GFI dan AGFI value

Model	RMR	GFI	AGFI	PGFI
Default model	.085	.775	.705	.591
Saturated model	.000	1.000		
Independence model	.214	.421	.360	.381

After GFI, value of the following data management which was taken AGFI (Adjusted Goodness of Fit Index), where the value was same or bigger than 0.90. On the managed data, showed that the AGFI value was as big as 0,705 such as in the table 4.

The other value was taken CMIN/DF value. CMIN/DF is the minimum Sample Discrepancy function which was divided with the degree of freedom, on the managed data, was found CMIN/DF is as big as 1.834 such in the table 5

Table 5. CMIN/DF VALUE

Model	NPAR	CMIN	DF	P	CMIN/DF
Default model	50	293.457	160	.000	1.834
Saturated model	210	.000	0		
Independence model	20	1039.385	190	.000	5.470

To compare a tested model though baseline model, taken TLI value. TLI value which is recommended as guideline to accept a model is ≥ 0. On the managed data, was found that TLI value was 0,813 such as in the table 6.

Table 6. TLI dan CFI Value

Model	NFI Delta1	RFI rho1	IFI Delta2	TLI rho2	CFI
Default model	.718	.665	.848	.813	.843
Saturated model	1.000		1.000		1.000
Independence model	.000	.000	.000	.000	.000

In the table 6 can also descripted the CFI value. Found CFI value is 0,843, where the highest fit is recommended for CFI is ≥ 0.90.

The results of this study are in line with the research taken by Suki & Ramayah. From their work, they identified the factors that determine users' acceptance of e-Government services in Malaysia and its causal relationships using a theoretical model based on the Technology Acceptance Model. Data relating to the constructs were collected from 200 respondents in Malaysia and subjected to Structural Equation Modeling analysis. The proposed model fits the data well. Results indicate that the important determinants of user acceptance of the e-Government services are perceived usefulness, ease of use, compatibility, interpersonal influence, external influence, self efficacy, facilitating conditions, attitude, subjective norms, perceived behavioral control, and intention to use e-Government services/system [8].

Other research results are in line was from Witkemper & Waldburger. While much of the current literature tends to focus on the direct effect of social influence on consumer online shopping behavior, the research drew heavily on social influence theory and argued for an alternative theory focusing on the moderating role of social influence. In particular, they explored how positive messages in online discussion forums may affect consumers' decisions to shop online. They used a laboratory experiment in which the treatment group was required to read positive messages about online shopping experience in an online discussion forum. Positive social influence was found to reinforce the relationship between beliefs about and attitude toward online shopping, as well as the relationship between attitude and intention to shop. They believe that their alternative theory provides new insight into the complex processes through which social influence is brought to bear on consumers' online shopping decisions [9].

4. Conclusion

From the data management, is found the value of seven indicators they are validation, reliability, RMSEA, GFI, AGFI, CMIN/DF, TLI and CFI. From seven indicators, the only one indicator which doesn't fulfill the regulations, it is AGFI. But it can really state that the result of the research is accepted. Because the other six indicators fulfill the regulations. It can be concluded that the media social have the important role and benefit in increasing the culinary product marketing, besides that based on the data management result can be concluded that all variables x have the positive influence to the variable y, so, the result of research can be the recommendation for the other culinary micro middle enterpriser to use the media social in selling the product.

5. References

[1] Hox, J. J., & Bechger, T. M. 2007. *An introduction to structural equation modeling*.

[2] Irianto, Aloysius Bagas Pradipta. 2015. *Pemanfaatan Social Media Untuk Meningkatkan Market Share UKM*, jurnal TEKNOMATIKA Vol. 8, No. 1, Juli 2015

[3] Iskandar, Wahyu, et al. 2015. *Analisis Kelayakan Usaha Steak Cafe Di Kota Pekanbaru*. Jurnal Online Institut Teknologi Nasional. Vol 1, No 3, Januari 2015.

[4] Munizu, Musran. 2010. *The influence of external and internal factors to small businessess performance in South Sulawesi*, Journal of Management and Enterpreneurship, vol.12, no. 1. ISSN: 1411-1438. Page 33-35.

[5] Priambada, Swasta. 2015. *Manfaat Penggunaan Media Sosial Pada Usaha Kecil Menengah (UKM)*. Seminar Nasional Sistem Informasi Indonesia, 2-3 November 2015.

[6] Sudaryanto, et al. 2015. *Strategi Pemberdayaan UMKM Menghadapi Pasar Bebas Asean*. Artikel dari http://fiskal.kemenkeu.go.id/. accessed 2 Mei 2016.

[7] Schermelleh-Engel, K., Moosbrugger, H., & Müller, H. 2003. *Evaluating the fit of structural equation models*: Tests of significance and descriptive goodness-of-fit measures. Methods of psychological research online, 8(2), 23-74.

[8] Suki, N. M., & Ramayah, T. 2010. *User acceptance of the e-government services in Malaysia: structural equation modelling approach*. Interdisciplinary Journal of Information, Knowledge, and Management, 5(1), 395-413.

[9] Witkemper, C., Lim, C. H., & Waldburger, A. 2012. *Social media and sports marketing: Examining the motivations and constraints of Twitter users*. Sport Marketing Quarterly, 21(3), 170.

IOP Conf. Series: Earth and Environmental Science **97** (2017) 012003 doi:10.1088/1755-1315/97/1/012003

The Implementation of Analytical Hierarchy Process Method for Outstanding Achievement Scholarship Reception Selection at Universal University of Batam

Marfuah [1]**, and Suryo Widiantoro** [2]
Universitas Universal, Batam, Indonesia
E-mail: marfuah@uvers.ac.id

Abstract: Universal University of Batam offers outstanding achievement scholarship to the current students to be each year of new academic year, seeing the large number of new Students who are interested to get it then the selection team should be able to filter and choose the eligible ones.The selection process starting with evaluation and judgement made by the experts. There were five criteria as the basic of selection and each had three alternatives that must be considered. Based on the policy of University the maximum number of recipients are five for each of six study programs. Those programs are art of music, dance, industrial engineering, environmental engineering, telecommunication engineering, and software engineering. The expert choice was subjective that AHP method was used to help in making decision consistently by doing pairwise comparison matrix process between criteria based on selected alternatives, by determining the priority order of criteria and alternatives used. The results of these calculations were used as supporting decision-making to determine the eligible students receiving scholarships based on alternatives of selected criteria determined by the final results of AHP method calculation with the priority criterion A (0.37%), C (0.23%), E (0.21%) , D (0.14%) and B (0.06%), value of consistency ratio 0.05. Then the alternative priorities 1 (0.63), 2 (0.26) and 3 (0.11) the consistency ratio values 0.03, where each CR $\leq$ 0.1 or consistent weighting preference.

Keywords : Analytical Hierarchy Process (AHP); Criteria; Alternative; Outstanding Scholarship.

1. Introduction

Universal University is one of the universities located in Batam City, Riau Islands Province, newly developed and able to open eleven undergraduate study programs, namely Dance Art, Music Art, Accounting, Management, Information Systems, Informatics Engineering, Software Engineering, Engineering Industry, Environmental Engineering, Telecommunication Engineering and Mandarin Language Education. To support the lecture process, Universal University provides scholarship for outstanding students. It is applied to six study programs of Music Art, Dance Art, Industrial Engineering, Environmental Engineering, Telecommunication Engineering, and Software Engineering. It aims to motivate students, alleviate the cost of lectures as well as improve students learning performance in achieving the vision of the studies taken by them.

The scholarship recepients are those who eligible. It involves the analysis of multi-criteria decision-making by prioritizing which criterion was the most important for students in the selection process [1]. The scholarship distribution process for the students was conducted by the selection team as the expert and with the scholarship budget which has been determined by Universal University. In the selection process, some criteria of assessments are need to be considered in order to know the students who deserve schoarship, with many enthusiastic scholarship enthusiasts it takes time and excessive expert. Thus one of the solutions offered is to find the best alternative based on the criteria that have been determined using Analytical Hierarcy Process (AHP) method by finding the weight value for each attribute and sorting the candidates that will determine the optimal alternative is that the eiligible recipients. Because the expert choice is subjective so AHP method assists in making decisions consistently by performing pairwise matrix comparison process between criteria based on selected alternatives, by determining the priority order of the criteria and the alternatives used. By that alternative, the system gives information needed to know which students deserve the scholarship.

2. Literatur Review and Method

2.1 *Literatur Review*

AHP method is one of analysis technique in decision making involving multi criteria and alternative criterion, build hierarchy with different level of importance so as to produce weighting with acceptable level of consistency.

The previous study used AHP method as a simple methodology for selecting suppliers in the automotive industry in Pakistan, identifying criteria in terms of price, quality, delivery and service determined by expert opinion. The use of AHP helps in prioritizing criteria and gives the impression of consistency throughout the decision making process, thus providing insight into the various factors that managers need to consider when making decisions [2].

The model of decision making in construction management is multi criteria, one of the optimization methods used is AHP in the selection of contractor for swimming pool construction. The multi-criteria aspect is important when making construction management decisions based on 9 criteria according to experts namely technical experience, resource performance, financial stability, performance management and employee qualifications, capacity, safety, operation and equipment. The results of this study were found to be the best contrast factor with a total score of 0.551 as the preferred constractor [3].

The AHP method was also used in housing renovation priorities combined with geographic information systems. This study took into account one of the basic human needs, society and civilization to ensure the quality of living standards. The city planning and management institutions were doing housing renovations because of the high occupancy rate, so the low economic people find it too expensive. Priority and preparation of the renovation plan required consideration relative to existing resources. This study considered all units with field study and census data collection techniques. The first priority result was 17.86%, the second priority was 77.24% and the remainder included priority 3, 4 and 5 [4].

2.2 *Analytical Hierarchy Process Method*

AHP is one of the proven techniques in solving a multi-criteria problem or sub criteria in which each criterion has a different level of importance in decision-making [1].

To get a decision using AHP method, there are 3 basic stages as shown in the following figure 1.

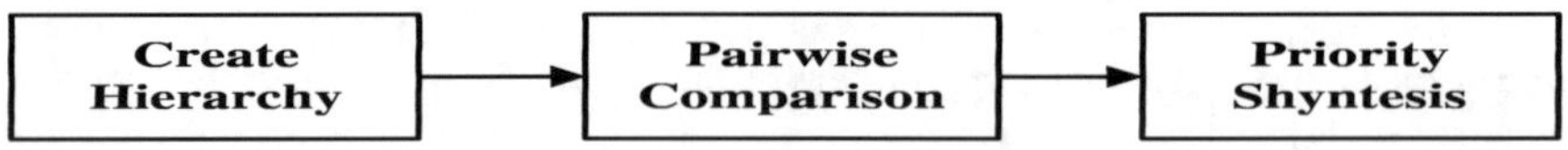

Figure 1. Basic AHP Analysis [1]

Figure 1 explains that the main basic stage in performing analysis of decision making involving multiple criteria or sub criteria is to build and develop a hierarchical criterion model which is relatively more important than the awardees of outstanding scholarship. Pairwise comparison performs a comparative assessment or defining and implementing data collection to obtain pairwise comparison data on hierarchical structure elements. Priority synthesis builds priority ranking overall.

The Criteria and Alternative Process are done by having pairwise comparison. According to Saaty for various issues, the scale 1 to 9 is the best scale for expressing opinions. In determining the priority of criterion, AHP performs the element priority analysis by applying pairwise comparison method between the two elements so that all elements are met. Priorities are determined on the basis of an expert's view of interest in decision-making, as shown in the following tabel 1.

Table 1. AHP Importance Scale [5]

Score	Relative Importance
1	Objectives *A* and B are of equal importance
3	Objective A is weakly more important than B
5	Objective A is strongly more important than B
7	Objective A is very strongly more important than B
9	Objective A is absolutely more important than B
	2,4,6,8 are intermediate values

The application of AHP method for the selection processof outstanding scholarship was done through a procedure, as shown in the following figure 2.

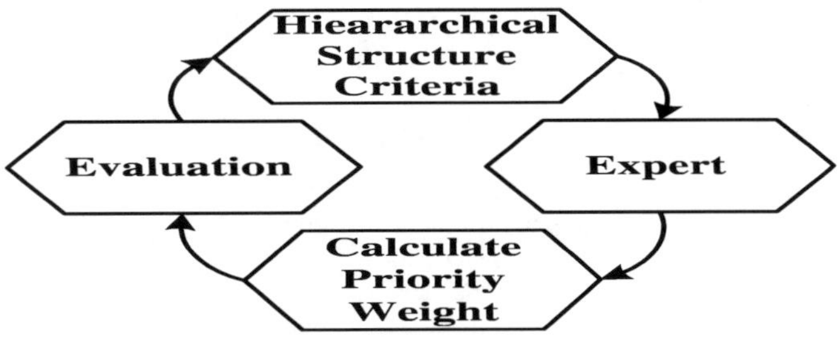

Figure 2. AHP procedure [2]

Figure 2 illustrates that the hierarchical criteria structure was the scholarship recipients that had been set by the university, the next step was to collect data by expert in pairwise matrix matrices, then calculated the criteria and alternative priority weighting. The next step was to evaluate whetherthe result of the calculation is acceptable or not measured by the consistency ratio where CR <0.1 if not then do the improvement [6].

3. Analysis

3.1 *Needs Analysis*

Analysis of needs related to input data that will be processed into output on the selection process of outstanding achievement scholarship obtained from the administration of the University and the selection team as the experts, the needs analysis obtained as follows:

a. Needs input. The required input data is student data consisting of student's number, name, study program, class information, history of achievement, average score of semester 1 to 5, recommendation of principal, score of achievement scholarship test and interview result. The data is the consideration of an expert that is formed in a hierarchy of criteria and alternative criteria.
b. Process Requirement. Criteria and alternative criteria data are processed by an expert using a pairwise matrix scale and AHP Random Index Consistency list.
c. Output Needs. The resulting output is a priority weighting sequence of criteria and alternative criteria that assist in consistent decision making.

3.2 *Data Collecting Technique*

Data collectiing techniques used to help obtaining the appropriate data are:

a. *Interview.* Collecting data through face-to-face interview directly to the interviewee, namely the administration department along with the scholarship selection team as the experts, related to data, process and criteria comparison scale and alternative criteria.
b. *Library.* Collecting data through journal literature, books and documents related to scholarship data.

3.3 *Analysis of AHP*

The next stage of analysis was to model business processes to map out the process flow on which development was based, as shown in the following figure 3.

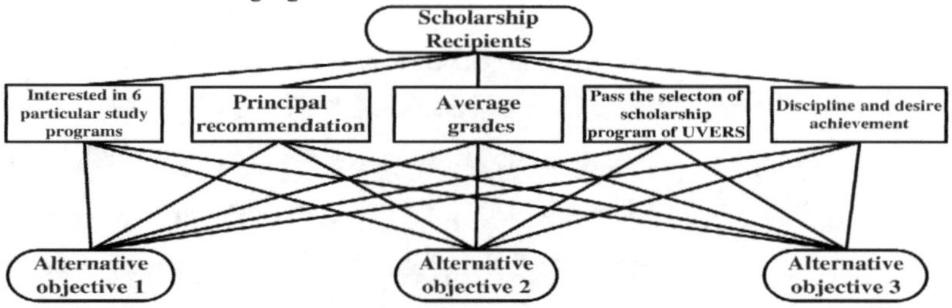

Figure 3. General framework of the AHP model [5]

Based on Figure 3 there were 5 criteria used in the selection process of outstanding scholarship, the existing criteria as shown in the following table 2.

Table 2. Criteria scale and its definition

No	Definition	Criteria
1	Interested in the study programs of Art Music Art, Dance Art, Industrial Engineering, Environmental Engineering, Telecommunication Engineering and Software Engineering	A
2	Principal recommendation	B
3	Average grades in semester 1 - 5 min 8 (Scale 10)	C
4	Pass the scholarship program selection of Universal University min 7 (scale 10)	D
5	Have a good discipline and great desire for college and achievement	E

From some criteria there was a need to be tested whether the consistency ratio was acceptable or not. In particular, a matrix (i, j) is said to be consistent if all elements follow some rule rules. As shown in equations (1) and (2) [6]:

$$a_{i,j} = a_{i,k} \cdot a_{k,j} \quad (1)$$

$$a_{i,j} = \frac{1}{a_{j,i}} \quad (2)$$

Where *i, j* and *k* represent any alternative of the matrix. If "A" is 9 times more important than "B", then "B" should be 1/9 times more important than "A". The relational scale used in the rankings as shown in the following table 3.

Table 3. Criteria Comparison Matrix

Criteria	A	B	C	D	E
A	1	9	2	3	1
B	0.11	1	0.33	0.33	0.33
C	0.5	3	1	3	1
D	0.33	3	0.33	1	1
E	1	3	1	1	1
Total	2.94	19	4.67	8.33	4.33

Based on Table 5 the matrix element *a [i, i]* = 1 where *i* = 1,2, ... *n*. where each element of the upper triangular matrix was the input and the lower triangle with the formula for *i j*.

The pairwise comparison matrix can also be shown in equation (3) [8].

$$A = \begin{bmatrix} a_{11} & \cdots & a_{1n} \\ \vdots & \vdots & \vdots \\ a_{n1} & \cdots & a_{nn} \end{bmatrix} = \begin{bmatrix} w_1/w_1 & \cdots & w_1/w_n \\ \vdots & \vdots & \vdots \\ w_n/w_1 & \cdots & w_n/w_n \end{bmatrix} \quad (3)$$

Then the next step summited each column to determine the value of the element in each criteria column by dividing the value of each cell with each number of columns as shown in the following table 4.

Table 4. Normalized Relative Weights of Criteria Comparison Matrix

Criteria	A	B	C	D	E	Total
A	0.34	0.47	0.43	0.36	0.23	1.83
B	0.04	0.05	0.07	0.04	0.08	0.28
C	0.17	0.16	0.21	0.36	0.23	1.13
D	0.11	0.16	0.07	0.12	0.23	0.69
E	0.34	0.16	0.21	0.12	0.23	1.06
Total	1	1	1	1	1	5

After obtaining the element value in each column, the next step was determining the average value of the row of criteria for each row in Table 4 where the number of rows was divided by the number of criteria as shown in the following table 5.

Table 5. Value of Eigen Vector

A	B	C	D	E
0.37	0.06	0.23	0.14	0.21

Based on the result of the priority criteria calculation, it was found the priority results with the following criteria as A, C, E, D and B.

Create a matrix of sums for each row, for a consistent matrix, as shown in equations (4), (5) and table 6 [8].

$$A = \begin{bmatrix} w_1/w_1 & \cdots & w_1/w_n \\ \vdots & \vdots & \vdots \\ w_n/w_1 & \cdots & w_n/w_n \end{bmatrix} \times \begin{bmatrix} w_1 \\ \vdots \\ w_n \end{bmatrix} = n \begin{bmatrix} w_1 \\ \vdots \\ w_n \end{bmatrix} \quad (4)$$

or in matrix form as follow:

$$A \cdot w = nw \quad (5)$$

where :
A = Comparison Matrix
W = Vector eigen
N = Dimensions of the matrix

Table 6. Matrix of the Sum of each row

Criteria	A	B	C	D	E	Total
A	0.37	0.50	0.45	0.42	0.21	1.95
B	0.04	0.06	0.08	0.05	0.07	0.29
C	0.18	0.17	0.23	0.42	0.21	1.21
D	0.12	0.17	0.08	0.14	0.21	0.72
E	0.37	0.17	0.23	0.14	0.21	1.11

The next step was to obtain the lamda (λ) value used to determine the Maximum Eigen value (λ_{max}) obtained from the sum of the total matrix of the sum of each row divided by the Eigen vector as shown in the following table 7.

Table 7. Calculation of Consistency Ratio

Criteria	Number of rows	*Eigen Vector*	Results
A	1.95	0.37	5.32
B	0.29	0.06	5.19
C	1.21	0.23	5.32
D	0.72	0.14	5.16
E	1.11	0.21	5.23
	Total		26.22

Then calculate the consistency value as a deviation or degree of consistency as shown in equations (6) [9]:

$$CI = \frac{\lambda_{max} - n}{n - 1} \quad (6)$$

The index was used by comparing it with a random consistency index as shown in the following table 8.

Table 8. List of Random Index Consistency (RI) [4]

n	*RI*	*n*	*RI*
2	0	9	1.45
3	0.58	10	1.49
4	0.9	11	1.51
5	1.12	12	1.48
6	1.24	13	1.56
7	1.32	14	1.57
8	1.41	15	1.59

After finding the CI value, the next step was calculating the value of CR as shown in equations (7) [9].

$$CR = \frac{CI}{RI} \quad (7)$$

The calculation of consistency ratio was to know whether the criterion comparative assessment is consistent or not. Where (CR) <= 0.1. If the CR value was greater than 0.1, then the pairwise comparison matrix must be corrected.

Amount (sum of result values): 26.22

N (number of criteria): 5

Λ max (number / n): 5.24

Consistency Index (CI) = CI ((λ_{max}-n) / (n-1): 0.06

Consistency Ratio = CI / RI, the RI value for n = 5 is 1.12 CR (CI / IR): 0.05

CR <0.1, then the consistency ratio of the calculation was acceptable (consistent weighting preference).

Alternative criteria as a material comparison of existing criteria as shown in the following table 9.

Table 9. Alternative Criteria

No	Criteria	Alternative
1	A	A1. Very interested A2. Interested A3. Quite Interested
2	B	B1. Very whorty B2. Worthy B3. Quite whorty

3	C	C1. Excellent>=9 C2. Very good=8.5<9 C3. Good =8<8.5
4	D	D1. Excellent>=9 D2. Very good =8<9 D3. Average =7<8
5	E	E1. Excellent E2. Very good E3. Average

Based on Table 9 the criterion alternative was calculated in the same way as the criteria process modeling, as shown in the following table 10.

Table 10. Pairwise Comparison Matrix of Alternative Criteria and Consistent Results

Alternative	1	2	3	*Priority*	*Sub- Criteria Priority*
1	1	3	5	0.63	1
2	0.33	1	3	0.26	0.41
3	0.2	0.33	1	0.11	0.17
Total	1.53	4.33	9	1	1.58
	Λ				9.12
	λ_{max}				3.04
	CI				0.02
	CR				0.03

CR <0.1, so the consistency ratio of the calculation is acceptable. The results of the scholarship selection process, as shown in the following table 11.

Table 11. AHP Implementation

Code	A	B	C	D	E	Priority
1	Interested 0.10	Worthy 0.01	Excellent >=9 0.14	Average =7<8 0.01	Averag e 0.06	3 (0.32)
2	Very Interested 0.23	Worthy 0.01	Very Good =8.5<9 0.06	Average =7<8 0.01	Very good 0.13	1 (0.45)
3	Interested 0.10	Very Worthy 0.04	Very Good =8.5<9 0.06	Average =7<8 0.01	Very good 0.13	2 (0.34)

And so on…

Based on the implementation of the student data in table 11, the priority was obtained from the results of criteria and alternative calculations used with the results of student 2wasmore eligible to receive thescholarship compared to students 1 and 2, while student 3 was more eligible for the scholarships compared to student 1.

4. Results and Discussion

AHP plays an important role to help solving problems in choosing the best alternative to prioritize based on multi-criteria data of awardees of outstanding scholarship at university. The results obtained by the AHP method in the selection process of outstanding scholarship at Universal University with the priority criteria A, C, E, D and B, the consistency ratio value 0.05, based on the order of priority the main factor in the selection of awardees of scholarship was interest in 6 study programs related to the priority value 0.37%, followed by the average value of report cards 0.23%, having good discipline and great desire to lecture and achievement 0.21%, pass the selection of universal university education scholarship program 0.14% and the recommendation of the principal 0.14%. The recommendation of the principal had a relatively low importance although it strengthened achieving conformity between the criteria fulfilled with the responsible party in previous education so that this criterion should not be ignored. Alternative priorities 1 (0.63), 2 (0.26) and 3 (0.11) had a consistency ratio of 0.03 in which each CR < 0.1.

The AHP method could be used to avoid contradictions, because AHP could only be used for consistent decisions. For decisions in the selection ofawardees were made by experts who evaluated and assessedthem carefully. Expert choices were subjective but the AHP structure assisted in consistent decision making [10].

In the previous study the AHP method had been used for decision making in selection of logistics service providers for a company in Istanbul, Turkey. The most important factor in the selection of logistic service provider companies was compatibility 47.15% and factors with relatively low importance, long term relationship 10.82% [9]. AHP was also applied in construction management decisions to select the best swimming pool contractors based on the six criteria used [3].

5. Conclusions

The AHP method provides the ease of decision making for the selection process of scholarship awardees in Universal University, through the multi criteria and alternatives used this method is able to prioritize each criteria and alternatives based on the level of expert interest so as to generate weight value that determines the outcomes of students who deserve to receive outstanding scholarship.

6. Suggestions

Suggestions for the development of next research are as follows:

1. This research can be developed into a new one on the selection of dedicated scholarship at Universal University of Batam.
2. The research data can be processed by using different methods such as Topsis or combination of AHP - Topsis and Fuzzy Logic methods, to produce the development method with the comparison of methods applied.
3. This research can be developed by building applications that can assist stakeholders in work, research and development
4. Melakukan integrasi sistem dengan aplikasi penyeleksian penerimaan calon mahasiswa baru secara online, sehingga proses bisnis untuk seleksi penerima beasiswa berprestasi berjalan beriringan dengan seleksi penerimaan calon mahasiswa baru sehingga lebih optimal.
5. Having Iintegrated online the system with the application of new student candidates selection, so that the business process for selection of scholarship achievers goes hand in hand with the selection of new students that will be more optimal.

7. References

[1] Kim, N., Joungkoo, P and Jeong-Ja, C. *Perceptual differences in core competencies between tourism industry practitioners and students using Analytic Hierarchy Process (AHP).* Journal of Hospitality, Leisure, Sport & Tourism Education. **20**, pp. 76-86, 2017.

[2] Dweiri, F., Sameer, K., Sharfuddin, A.K and Vipul, J. *Designing an integrated AHP based decision support system for supplier selection in automotive industry.* Expert Systems With Applications, **62**, pp. 273–283, 2016.

[3] Erdogan, S.A., Jonas, S., and Zenonas, T. *Decision Making in Construction Management: AHP and Expert Choice Approach.* Procedia Engineering, **172** , pp. 270 – 276, 2017.

[4] Foroughi, S and Mezgeen, A, R. *Housing renovation priority in the fabric texture of the city using the analytic hierarchy model (AHP) and geographic information system (GIS): A case study of Zanjan City, Iran.* The Egyptian Journal of Remote Sensing and Space Sciences. pp. 1-10, 2016.

IOP Conf. Series: Earth and Environmental Science **97** (2017) 012003 doi:10.1088/1755-1315/97/1/012003

[5] Dalalah, D., Faris, A and Mohammed, H. *Application of the Analytic Hierarchy Process (AHP) in Multi-Criteria Analysis of the Selection of Cranes*. Jordan Journal of Mechanical and Industrial Engineering, **4** (5), pp. 567 – 578, 2010.

[6] Lai, P.L., Andrew, P., Malcolm, B and Anthony, B. *Evaluating the efficiency performance of airports using an integrated AHP/DEA-AR technique*. Transport Policy, **42** , pp. 75–85, 2015.

[7] Albert, g., Oren, M., Ilit, O and Tsippy, L. *Which smartphone's apps may contribute to road safety ? An AHP model to evaluate experts' opinions*. Transport Policy, **50**, pp. 54–62, 2016.

[8] Chiu, W.Y., Lee, T.D and Lin, T.Y. *Performance Evaluation Criteria For Personal Trainers : An Analytical Hierarchy Process Approach.* Social Behavior and Personality, **38** (7), pp. 895-905, 2010.

[9] Gurcan, O.F., Ibrahim, Y., Omer, F.B., Cigdem, Y.A. and Fahrettin, E. *Third Party Logistics (3PL) Provider Selection with AHP Application*. Procedia - Social and Behavioral Sciences, **235**, pp. 226 – 234, 2016.

[10] Petruni, A., Evanthia, G., Ewan, D., Jie, G., Maria, C.L and Micaela, D. *Applying Analytic Hierarchy Process (AHP) to choose a human factors technique: Choosing the suitable Human Reliability Analysis technique for the automotive industry*. Safety Science. pp.1-11, 2017.

IOP Conf. Series: Earth and Environmental Science **97** (2017) 012004 doi:10.1088/1755-1315/97/1/012004

Implementation Of Haversine Formula And Best First Search Method In Searching Of Tsunami Evacuation Route

Anisya, Ganda Yoga Swara
[1,2]Institut Teknologi Padang, Padang, 25143, Indonesia
Email : anisya@itp.ac.id[1],gandayogaswara@yahoo.com[2]

Abstract : Padang is one of the cities prone to earthquake disaster with tsunami due to its position at the meeting of two active plates, this is, a source of potentially powerful earthquake and tsunami. Central government and most offices are located in the red zone (vulnerable areas), it will also affect the evacuation of the population during the earthquake and tsunami disaster. In this study, researchers produced a system of search nearest shelter using best-first-search method. This method uses the heuristic function, the amount of cost taken and the estimated value or travel time, path length and population density. To calculate the length of the path, researchers used method of haversine formula. The value obtained from the calculation process is implemented on a web-based system. Some alternative paths and some of the closest shelters will be displayed in the system.

Keywords : best first search, haversine formula, tsunami evacuation route

1. Introduction

Padang is one of the cities prone to earthquake disaster with tsunami due to its position at the meeting of two active plates. Spreading of population in the coastal area of coastal area also with central government and office. In the earthquake disaster experience in previous years, the city of Padang has a problem in the evacuation of the population.Best-First Search is a graphical search algorithm, which means the search space can be represented as a node connected by a path. In this case, the distance calculations of each point use the haversine method of the formula.

The haversine formula is an equation important in navigation, giving great-circle distances between two points on a sphere from their longitudes and latitudes.

Based on the above background then the formulation of the problem in this research is how to use haversine formula in designing an android application that can help the community in determining the location of the nearest shelter.

So that the author does not come out of the existing problems and research results can be obtained with good, perfect and directed, the authors limit the scope of the discussion. In addition, given the limited time, cost and the ability of the author, the research takes the restrictions of the problem as follows:

1) This research is only intended to address the problems that exist within the community about the evacuation site.
2) Restricted only to overcome the difficulties of the people in the selection of the closest evacuation routes specifically red zone (coastal territory).
3) Restricted to data obtained in vulnerable areas or 2 km from coastline.

IOP Conf. Series: Earth and Environmental Science **97** (2017) 012004 doi:10.1088/1755-1315/97/1/012004

Through the results of this study is expected to produce a capable:

1) Produce an application system that can assist the community in getting information about tsunami shelter nearest to Padang City by using artificial intelligence application of heuristic method.
2) Scientific Publication.

2. Theoretical Basis

PHP: Hypertext Preprocessor Is a scripting language that can be embedded or inserted into HTML. PHP is widely used to program dynamic websites. PHP can be used to build a CMS.
MySQL (Can be read by mai-es-ki-el or it could be mai-se-kuel) is a database software relation (*Relational Database Management System* atau *DBMS*),as well as ORACLE, POSTGRESQL, MSSQL, etc. SQL stands for *Structure Query Language*, defined as a syntax of certain commands or program languages used to manage a database. So MySQL is the software and SQL is the command language. Heuristic search is a strategy to conduct a selective search process and can guide the search process that has the greatest chance of success, but with the possibility of sacrificing completeness. To apply the heuristic search required a heuristic function. The heuristic function is the rules used to obtain the desired solution. This heuristic search can overcome some obstacles that arise in previous methods, such as solving problems that have high complexity.

The heuristic search consists of:

1) Generate and test
2) Hill Climbing
3) Best-First-Search
4) Algorithm A*
5) Simulated Annelaing

In this research, the algorithm used will use Best-First Search heuristic.

The Best-First Search algorithm is a search algorithm that discusses graphs by extending the most promising nodes selected according to the prescribed rules.
Algorithm (T.Sutojo: 2011).

1) Create a stack, initialization, as the root node (first node).
2) When the first node ≠ goal, the node is removed and replaced with its children.
3) Furthermore, the entire nodes in the Stack are sorted by Ascendingg based on the heuristic function used.
4) When the first node ≠ GOAL, repeat step point b.
5) When the first node = GOAL, find the solution by tracing the path from GOAL to the root node.
6) Done.

The haversine formula is an equation important in navigation, giving great-circle distances between two points on a sphere from their longitudes and latitudes.

$$Jarak = 2.R.arcsin\left\{\sqrt{\sin^2\left(\frac{Lat_1 - Lat_2}{2}\right) + \cos(Lat_1).\cos(Lat_2).\sin^2\left(\frac{Long_1 - Long_2}{2}\right)}\right\}$$

Information : Lat is latitude, Long is longitude, R is radius of earth (radius = 6,371km).

This study used several references from pre-existing studies. Among them is, Shanty Dewi, 2010 with the title research Design Information Application Shelter Transit Transjakarta Search with Breadth First Search Method. Ifatul Faizah, 2010, with Design Router Software Build. Traveling in Malang Using Dijkstra's Algorithm. Research conducted by Rike Nur Setiyani, 2015, Implementation of Best-First Search (BEFS) Algorithm on Salesman Problem Traveling (Case Study : Tourism Travel In Yogyakarta City), in the study researchers ignore the density of traffic, traffic lights, road width, and obstacles. While research on this paper use these obstacles as a determinant variable heuristic value.

IOP Conf. Series: Earth and Environmental Science **97** (2017) 012004 doi:10.1088/1755-1315/97/1/012004

3. Methods

3.1 Types of Research

The type of research that the authors do is research on making the search system of the nearest tsunami shelter search path using heuristic method - best first search. Aims to generate and develop new and useful innovations to help communities in the event of a tsunami.

3.2 Research Sites

This research was conducted at ITP, by analyzing, designing, and building database.

3.3 Desain of Context Diagram

Figure 1. *Context Diagram*

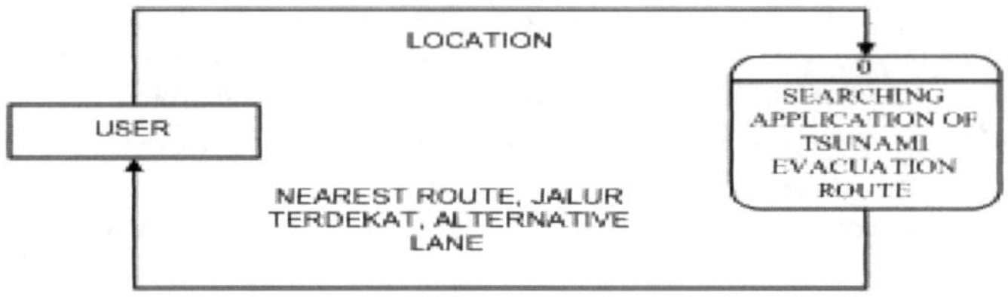

In Figure 1 above illustrates the outline of the nearest path search application by involving 1 entity only, is the user.

3.4 Input / Output Design

Here is the design of the main form view, where this form will be run in the smartphone browser or directly from user who using android.

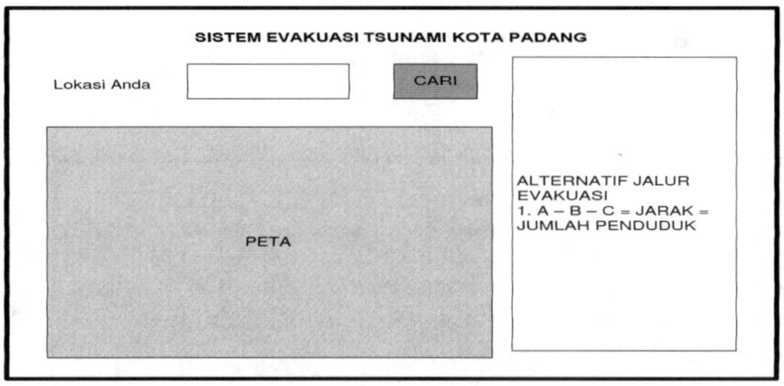

Figure 2. System View Design

4. Results and Discussion

4.1 Entity Relationship Diagram

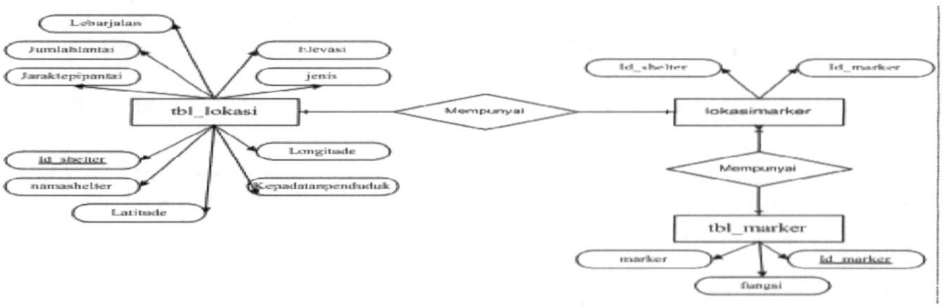

Figure 3. Entity Relationship Diagram

4.2 Database Design

1. Database : shelterpadang
   Table : tbl_lokasi

Table 1. Desain file of tbl

No.	Field	Type	Length	
1	Id_shelter	Varchar	12	Primary Key
2	namashelter	varchar	25	
3	Latitude	varchar	10	
4	Longitude	varchar	10	
5	Jumlahlantai	varchar	5	
6	Elevasi	varchar	5	
7	Jaraktepipantai	varchar	5	
8	Areaevakuasi	varchar	5	
9	Kepadatanpenduduk	Int		
10	Lebarjalan	int		
11	jenis	varchar	20	

3. Database : shelterpadang
   Table : tbl_marker

Table 2. Desain file of tbl_marker

No.	Field	Type	Length	
1	Id_marker	Varchar	12	Primary Key
2	fungsi	varchar	25	
3	marker	varchar	25	

Table 3. Nodes

Nodes In Map Search			
A	SDN 15	F	JL. JAYAPURA II ASRATEK
B	SMK N 5 PADANG	G	TELKOMSEL
C	SMP N 7 PADANG	H	STIE KBP
D	SMP N 25 PADANG	I	PLN BELANTI
E	SMA N 1 PADANG	J	STIKES ALIFAH PADANG

Variable =

a. Population density

Table 4. Population Density

Population density	Value
1 – 5000	1
5000 ≤ P.D ≤ 7500	2
7501 ≤ P.D ≤ 12.500	3
12.501 ≤ P.D ≤ 20.000	4
P.D ≥ 20.001	5

b. The width of the road

Table 5. The width of the road

The width of the road	Value
W.R ≤ 5 m	1
W.R ≤ 7 m	2
W.R ≤ 10 m	3

Population Density in node

Table 6. The width of the road

Node	Population Density
A	10000
B	12000
C	12000
D	7000
E	12000
F	4000
G	45000*
H	45000*
I	45000*
J	45000*

* If the location is 1 subdistrict & 1 lane, the population density is the same.

F → A = 1.5 km
F → B = 1.7 km
F → C = 1.4 km
F → D = 0.95 km
F → E = 1.3 km
F → G = 0.4 km
F → H = 0.65 km
F → I =1.9 km
F → J =0.75 m

heuristic formula of Best First Search

$$f(n) = g(n) + h(n) \text{ ....(2)}$$

f(n) = Total Cost needed to reach the node
g(n) = mileage
h(n) = approximate value

Table 7. Result of approximate value

Node	Variable Density of Population Values	Distance to F	Width of Road		approximate value (Density of Society + Width of Road)/10	$f(n)=g(n)+h(n)$
		$g(n)$	Width of Road	Variable Value	$h(n)$	
A	3	1.5	5 m	1	0.4	1.9
B	3	1.7	5 m	1	0.4	2.1
C	3	1.4	5 m	1	0.4	1.8
D	2	1.0	5 m	1	0.3	1.3
E	3	1.3	7 m	2	0.5	1.8
G	5	0.65	8 m	3	0.8	1.45
H	5	0.4	8 m	3	0.8	1.2
I	5	1.9	8 m	3	0.8	2.7
J	5	0.75	8 m	3	0.8	1.55

Table 8. Case

Shelter	Address	latitude	longitude
St. JAYAPURA II	Lolong	-0.912987	100.356051
PLN Rayon Belanti	Khatib Sulaiman	-0.916874	100.360254

Result using Haversine Formula :

1 derajat = 0,0174532925 radian
Lat_1 = --0.912987 * 0,0174532925 = -0.015934629 derajat
Lat_2 = --0.916874* 0,0174532925 = -0.01600247 derajat
$Long_1$ = 100.356051 * 0,0174532925 = 1.751544 derajat
$Long_2$ = 100.360254 * 0,0174532925 = 1.751617 derajat
ΔLat = - 0.015934629 – (- 0.01600247) = 0.000068
ΔLong = 1.751544 – 1.751617 = -0.000073
$x = \sin^2(0.000068 /2) = 1.1506E-09$
y = cos(-0.015934629)*cos(-0.01600247) = 0.999745026
$z = \sin^2(-0.000073 /2) = 1.34528E-09$
a = x+(y*z)= 1.1506E-09 + (0.999745026 * 1.34528E-09) = 2.49554E-09
R = 6371
Jarak = 2*6371*arcsin(√ 2.49554E-09) = 0.636531216 km

From the table above, so the node to which node H.

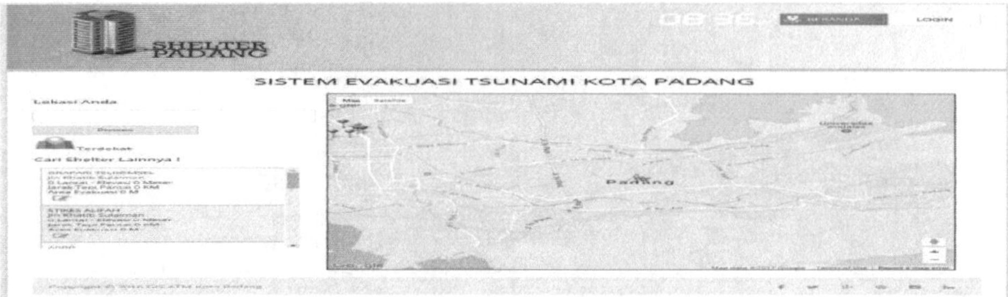

Figure 4. Main View of System

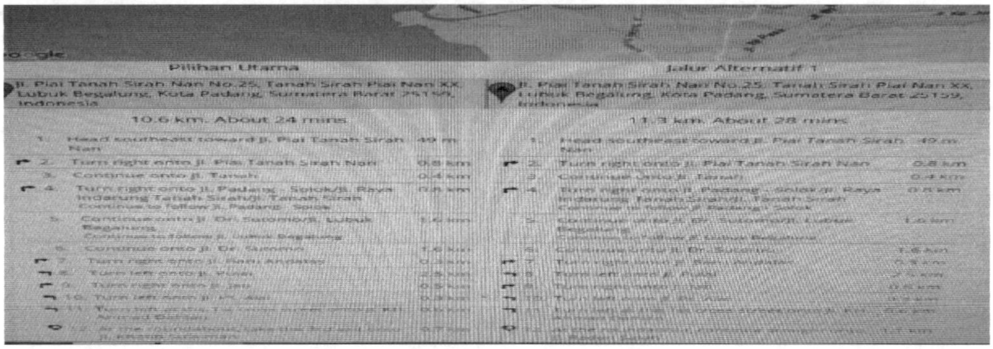

Figure 5. Alternate Path View

5. Conclusions

In this paper use haversine formula to calculate the distance of each node. The heuristic function used in this case takes into account the population around the node as well as the width of the road to be traversed. And also the optimum time obtained based on the density of the population that will evacuate, and the extension of the lane. And numerical calculations are applied to web-based systems. Where users will be able to directly access the system. For the next stage, this application will be developed in other sub-districts and socialized with related parties so as to be able to be implemented in the community. For suggestions from this paper, in order for further research, whether done by the researcher himself or any other researcher, can develop this application for a wider scope such as 1 city of Padang. And made more flexible for other cities.

6. References

[1] Anisya. 2013. *Aplikasi Sistem Database Rumah Sakit Terpusat Pada Rumah Sakit Umum (Rsu)'Aisyiyah Padang Dengan Menerapkan Open Source (Php-Mysql).* 2, Padang : Jurnal Momentum ISSN: 1693-752X, 2013, Vol. 15.

[2] Alamsyah. 2010. *Pemanfaatan Metode Heuristik dalam Pencarian Minimum Spanning Tree Dengan Algoritma Semut*. MEKTEK Tahun XII NO. 3, September 2010.

[3] Alamsyah. 2010. *Pemanfaatan Metode Heuristik pada Pencarian Jalur Terpendek dengan Algoritma Genetika*. Jurnal SMARTek, Vol. 8 No. 4. Nopember 2010. Hal 307 – 316.

[4] A.K.Balirsingh. 2011. *Robust FACTS Controller Design Employing Modern Heuristic Optimization Techniques.* World Academy of Science, Engineering and Technology 56 2011.

[5] Dewi, Shanty. 2010. *Perancangan Aplikasi Informasi Pencarian Shelter Transit Transjakarta Dengan Metode Breadth First Search.* s.l. : Universitas Gunadarma, 2010.

[6] Fauzan. *Evaluasi Existing Building dan Pembuatan Peta Evakuasi Vertikal Terhadap Tsunami Di Kota Padang*. Universitas Andalas.

Genetic Algorithms to Optimizatize Lecturer Assessment's Criteria

Deny Jollyta[1], Johan[2] and Alyauma Hajjah[3]
[1, 2, 3] STIKOM Pelita Indonesia Pekanbaru, Riau, Indonesia
Telp: 076124418, Fax: 076135508
E-mail: deny.jollyta@lecturer.pelitaindonesia.ac.id

Abstract: The lecturer assessment criteria is used as a measurement of the lecturer's performance in a college environment.To determine the value for a criteriais complicated and often leads to doubt. The absence of a standard valuefor each assessment criteria will affect the final results of the assessment and become less presentational data for the leader of college in taking various policies relate to reward and punishment. The Genetic Algorithm comes as an algorithm capable of solving non-linear problems. Using chromosomes in the random initial population, one of the presentations is binary, evaluates the fitness function and uses crossover genetic operator and mutation to obtain the desired crossbreed. It aims to obtain the most optimum criteria values in terms of the fitness function of each chromosome. The training results show that Genetic Algorithm able to produce the optimal values of lecturer assessment criteria so that can be usedby the college as a standard value for lecturer assessment criteria.

Keywords:Lecturer Assessment Criteria, Binary, Genetic Algorithm, College

1. Introduction

The College with its products in the form of education services is an institution that serves as a place to organize education or teaching, research and community service. The quality of college is strongly influenced by the presence of students, lecturers, facilities and infrastructure. Lecturers are professional educators with the main task of transforming, developing, and disseminating science, technology and the arts through education, research and devotion.

Education quality of the college is determined by the lecturer's eligibility to teach in accordance with his competency. In the college, the performance of lecturers is measured through questionnaires filled by students based on predetermined criteria. The problem that occurs is the difficulty in determining the optimal value standard for each criteria. The standard of value needs to be determined given the difficulty of providing value to pedagogic competence, professional competence, personality competence, and social competence of a lecturer. In addition, the assessment results can be used by the head of university as a consideration in taking various policies relating to the reward and punishment of lecturer.

Genetic Algorithm is a Heuristic algorithm based on the mechanism of biological evolution (Kusumadewi, 2003). Genetic algorithms can be applied to some types of non-linear objective functions and have the flexibility to be implemented efficiently on a particular problem (Zheng, etl, 1999). The determination of the standard value of each lecturer's assessment criteria can be calculated through criteria organized as a gene in each lecturer grouped into chromosomes. Through the stages of crosses and mutations, then the optimal value can be reviewed from the fitness function of each chromosome.

2. Metodology

The research methodology outlines all the activities carried out during the research. The stages passed in this research are shown in the following figure.

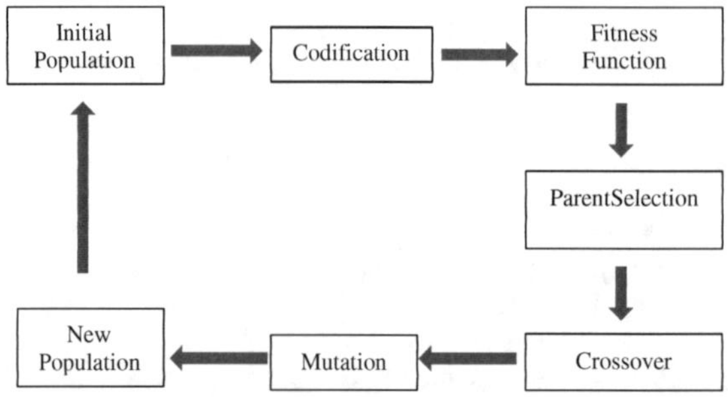

Figure. 1 Research Stages of Optimization of Lecturer Assessment Criteria

2.1 The Initial Population

This stage is the beginning of the training process using Genetic Algorithm. The initial population is to determine the number of samples to be a population in studied. In this study, 24 samples were taken.

2.2 Codification

The coding technique is a technique that states the initial population as a potential solution of a problem to the chromosome (Gen and Cheng, 2000) as a key issue when using the Genetic Algorithm. Genes can be presented in binary form. Codification is neded to be processed using Genetic Algorithm. The lecturer to be assessed is encoded as a Chromosome. The criteria that lecturer asessment are coded as genes to be arranged in binary form.

2.3 Fitness Function

Each chromosome is assigned a fitness value. The chromosome with the highest fitness value will survive, while the chromosome with low fitness value will die.

2.4 Roulettewheel (Parent Selection)

Parent selection is performed to obtain a pair of chromosomes that are used as parent. To get it generated random value as much as the number of chromosomes. If the random value of a chromosome is smaller than the predetermined probability (pc), then the chromosome will be selected as the parent.

2.5 Crossover

This process is done after obtaining the parent chromosome. The parent chromosomes are then crossed with each other to get the off-spring.

2.6 Mutation

Mutation is done to replace the missing genes caused by selection. Mutation of genes are performed by generating a random number of each gene. Random number less than mutation probability (pm) will be mutated, 0 to 1 and 1 to 0.

2.7 New Population

The mutation result is become an initial population for the next generation. The new population will be reprocessed as well as the initial population. This process is done until the fitness values of all individuals are equal.

3. Experiment

The search for optimal values for each lecturer's assessment criteria, starting with codification in the first generation population.Codification performed on the data that has been prepared. These data consist of 24 lecturers at STIKOM college as shown in Table 3 were coded as individual chromosome L1-L24. Each individual has four criteria and each them has 4 sub criteria as shown in table 1. GA process performed on all lecturers.

Table 1. Lecturer Code

Number	Lecturer Code
1	L 1
2	L 2
3	L 3
-	-
23	L 23
24	L 24

Table 2. Codification of Lecturer Asessment Criteria

Lecturer Code	**X1**				**X2**			
	Pedagogic Competency				Professional Competency			
	G1	**G2**	**G3**	**G4**	**G1**	**G2**	**G3**	**G4**
Lecturer Code	**X3**				**X4**			
	Personality Competency				Sociality Competency			
	G1	**G2**	**G3**	**G4**	**G1**	**G2**	**G3**	**G4**

The Genetic Algorithm began by assigning binary bits to each gene of a chromosome and calculating its by the formula :

$$\text{Fitness }(Pos)=((2-SP)+2(SP-1)(Pos-1))/(N-1)\ldots\ldots\ldots\ldots\quad (1)$$

Value Interval $SP \in [1,2]$
Following the calculation of the fitness value of each chromosome :

$$F(1) = ((2-1.5)+2*(1.5-1)*(1-1))/(24-1) = 0.02173913$$

$$F(2) = ((2-1.5)+2*(1.5-1)*(2-1))/(24-1) = 0.065217391$$

$$\vdots$$

$$F(24) = ((2-1.5)+2*(1.5-1)*(24-1))/(24-1) = 1.02173913$$

The next step is to select the chromosome to become the parent chromosome. The chromosome is chosen proportionately according to its fitness value. The Roulettewheel method places each chromosome on a piece of the circle proportionately according to its fitness value. Chromosomes that have greater fitness values occupy larger circle cuts than low-value chromosomes.

To be able to perform chromosome selection, the steps that must be done is as follows ::

- Determine *fitness relative* (p_k) of each chromosome
- p1 = F1 / Total of *Fitness Value* = 0.02173913 / 12.56521739 = 0.002
- Determine *fitness* cumulative (q_k) of each chromosome q1 = p1 = 0.002
- Generate *random* numbers as much as the number of chromosomes.
- Create a cumulative fitness value interval of each chromosome.

Next is the crossover process by generating random numbers. Crossover probability (pc) is one of the parameters that shows the ratio of off-spring in each generation. Pc used is 0.25 by means of one-point crossover. The chromosome chosen as the parent is that has a random number smaller than pc.

Table 3. Random Numbers

Lecturer Code	Chromosome	Random Value
L 7	1*	0.04414
L 15	2*	0.45456
L 22	3*	0.03923
L20	4*	0.67293
L18	5*	0.07585
L20	6*	0.34349
L 13	7*	0.16861
L 18	8*	0.58825
L 12	9*	0.46987
L 9	10*	0.25414
L 20	11*	0.67622
L 7	12*	0.95616
L 23	13*	0.61232
L 24	14*	0.76767
L 21	15*	0.90898
L 17	16*	0.34174
L 19	17*	0.35652
L 15	18*	0.19751
L 22	19*	0.87489
L 24	20*	0.47587
L 20	21*	0.56944
L 9	22*	0.52767
L 10	23*	0.5182
L 18	24*	0.11539

Table 4. Parent Chromosome and Off-Spring

Parent Chromosome																		
L7	1'	1	0	1	0	1	0	1	1	1	0	0	0	1	0	0	1	0,282608696
L22	3'	0	0	0	0	0	1	0	0	1	0	0	0	1	1	0	0	0,934782609
L18	5'	1	0	0	0	1	1	0	0	0	0	0	0	0	1	0	0	0,760869565
L13	7'	1	0	1	1	1	0	1	0	1	0	0	1	1	0	0	0	0,543478261
L5	18'	1	0	0	1	1	0	1	0	1	0	1	1	1	0	0	0	0,630434783
L18	24'	1	0	0	0	1	1	0	0	0	0	0	0	0	1	0	0	0,760869565
Off-Spring																		
L7	1'	1	0	1	0	0	1	0	0	1	0	0	0	1	1	0	0	0,282608696
L22	3'	0	0	0	0	1	0	1	1	1	0	0	0	1	0	0	1	0,934782609
L18	5'	1	0	0	1	1	0	1	0	1	0	0	1	1	0	0	0	0,760869565
L13	7'	1	0	1	0	1	1	0	0	0	0	0	0	0	1	0	0	0,543478261
L5	18'	1	0	0	0	1	1	0	0	0	0	0	0	0	1	0	0	0,630434783
L18	24'	1	0	0	1	1	0	1	0	1	0	1	1	1	0	0	0	0,760869565

From table 3 above there are 6 random numbers that smaller than pc. Chromosomes which has that random numbers, will be arrange in pairs as parent. One-point crossover is a method by determining the point of crossing. In this case, crossing point lies in the fourth bit. Any different genes from the parent pair, will be crossed. The crosses then become off-spring. The next process is a mutation, which starts by counting the number of bits in the population where :

Number of bits = popsize * Number of Genes................(2)
= 24 * 16
= 384

The mutation (pm) used is 0.05, meaning that there will be a 5% chance of the number of bits to be mutated, ie 20 bits. To specify the bits must be generated random numbers for each bit in the population (as many as 384 random numbers). The selected bit is a bit whose random number is less than pm. The selected bit will be mutated, 0 to 1 and 1 to 0. The mutation result is the final population on the whole of the first generation process. This final population became the initial population in the next generation.

4. Result

Every generation will do the same steps. The training stops when fitness value of each individual is the same, means the number of generation can not be predicted. Some cases proved that the number of generation can reach up to 1000 even more. The articles about optimization using genetic algorithms showing up to 1000 generations [3]. Others showing that the generation up to 2500 [8]. It coused by random numbers and genes which influenced the training results.
Experiments of this research showed that the results of training on the data yield varying values for each assessment criterion. Training stopped at the 14th generation, where all individuals display the same fitness value of 1.02174.

To optimize the bit value of each gene, will be achieved by using the formula :

$$x = r_b + (r_a - r_b)(g_1 x 2^{-1} + g_2 x 2^{-2} + \ldots + g_n x 2^{-n}) \ldots\ldots \quad (3)$$

Note :

x = variable (criteria)
r_a = upper limit of interval value
r_b = lower limit of intervalvalue
g = gen of variable
2^{-n} = numbers of genin variable

The following calculation is to get x value (criteria) of first chromosome of first generation.

$$x_1 = 1 + (16-1)((1(2^{-1}) + 0(2^{-2}) + 1(2^{-3}) + 0(2^{-4}))$$
$$= 10{,}275$$
$$x_2 = 1 + (16-1)((0(2^{-1}) + 1(2^{-2}) + 0(2^{-3}) + 0(2^{-4}))$$
$$= 4{,}75$$
$$x_3 = 1 + (16-1)((1(2^{-1}) + 0(2^{-2}) + 0(2^{-3}) + 0(2^{-4}))$$
$$= 8{,}5$$
$$x_4 = 1 + (16-1)((1(2^{-1}) + 1(2^{-2}) + 0(2^{-3}) + 0(2^{-4}))$$
$$= 12{,}25$$
$$\sum x = 35{,}875$$

The optimization of bits applies to all genes of each generation. Optimization aims to find the best value of each criterion of each generation taking into account the highest total value of criteria (x). The highest total value of x of each generation, will be decisive to determine the optimal value of x .

The selection of the most optimal value of x can be determined by several reasons :

- Towards the end of generation, the value of x and the total value of x denotes the same value. The generation according to the reason was selected as the most optimal value.
- All generations showed different results. The generation with the highest value was selected as the most optimal value.

In this research, the optimal value shown by 3rd generation. The optimal value is taken from the individual that has the best criterion value and total value which is indicated by the total value of 47.125. This means, every criterion has obtained the most optimal value as follows :

Table 5. The Optimal Value of Lecturer Asessment Criteria

Variabel x	**Name of Criteria**	**Optimal Value**
x_1	Pedagogic Competency	**14,125**
x_2	Professional Competency	**9,4375**
x_3	Personality Competency	**13,1875**
x_4	Sociality Competency	**10,375**

5. Conclusion

The result of this research proved that Genetic Algorithm can achieve the optimize value of lecturer assessment criteria as follow :

Pedagogic Competency : 14,125
Professional Competency : 9,4375
Personality Competency : 13,1875
Sociality Competency : 10,375

Each criterion has the highest score is 16. Achievement of value through Genetic Algorithm process is the most optimal value that can be achieved by a lecturer, so it can be assessment measure for lecturer assessment criteria. The optimal values largely determined by the generation of random number and genes number because these influence the training result. The Optimal value is expected to facilitate universities in assessing the performance of lecturers through a questionnaire that will be filled by students so that the assessment becomes more appropriate and objective.

6. References

[1] Aristoteles. 2013. "Penerapan Algoritma Genetika Pada Peringkasan TeksDokumen Bahasa Indonesia", *Prosiding Semirata FMIPA Universitas Lampung.*
[2] Gen and Cheng. 1997. *Genetic Algorithm and Engeenering Design* : John Wiley & Son, Inc, New York
[3] Indrianingsih Yuliani. 2010. "Algoritma Genetika Untuk Menyelesaikan Masalah Optimasi Fungsi Berkendala Dengan Pengkodean Bilangan Bulat", *Jurnal Angkasa,* vol. 2, no. 1.
[4] Jollyta Deny. 2017. "Genetic Algorithm To Optimize The Value Of Customer Satisfaction For Motorcycle User", *Abstract Book ADRI Sumbar 7th-Part 03*
[5] Jollyta Deny. 2011 "Penggunaan Genetic Algorithms dalam Mengoptimalkan Akurasi Penilaian SCL Untuk Meningkatkan Motivasi dan Hasil Belajar", *IEEE Trans. on General Writing*, vol. 1, no. 2, pp.1-10.
[6] Kusumadewi. 2003. *Artificial Intelligence (Teknik dan Aplikasinya)* : Penerbit Graha Ilmu
[7] Saptono F, Hidayat T. 2007. "Perancangan Algoritma Genetika Untuk Menentukan Jalur Terpendek". *Seminar Nasional Aplikasi Teknologi Informasi 2007 (SNATI 2007).*
[8] Sundarningsih D, Mamudy WF, Sutrisno. 2017. "Penerapan Algoritma Genetika untuk Optimasi Vehicle Routing Problem with Time Window (VRPTW)", *Jurnal Pengembangan Teknologi Informasi dan Ilmu Komputer*, vol.1, no. 2, pp.100-107
[9] Suyanto. 2005. *Algoritma Genetika dalam Matlab.* Penerbit Andi
[10] Zheng, et ll. 1999. *Genetic Algorithm Applications: Assigment* #2 for Dr. Z.Dong
[11] Zukhri Z. 2013. *Algoritma Genetika : Metode Komputasi Evolusioner untuk Menyelesaikan Masalah Optimasi* : Penerbit Andi

Design Of Computer Based Test Using The Unified Modeling Language

Agus Tedyyana, Danuri and Lidyawati
[1] Politeknik Negeri Bengkalis, Riau, Indonesia
E-mail: agustedyyana@polbeng.ac.id, danuri@polbeng.ac.id, lidyawati@polbeng.ac.id

Abstract: The Admission selection of Politeknik Negeri Bengkalis through interest and talent search (PMDK), Joint Selection of admission test for state Polytechnics (SB-UMPN) and Independent (UM-Polbeng) were conducted by using paper-based Test (PBT). Paper Based Test model has some weaknesses. They are wasting too much paper, the leaking of the questios to the public, and data manipulation of the test result. This reasearch was Aimed to create a Computer-based Test (CBT) models by using Unified Modeling Language (UML) the which consists of Use Case diagrams, Activity diagram and sequence diagrams. During the designing process of the application, it is important to pay attention on the process of giving the password for the test questions before they were shown through encryption and description process. RSA cryptography algorithm was used in this process. Then, the questions shown in the questions banks were randomized by using the Fisher-Yates Shuffle method. The network architecture used in Computer Based test application was a client-server network models and Local Area Network (LAN). The result of the design was the Computer Based Test application for admission to the selection of Politeknik Negeri Bengkalis.

Keywords: Unified Modeling Language, Paper Based Test, Computer Based Test, Fisher-Yates Shuffle, Criptography, Local Area Network

1. Introduction

Politeknik Negeri Bengkalisis one of the vocational colleges and the only State polytechnic in Province of Riau, There are three ways of admissions namely through the search of Interests and Talents, Joint selection admission test Polytechnic and path Self examination conducted Paper Based Test (PBT).

Admission tests of Politeknik Negeri Bengkalis through Joint Selection Examination Polytechnic and independent are using Paper Based Test (PBT), which is through the written test by distributing a booklet and answer sheet to the test takers then answering the test that have been provided by the committee. Paper Based Test (PBT) Model also has some disadvantages, They are the use of too much paper, the possibility of questions leaking to the public, manipulation of test data result and other fraud during the test.

This study is designing Computer Based Test (CBT) application using Unified Modeling Language (UML) which consists of *Usecase diagrams, Activity* and *Sequence diagrams*. Computer Based Test, is defined as a test or assessment given by the computer whether they are *stand-alone* or dedicated network, or with other technologies devices which connected to the Internet or *World Wide Web (WWW)* and mostly use the

Multiple Choice Questions (*MCQS*), Computer Based Test itself has been applied since 1960 for the knowledge test and problem-solving test, Recent Computer Based Test has been adopted in some schools in Indonesia for the national exam High School level. In this study using the Fisher-Yates Shuffle method into computer-based exam application in randomize the problems so that the problems raised are different. And using the RSA cryptographic algorithm for security exam questions.

2. Method

2.1. Unified Modeling Language Diagrams

graph shaped Diagram showing the symbol of model elements arranged to illustrate certain parts or aspects of the system. A diagram is part of a particular view and generally described to a particular view. The type of diagram such as:

a) Use Case Diagram describes a number of external actors and their relation to the *use case* provided by the system. *Use case* is a description of the functions provided by the system in the form of text as documentation of the use case symbol but it can also be done in diagrams activity. Use case described only seen from the outside of an actor (the state of the system environment the user visits) and not how to function in the system
b) Class diagram described the static structure of classes in the system. Class represents something that is handled by the system. Class can be connected to one another through a variety of ways: associated (connected to each other), dependent (a dependent class / using other classes), specialed (one class is a specialization of other class), or package (a unit group), a system usually has several class diagrams.
c) State Diagram describes all states (state) owned by an object of a class and the circumstances which led the state change. The events can be another object that sent the message. State class is not described for all classes, only the state has a number of well-defined and class conditions changed by a different state.
d) Sequence Diagram Illustrated the dynamic collaboration among a number of objects. Their role is to show a series of messages sent between objects is also interaction between object, something that happens at a certain point in the execution system.
e) Collaboration Diagram describes dynamic collaboration such as sequence diagrams. In order to show the message exchange, collaboration and connection diagrams describe object (refer to the context). If the emphasis is on time or sequence, sequence diagrams are used, but if the emphasis on the context collaboration diagrams are used.
f) Activity Diagram Describes a series of flow of activity, it is used to describe activities that are formed in an operation that can also be used for other activities such as the *use case* or interaction
Component Diagram describes the physical structure of the code's component. The compoenent may include source code, binary component, or executable component. A logic's component contains information about the class or classes that are implemented to make the mapping of the logical view to the component view.
g) Deployment diagrams describe the physical architecture of the hardware and software show the relationship between computer and the devices (nodes) to one another and the type of the relationship. In the nodes, the executable component and object allocated for showing units of software executed by the particular node and dependence components.

2.2. Computer Based Test (CBT)

Computer Based Test (CBT), is defined as a test or assessment given by the computer whether they are stand-alone or dedicated network, or with other technologies devices that are connected to the Internet or World Wide Web (WWW) and mostly use the Multiple Choice Question (MCQs). Computer Based Test itself has been applied since 1960 to test the knowledge and problem-solving test.

2.3. Fisher Method Yates Shuffle

Fisher-Yates Shuffle (named after Ronald Fisher and Frank Yates), also known as the Knuth Shuffle (named after Donald Knuth), is an algorithm to generate a random permutation of a finite set, in other words, it is used to shuffle some of the set. A variant of the Fisher-Yates shuffle, known as algorithms Sattolo it can be used to generate a random cycle of length instead. The basic process of Fisher-Yates shuffling is similar to randomly choose a numbered ticket out of the cab, or a card from a deck, The use of the Fisher-Yates Shuffle can be in two ways: original and modern methods, the basic method used to generate a random permutation of number 1 to N is as follows (Ibijola and Olu, 2012):

a) Write down the numbers from 1 to N.
b) Choose a K random numbers between 1 to the number of digits that have not been crossed out.
c) Calculated from below, strike figure K that has not been crossed out, and write that number in other places.
d) Repeat step 2 and 3 until all numbers already crossed.
e) The order of the numbers written in step 3 is a random permutation of the previous numbers

In the modern version, the numbers chosen are not crossed, but its position has exchanged with the last digit of the numbers that have not been selected. The following are examples of the modern version. Range is the number of digits that have not been selected, the roll is the random numbers, scratch is a list of numbers that have not been selected, and the result is a permutation results to be obtained.

2.4. *RSA Cryptography Algorithm*

Cryptography prior to the modern age was effectively synonymous with encryption, the conversion of information from a readable state to apparent nonsense. The originator of an encrypted message (Alice) shared the decoding technique needed to recover the original information only with intended recipients (Bob), thereby precluding unwanted persons (Eve) from doing the same. The cryptography literature often uses Alice ("A") for the sender, Bob ("B") for the intended recipient, and Eve ("eavesdropper") for the adversary.[5] Since the development of rotor cipher machines in World War I and the advent of computers in World War II, the methods used to carry out cryptology have become increasingly complex and its application more widespread

Cryptography is derived from the Greek, crypto and Graphia. Crypto means “to hide”, while Graphia means “writing”. In general, Cryptography is the study of mathematical techniques related to aspects of information security, such as data confidentiality, data authenticity, data integrity, and authentication of data. However, it is not all aspects of information security can be solved by cryptography. Additionally, cryptography can also be interpreted as a science or art to maintain the security of the message. RSA is an algorithm based on public-key cryptography scheme. RSA stands for the initials of the inventors: Ron Rivest, Adi Shamir, and Leonard Adleman. The security of the RSA algorithm lies on the difficulty of factoring large numbers into prime factors. Factoring is done to obtain the private key.

2.5. *Web Server*

web Server is a software that provides data-based services and functions to accept the requests from HTTP or HTTPS from the client and usually known as web browser which then sends the results back in the form of multiple web pages and generally will be in the form of HTML documents. In the simplest form, web server will send the HTML data to the web browser, so the request will look like in general, ie a website display. The most common use is the web server for placing the web site, but in practice its use expanded as a data storage or to run a number of business-class applications, Web servers are not only used for serving the World Wide Web. They can also be found embedded in devices such as printers, routers, webcams and serving only a local network. The web server may then be used as a part of a system for monitoring or administering the device in question. This usually means that no additional software has to be installed on the client computer, since only a web browser is required (which now is included with most operating systems)

3. Results and Discussion

UML (Unified Modeling Language) is a modeling language for systems or software that the object-oriented paradigm. Modeling actually used for the simplification of complex issues in a way that is easier to learn and understand. Types of diagrams in UML is the Use Case Diagram, activity diagrams and sequence diagrams, UML is a language to specify, visualize, construct and document the artifacts (part of the information that used to be produced by the process of making software, the artifact can be a model, a description or software) of the software system, such as the modeling business and non-system devices other software. Besides UML is a modeling language that uses the concept of object orientation. UML made by Grady Booch, James Rumbaugh, and Ivar Jacobson under the banner of Rational Software Corps. UML provides a notation that help model the system from various prespetktif. UML is not only used in the modeling software, but almost in all fields that require modeling.

3.1. Use Case Diagrams

Iin the use case diagram the actors are the admin, prospective students and supervisors. Administration staffhas 13 use case, the supervisor has four use cases and new students have 7 case. use case scenario described the sequence of steps describe between the user and the system.

IOP Conf. Series: Earth and Environmental Science **97** (2017) 012006 doi:10.1088/1755-1315/97/1/012006

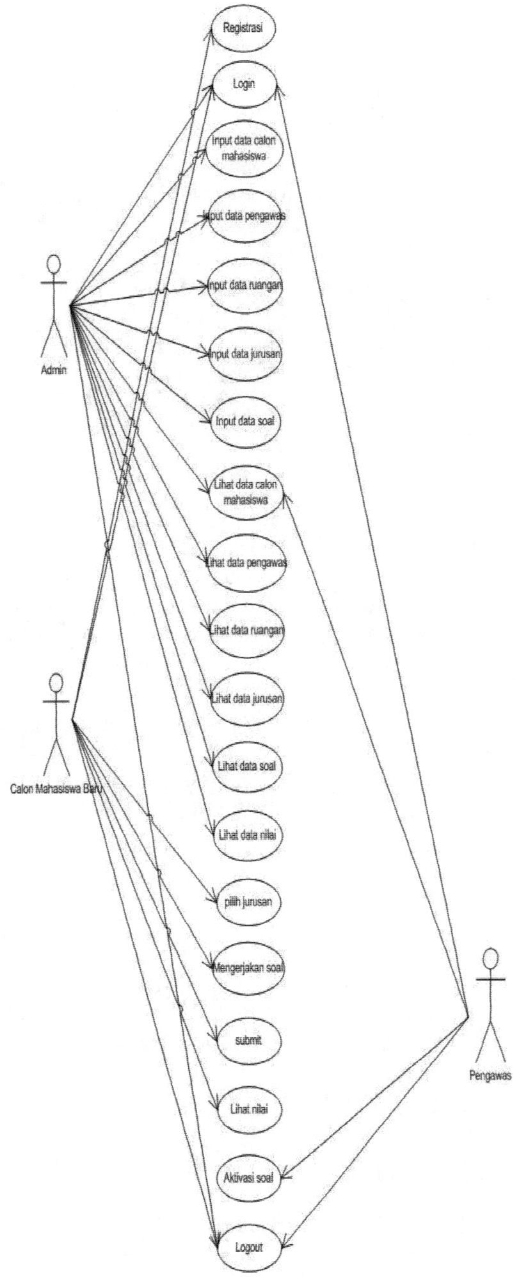

Figure 1. Use Case Diagram

a. Use case scenario login

table 1, Use case scenario login

Use case name	Login	
Use case id	1	
Actor	Administration staff, supervisors and new student candidate.	
pre condition	Actor must fill the username and password first	
triger	Actor can sign in and access the system	
Description	This use case describes the actor that fills your username and password to sign in and access the system	
Typical cource of events	Actor action • Input your username and password	system response • Check the username and password • Showing the page based on the authentication
Alternate courses	If the username and password are correct so the actor would enter into the system if false so the actor would reenter the username and password	
conclution	Actor has successful entered into the system	
Post condition	Showing the main page into the system based on each actors.	

b. Use case scenario registration

c.

table 2, Use case scenario registration

Use case name	Registration	
Use case id	2	
Actor	new student candidate	
pre condition	Actor must fill out and input the registration form for being a new student	
triger	Actor can sign up to be a new student	
Description	This use case describes the actor that fills and input the registration form for being a new student	
Typical cource of events	Actor action	system response
	Select a menu list of students	Showing the form for terms and conditions of registration
	select list	Displays form a list of new students
	Input data registration	
lternate courses	If the students' input do not complete the students can't register	
conclution	Actors can register	
Post condition	Data will be stored into the student registration data	

3.2. Activity diagram

a. Login
In the diagram activity that must be done by administration staff, supervisor and the student is entering usename and password. If the username and password is entered incorrectly then the system will display the username and password is entered incorrectly then do repetition to input a username and password. If the password and user name is entered correctly then it will display the main page or go to the system

b. Registration for new student candidates
In the activity diagram of registration for new students to the things that they must do are logging on, select a menu list of prospective new students on the menu page list of students fill out the form provided to enter data for new students if it is complete then stored.

c. Input the question data
In the activity of inputing data that done by the administration staff is logged in, then select the menu, the menu pages about, administration staff inputing the question data if it is suitable will be saved

3.3. Sequence Diagrams

The Administration staff inputs and adds the question data of the supervisor with taking the steps like log in to enter to the main page that contained the main menu, in the main menu administration staff will select one of the menu like question data, supervisors or room. If you want to add and input the question data so enter into the question menu.. , save and view the question data. If you want to add and input the supervisors data so enter into the supervisors menu. , then input the supervisor data, store and view the supervisor data. If you want to add and enter the data into the room the room menu, then input the room data, stored and viewed the room data.

4. Implementation

The design of the interface is done to allow users to interact with the system. The design interface is done using hypertext markup language (HTML5), Cascading Style Sheet (CSS 3) and jQuery for the system implemented a web-based server.

4.1. Design Interface Login

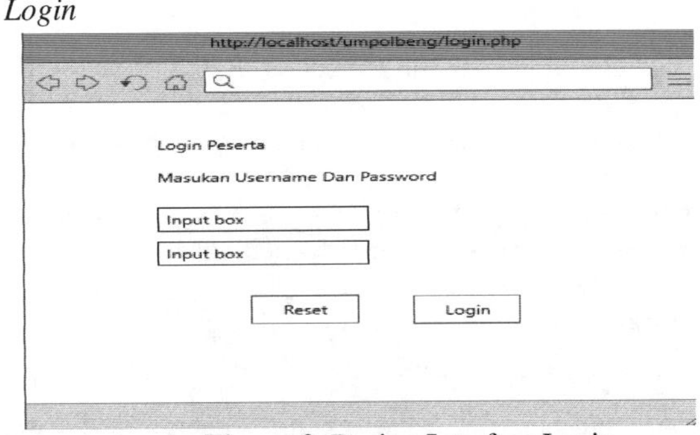

Figure 2. Design Interface Login

4.2. Administration staffInterface Design

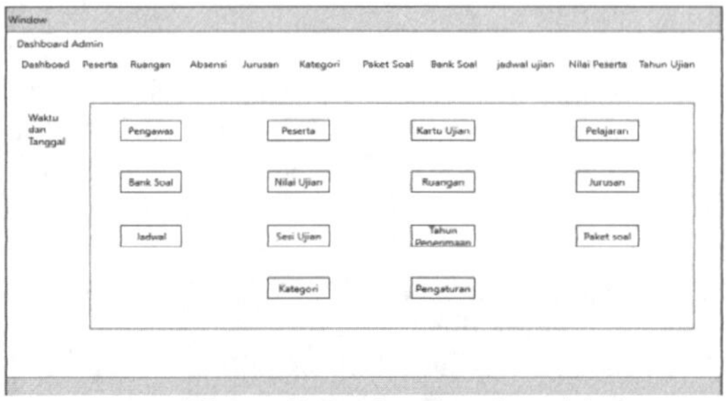

Figure 3. Design Interface Admin

4.3. Exam Supervisor Interface Design

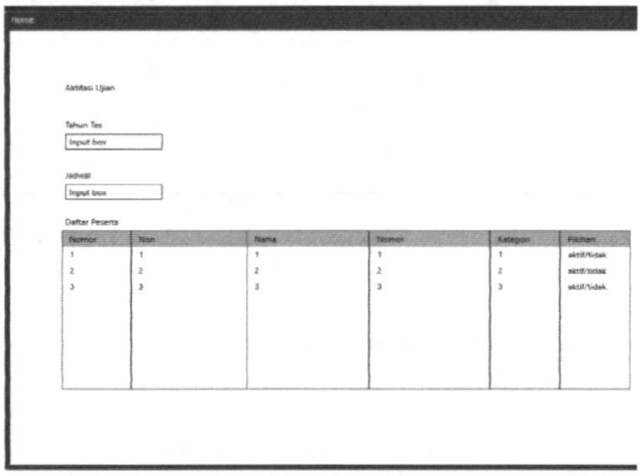

Figure 4. Interface Design Supervisor Exam

5. Conclusion

To undertake the designing application of Computer Based Test for admission of Politeknik Negeri Bengkalis. The first step that must be done is analyzing the needs of the system, then making the system design, next is making the system and the last one is testing of the system.

The system will be made with a procedural programming model and a web-based using the PHP programming code and MySQL database. Implementation environment using Apache web server. the network architecture that is used by Computer Based Test Application is a client-server network model with a Local Area Network (LAN). The results of designing in the form of application design Computer Based Test for admission Polytechnic of Bengkalis.

The design of this research using Fisher-Yates Shuffle method in randomize the problems so that the problems raised are different. As a further development suggestion, other randomisation methods can be used to randomize answers that arise so that in addition to raising different questions can bring up different answers

6. References

[1] Sorana-Daniela, B. and Lorentz, J. (2007). *Computer-based testing on physical chemistry topic: A case study. International Journal of Education and Development using Information and Communication Technology*, 3 (1), 94-95

[2] Ibijola, A., and Olu, A., 2012, *A Simulated Enhancement of the Fisher-Yates algorithm for Shuffling in Virtual Card Games using Domain-Specific Data Structures, International Journal of Computer Applications*, Vol.54, No.11 0975-8887 , page24-28.

[3] Arifin, 2009, *A Case Study of Use of Cryptographic Algorithms RSA algorithm As Safe, Mulawarman Informatics Journal*, Vol. 4, No. 3

[4] Maitimu, TR, 2008, *Design and Implementation of Web Server Load Balance Clustring and Scheme Using Linux Virtual Server via NAT*, Journal of Information Technology - AITI, Vol.5, No.1, p. 14-27

[5] Nugroho Adi. 2010. *Object-Oriented Software Engineering method USDP*. Andi offset: Yogyakarta

Effectiveness of Smartphone Application for the Development of Youth Anthusiasm to Malay Culture

Elvira Asril[1], Fajrizal[2], Fana Wiza[3]
Universitas Lancang Kuning, Pekanbaru, 28265, Indonesia
[1]Hp. 081275381981
E-mail : elvira@unilak.ac.id
[2]Hp. 08127602191
E-mail : fajrizal@unilak.ac.id
[3]Hp. 082284699447
E-mail : fana@unilak.ac.id

Abstract: This study will measure the effectiveness of Malay cultural applications, by socializing Melayu.com web, then distributing questionnaires to them (young people / high school students), and will be able to find out what features are of interest to them. With this smartphone introduction application of Malay culture, it is expected to increase young enthusiasm towards Malay culture which is really beautiful if it is known and understood. After the socialization of 30 high school students, the results obtained that they are less interested in the application. Because it is not user friendly, not interactive and rigid. Eventhough they have interest in this Malay culture,, they have not found an app or media that can attract attention. Thus, they ask if the application of Malay culture can add game content later, eg War games using background and other elements related to Malay culture.

Keywords: Efficiency, youth, Malay culture.

1. Introduction

Today, young people are more likely to like the outward cultural arts that seem to have globalized, as well as the influx of art from the media for example television. Some young people in some big cities have not studied or may not know the culture and art of the region than young people who live in small towns or villages that have been from small to learn the traditional arts of their respective regions. This study will measure the effectiveness of Malay cultural applications, then spread the questionnaires to students and students (young people), and later be able to find out what features are of interest to them. This media can also be used as a teaching material for teachers in schools. With this smartphone introduction application of Malay culture, it is expected to increase young enthusiasm towards Malay culture which is really beautiful if known and understood.

For some people, a smart phone is a phone that works with all operating system software that provides standard and fundamental relationships for app developers. For others, the smart phone is just a phone that features advanced features such as email, internet and e-book reading or a keyboard (both as finished and outgoing) and a VGA connector. In other words, the smart phone is a small computer that has the capability of a phone. Culture is a legacy of our ancestors or ancestors that are priceless. The Indonesian state is called a maritime country because it is surrounded by many islands, Indonesian culture is very

IOP Conf. Series: Earth and Environmental Science **97** (2017) 012007 doi:10.1088/1755-1315/97/1/012007

much and diverse, that culture is what we should keep and we preserve to be extinct or even claimed by other countries. (I Made Narthana). Malay culture is a hereditary culture carried out by the community. Malay culture is one of the pillars supporting the Indonesian national culture in particular and the culture of the world generally, in addition to various other cultures (Isjoni, 2007). Malay culture thrives and is thick in the midst of Indonesian society. Sancin, Director of the Socio-Cultural Cross Sapir Institute (January 5, 2009) argues that Malay that is synonymous with religion, language and custom is a solid integrity. (Khaerul Sobar)

2. Methods

Type of research conducted in this study is experimental research and spreading the questionnaire. This study also conducts an evaluation, aims to examine the travel process of a program as well as elaborate facts that are complex and involved in the program. For data collection in the form of Observation Technique, Library Study Technique, Questionnaire, System Requirement Analysis, Creating System Modeling.

3. Results

Here is an app view containing Malay cultural content. Already many links and content, with features quite interesting from the website melayuonline.com.

Figure 1. melayuonline.com

After the socialization of 30 high school students, the results obtained that they are less interested in the application. Because it is not user friendly, not interactive and rigid. But actually according to them, there is interest to this Malay culture. However, they have not found an app or media that can attract attention. And they ask that if later made application of Malay culture, to add game content related to Malay culture. Eg War games that use background and things related to Malay culture.

Table 1. Recapitulation of questionnaire

NO	Question 1	Question 2	Question 3	Question 4	Question 5
1	4	3	4	2	1
2	4	2	4	2	1
3	3	3	4	3	1
4	3	3	4	2	1
5	4	3	4	2	1
6	4	3	4	2	1
7	4	3	4	2	1
8	4	3	4	2	1
9	3	3	4	2	1
10	4	2	4	2	1
11	3	3	4	2	1
12	3	3	4	2	1
13	4	2	4	2	1
14	4	3	4	3	1
15	4	3	4	2	1
16	4	3	4	2	1
17	3	3	4	2	1
18	3	2	4	2	1
19	4	3	4	2	1
20	4	3	4	2	1
21	3	3	4	2	1
22	3	3	4	2	1
23	3	3	4	2	1
24	3	3	4	2	1
25	3	3	4	2	1
26	3	3	4	3	1
27	3	2	4	3	1
28	3	2	4	2	1
29	4	3	4	2	1
30	3	3	4	2	1
Average	**3,47**	**2,80**	**4,00**	**2,13**	**1,00**

Based on the above recapitulation it can be seen as lies in the points below:

1) Appearance of online malay culture is not interesting.
2) Content on the application is sufficient.
3) The use of animation and multimedia is not very interesting.
4) With the online application of Malay culture, the youths become aware of a bit of Malay culture.
5) Additional game content is needed for them, in order to attract youth interest in Malay culture.

4. Design

Based on the results of socialization and questionnaires conducted, the researchers designed the use case diagram an application that adds game content and in use animation or multimedia that will attract the attention of young children.

4.1 Use Case Diagram

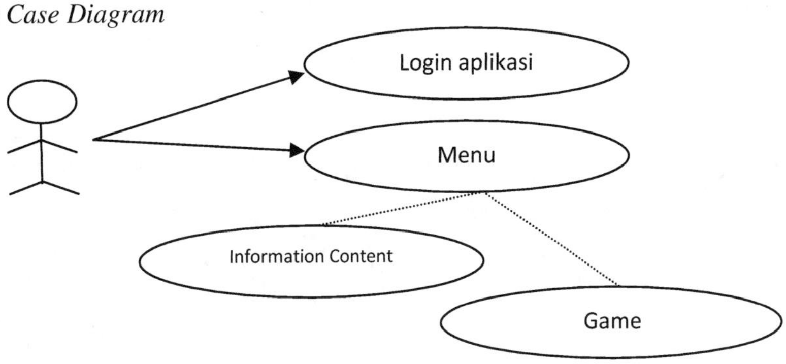

4.2 Class Diagram

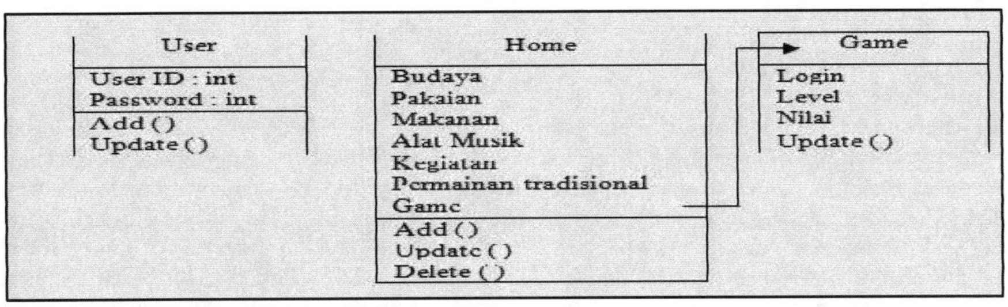

5. Conclusion and Suggestions

The paper would like to conclude into three major conclusions namely:

a. After introducing the application of Malay culture to young people (high school students) Pekanbaru, it can be seen that the application does not make them enthusiastic, because the application is rigid and not interactive.
b. Through questionnaires distributed, it is concluded that young people expect that applications containing Malay cultural content is more flexible and having interactive games.
c. The proposed application design can attract the attention of young people and hope that the application is immediately made.

Therefore this paper suggests the following items

a. Existing applications can now be developed (updated) to be more user friendly and there is game content in order to attract the attention and interest of young people.
b. With the vision of Unilak and Vision of Riau, which put forward the Malay culture it is expected that applications containing Malay culture (update) can be used as one of the teaching media in schools and campuses

6. Reference

[1] Andrew Campbell and Tanzeem Choudhury. 2012. *From Smart to Cognitive Phones Pervasive Computing.* IEEE C5 2015. 1536-1268/12/$31.00 © 2012 IEEE
[2] Didik Dwi Prasetyo, 2004. *Solusi Pemograman Berbasis Web Menggunakan PHP 5* Jakarta: Elex Media Komputindo.
[3] Jogiyanto H. .2005. *Analisis dan Desain Sistem Informasi*
[4] Munawar,2005,*Pemodelan Visual Dengan UML.*Graha Ilmu,Yogyakarta.
[5] Nugroho, Bunafit. 2004. *PHP dan MySQL* . Yogyakarta: Dreamweaver MX
[6] Riswanto, Eko.2007. *Analisa dan Perancangan Sistem Informasi.*Yogyakarta.
[7] Sibero, Alexander F.K. 2011. *Kitab Suci Web Programming.*Yogyakarta.
[8] Buku Aplikasi Berbasis Android, 2015, Nazruddin Safaat H in Android.

Natural Fiber Cut Machine Semi-Automatic Linear Motion System for Empty Fiber Bunches: Re-designing for Local Use

Asfarizal [1)], Anwar Kasim[2)], Gunawarman[3)], Santosa[4)]

[1)] *Padang Institute of Technology,Padang, 25142, Indonesia*
[2,3,4)] *Andalas University, Padang, 25163, Indonesia*
[1)]*e-mail: asfarizalsaad@yahoo.com*

Abstract: Empty Palm bunches of fiber is local ingredient in Indonesia that easy to obtain. Empty Palm bunches of fiber can be obtained from the palm oil industry such as in West Pasaman. The character of the empty Palm bunches of fiber that is strong and pliable has high-potential for particle board. To transform the large quantities of fiber become particles in size 0-10 mm requires a specially designed cut machine. Therefore, the machine is designed in two-stage system that is mechanical system, structure and cutting knife. Components that have been made, assembled and then tested to reveal the ability of the machine to cut. The results showed that the straight back and forth motion cut machine is able to cut out the empty oil palm bunches of fiber with a length 0-1 cm, 2 cm, 8 cm and the surface of the cut is not stringy. The cutting capacity is at a length of 2 cm in the result 24.4 (kg/h) and 8 cm obtained results of up to 84 (kg/h)

Keywords: straight motion, cutting knife, oil palm empty bunch fiber, particle

1. Introduction

Natural fibers for composite particle board requires the same fiber length uniformity and a high number of fibers needs to produce, but the amount available is very less. One of the raw fibers is the empty palm oil bunch (EFB) having a different length and diameter (0.4-0.6 mm), to obtain uniform length of fiber in large quantities can not be cut manually due to long and limited amounts. Therefore required a fiber cutting machines are semi-automatic long cut the fibers can be arranged as needed so that the retrieved long fibers are uniform. This study identifies the physical and mechanical properties of EFB fiber from the two methods of obtaining fibers i.e. boiling and evaporation to design the proper Fiber Cut Machine with Tensile strength 10.675 MPa and 10.7373 MPa. This design aims to produce a natural fiber cut machine especially EFB, for the application of composite materials cut machine with linear drive system using a knife cut made a spring leaf scars from the car.

2. Basic Design

The design basic principles of fiber cut machine works by utilizing a twist moment was moved by the rotating axis of roll chain mover blade and thruster. Fibers are vacuumed into the hover and impelled by thrusters beyond fiber will then cut by knifes. The cuts will take place continuously with the cutting length can be arranged. As for the details of principles fiber cut machine work is as follows:

1) Drive systems; Electric motor as a driving force that moves the blade shaft and thrusters. Using rond mover system transmission roller chains with the sprocket. Movers blade and impeller fiber twist your next exploit using the crankshaft which turn it into a linear motion in forward connecting rod which is directly connected with a knife cut and thruster
2) The process of use; The cut tool is quite simple in operation, because it is only driven an electric motor, the material entered into the hover, the material is pushed and then cut by the cutting blade.
3) Cutting Process; The cutting process begins with the encouragement of the thrusters and so on will be cut by knife cut. Working system between the motion of the blade and thruster motion of fiber that is when thrusters on the driving end position and will return to the starting position, at that time also a position of the knife is the beginning of the cuts.

The shaft; along with power forward rotation or twist your moment. The shaft is the main component in the transmission serves as the successor to power through a common round sectional round, attached to other machine elements also like fully, flywheel and other power mover elements. This gets kind of shaft load pure twist or twist and bending.

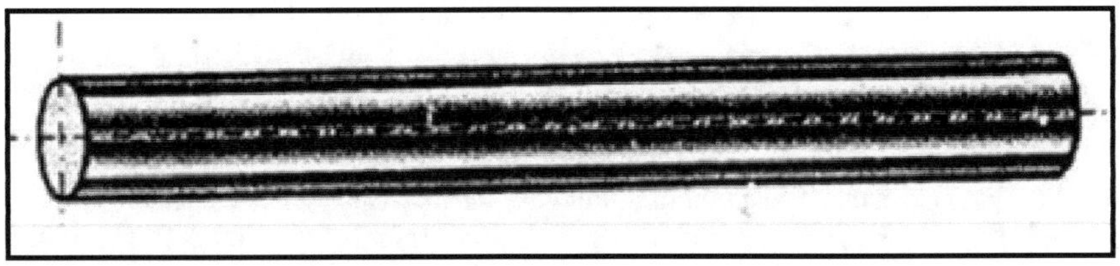

Figure 1. The transmission shaft

The Chains transmission and Sprocket; In its most simple form a chain drive consists of a chain and two sprockets – driver and driven. Drivers operating under severe load and at high velocity are encased in housing and provided with a lubrication system

The principal advantages of chain drives are the follows:

a. The effective transmission of power when the distance between the shaft centers is large (L_{max}. 8 m)
b. High efficiency reaching. ($\eta = 0.98$)
c. Less load on the shaft compared to a belt drive
d. The ability of one chain to transmit motion to several shafts.

Chain drives are used in agricultural machinery, bicycles, motor cycles.

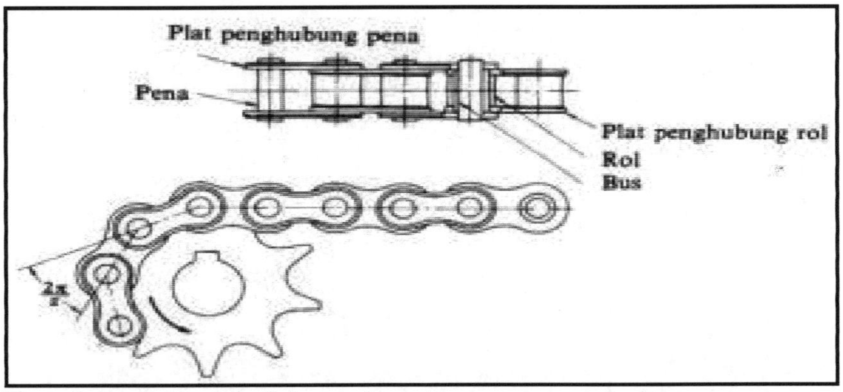

Figure 2. Transmisi rantai roll dan sprocket

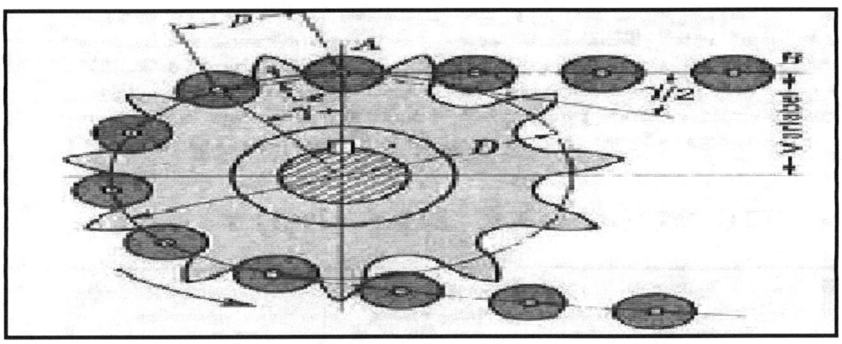

Figure 3. Chain roll dan sprocket (Dimension)

Chain speed v (m/s) can be calculated from

$v = (p.\ z_1.\ n_1)/60.1000$ 1)

The loads acting on one chain F (kgf) can be calculated as in the belt with the formula

$F = (102.\ Pd)/v$ 2)

3. Results

The Blade Cut; The materials used is medium carbon steel. the procedure of making a knife cut begins with warming up to 900 degree Ceclcius, the rolling until thick 3 mm, heat treatment (quench).
The composition of the medium carbon steel is, namely:0,457%C; 0,7304% Mn; 0.0240% S; 0,0257% P[4]. The tensile strength of leaf spring is 1332.5 kg/mm^2 [5]. The dimensions of the leaf springs with a length of 250 mm, wide: 72 mm and thick: 3 mm.

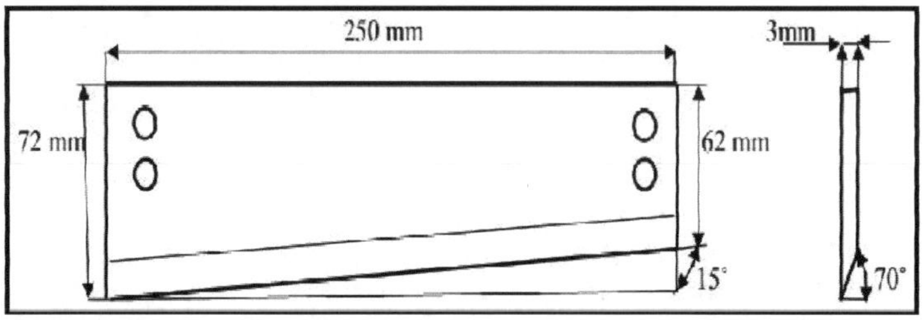

Figure 4. The blade cut-leaf spring

The style evoked the knife (F_{knife});

$F_{knife} = m\ V^2/(R)$ 8)

Analysis of the main components and motor power were, the style of cutting machine fiber consists of a few main components that are assembled into one main part. The components of this natural fiber cutting tools, namely: the style of cut fibers, electric motor, pull and belt, sprocket and chain rollers.

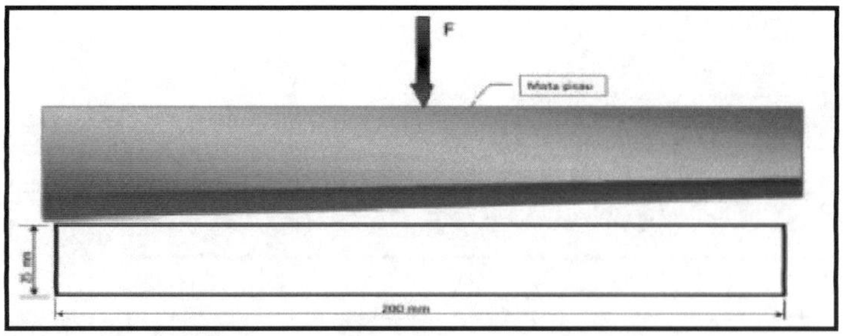

Figure 5. The total fiber cross-sectional area cut 25 x 200 mm

The tensile strength fibers is 10.7373 MPa [1], used to determine the strength to cut fiber:
$\tau_{shear} = (0.5 - 0.7)\ \sigma$; τ_{shear}, taken = 0.5 x 10.7373 = 5.3686 MPa or 5.3686 N/mm^2. Cut one strand of fiber force (F_{cut}) = τ shear . A = 5.3686 N/mm^2. (0,785 . 0.54^2) mm^2 = 1.23 N

The cutting is done with a knife cut fibers are reduced cut force of 50-75%, the average reduction in the style of cut 65%. cuts force: 0.35 x 12.3 N = 0.4305 N. Cut the effective cross-sectional area designed is Atotal = 20 mm x 100 mm = 2000 mm2, then cutting the total fiber forces will be obtained and if the cross section cut no cavities, Ftotal = 0.4305 N/mm2 x 2000 mm2 = 861 N

The speed of the belt and chain; the transmission from the motor is used belt V then proceed with the transmission Chain rollers, roller chain used for mountain bikes. Determine the reduction (i), on the design of this round of 1400 rpm motor, therefore needs to be reduced to ± 60 rpm at the crankshaft cutting fibers.

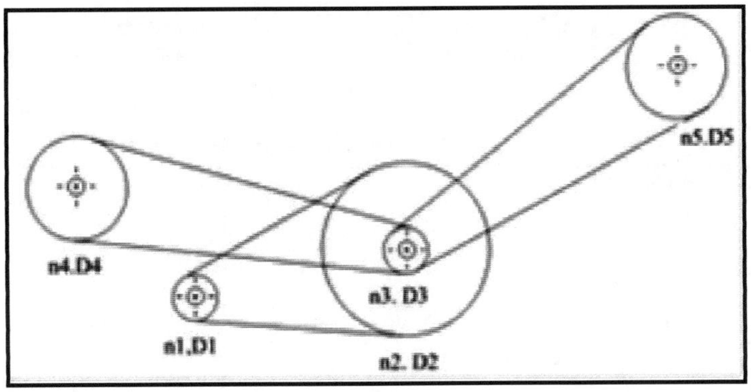

Figure 6. Chain and belt transmission system

Preliminary data are available:
n_1 = 1400 rpm, $D_1 = D_2$ = 2 in, 12 in, z_3 = 12, $D_4 = D_5 = Z_4 = Z_5$ = 48
Puli 1 and 2: $i_1 = D_2/D_1 = n_1/n_2$ = (12 in)/(2 in) = (1400 rpm)/(n_2 rpm), retrieved n_2 = 233 rpm
Sprocket 3 and 5: $i_2 = z_5/z_3 = n_3/n_5$ = (48)/(21) = (233 rpm)/(n_5 rpm), retrieved n_5 = 58 rpm
The movement of the blade cut and ascending: 58/minute
Then itotal = n_1/n_5 = (1400)/(58) = 24
Determine the linear Speed of the belt (v)
v = (π. d_p. n_1)/60.1000 = (π. 50,8.1400)/60.1000 = 3.71 m/s
Determine the speed of cut knife (v)
Known D_{shaft} Crankshaft = 60 mm, n = 58 rpm
v = (π. dp. n_1)/60.1000 = (π. 60 mm. 58 rpm)/60.1000 = 0,182 m/s

Twist moment Motor; selected motor low-power: Power = 0.5 Hp = 0.373 Kw rotation of the motor (n_1) = 1400 rpm, rotation of the blade cut (n_5) = 58 rpm, crankshaft Diameter d_5 = 60 mm, the force on the blade (F_{total}) = 861 N

Torque motor (T_m) = 9.74,10^5.0.373/1400 = 259 kgmm the torque on the shaft of the blade cut (T_5) are: T_m/T_5 = (n_{blades})/(n_{motor})
$T_5 = (T_m \cdot n_{motor})/(n_{blades})$ = (259 kgmm. 1400 rpm)/(58 rpm) = 6263.82 kgmm
So that resulting cut force of produced electric motors in the F_5 was

$$T_5 = F_5 \times \frac{D}{2}$$

F_5 = 6263.82 kgmm x 2/60 mm = 208.7 Kg. = 2047.347 N
So with ½ Hp motor capable of use for the powerplant and cutters because $F_5 \geqq F_{total}$

The result of the testing carried out to know the cut machine cutting capacity per hour, i.e.:

IOP Conf. Series: Earth and Environmental Science **97** (2017) 012008 doi:10.1088/1755-1315/97/1/012008

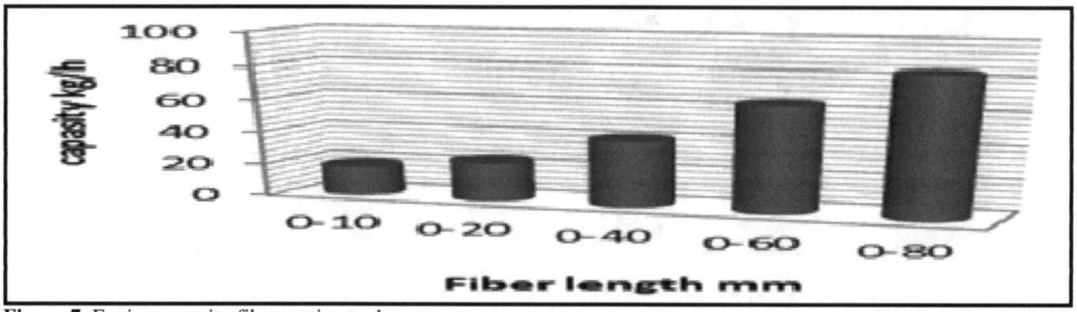

Figure 7. Engine capacity fiber cutting tools

Testing the Fiber Cutter machine respectively for 0.5 hours cutting results obtained with a length ranging from 0-10, 0-20, 0-40, 0-60 and 0-80 mm shown in Figure 7 Even though its capacity is increasing every hour, more dominant is caused by different fiber lengths. The cut surface of the fiber is not stringy and the knife cut able to survive during the cutting.

4. Conclusion

As for the conclusion of the spring fiber cut machine design this is the force of cut fibers of 861 N, motor power used enough 0.5 HP, with force of cutting 861 N. The cut surface of the fiber is not stringy and the knife cut able to survive during the cutting and do not need sharpening. The length of the fiber can be arranged according to the wishes and the shortage of this cut machine is the absence of clamp fibers that have an impact on the work of the blade cut. Therefore this study proposed this model of the natural fiber cut machine semi-automatic linear motion system for empty fiber bunches:

The Final Design

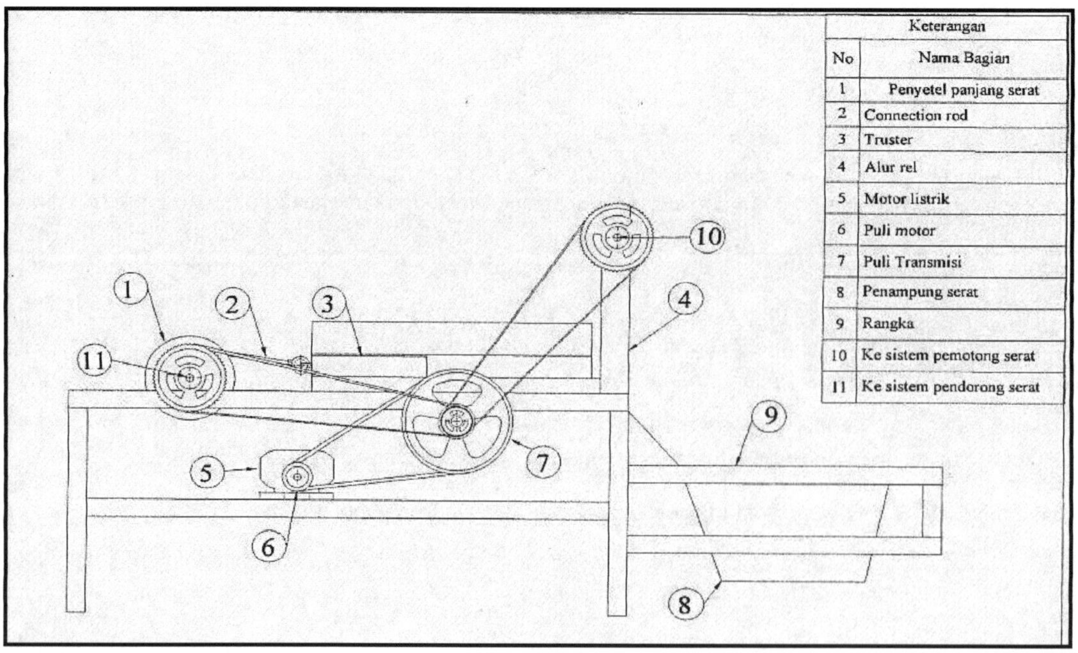

Figure 8. The Final design

Figure 8. The Final design after manufacturing

Reference

[1] L. U. A. H. Agustina, "Karakteristik serat tandan kosong kelapa sawit (tkks) dengan perlakuan perebusan dan pengukusan," ZIRAA'AH, vol. 41, pp. 97–102, 2016.

[2] J. E. Shigley, Mechanical Engineering Design, 4th ed. Michigan: University of Michigan, 1984.

[3] K. Suga Sularso., Dasar Perencanaan dan Pemilihan Elemen Mesin. Jakarta: Paradya Paramita, 1980.

[4] T. Hidayat, "Analisa Kegagalan Pegas Daun (leaf spring) pada Toyota Kijang Kapsul 7K-EI tahun 2000."

Gardens on the Arid Climate

Weldy Eka Saputra
University of Bahrain, Isa Town, Kingdom of Bahrain
Phone: +973 33135925
E-mail: wsaputra@uob.edu.bh

Abstract. Bahrain is located in the climate of the arid zone which rainfall is low and irregular. This paper discusses the approaches which response to the local context that has been implemented by the government of Bahrain to sustain the quality of the public garden in the arid climate, turning to green. Generally, the approach is an improvement in the central treatment of waste water system plant that used to irrigate the landscaping, agriculture as well as for industry use. These approaches are not the only technologically, but also involves the participation of community to achieve sustainable garden in this country.

Keywords: arid climate, Bahrain, desert garden, sustainable, wastewater

1. Introduction
Bahrain means 'two seas' consist of 33 islands which area about 780 km^2 is located off the coast of Saudi Arabia in the west and Qatar Peninsula in the east as shown in Figure 1. A long time ago Bahrain was recognised as a centre of pearl trading in the world, as well as the first country where the crude oil discovered in the Gulf region on 1932. Because of the strategic location in the Persian Gulf, Bahrain was under influence of some civilisations in the world such as Sumerians, Persians, Assyrians, Babylonians, Portuguese, Arabs and British.

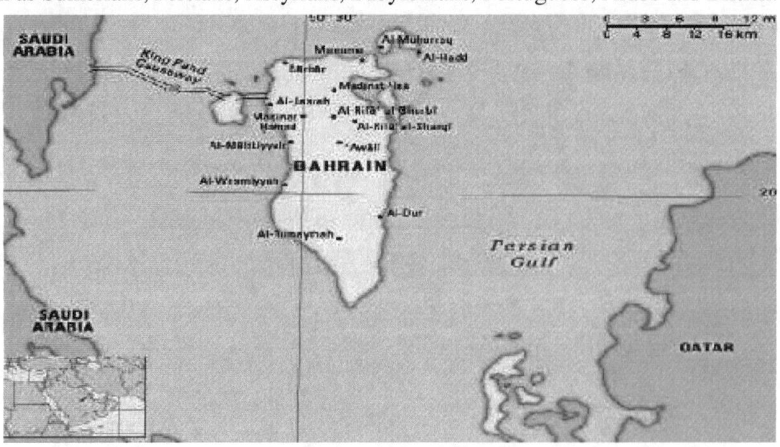

Figure 1. Location of Bahrain

The Bahrain's location also gives influence into the Bahrain climate which comes from Saudi Arabia, Iraq and Iran. According to the Bahrain's Meteorological Directorate, Bahrain's climate is categorised as the arid climate, the rain volume is less than 71mm and irregularly, as well as the average temperature is 17°C-36°C, but sometimes the temperature is extremely more than 40°C and the average of relative humidity is 39%-88%. This climate challenges for plants to sustain for long period. The good maintenance, as well as appropriate approaches, base on the local climate, needs to sustain the growth up and age of plantations.
The aim of this study is to understand the sustainable approaches in the garden (public gardens) according to the local context, the arid climate zone. This study has some objectives: (1) Investigate the sustainable approaches in the arid climate zone, (2) Contribute as a reference in a sustainable development in the arid climate, (3) Share information the current technology is using in the vegetation irrigation in arid climate zone.

Generally, the meaning of arid climate is insufficient rainfall volume (less than 250mm) or lack of water to support plantation grow up, a short word called "dry". This situation is raised because of some factors may be distinguished (Wescoat, 1996): (1) *Latitude of the region*, (2) *Effects of rain shadow*, (3) *Condition of continental*, (4) *Ocean temperature*, (5) *High altitudes and polar latitudes.* Excessive evaporation can cause drought on the soil that can inhibit plant growth. This situation becomes a disadvantage for plants to grow for long times. Some actions must be taken to break out of this unfortunate situation.

On the 1980s, the Xeriscape system introduced by Denver Water which means 'xeros' (dry) and 'scape' (landscape). This dry landscape system saves the water, the environment protection, and cost less in the landscape maintenance (Lyle, 1996). There are seven principles in Xeriscape system gardens:

(1) Plan and design. Inventory and analysis of existing site situation are necessary. All data must collect such as drainage flow, types of soil, existing plants, exposures (including soft scape, hardscape, building, etc). In this point also have to consider the plan and design the future garden such as pathway, hardscape and soft scape area, etc. These components, existing and new development, should be integrated into the whole design of the garden.

(2) Soil improvement. One of the important component in the garden design is soil. The analysis and evaluation of soil condition should be done to understanding the soil characteristics, otherwise, the whole design of garden will be useless. The weak soil or unfavourable to grow crops can mix fertiliser into the soil to improve the soil quality.

(3) Efficient irrigation. Well-planned of placement of sprinkler outlets to irrigate soil or plants directly as well as to avoid runoff waste of water. The grouping of plants with same characteristics also help effectiveness in irrigation network.

(4) Proper plant selection. The low water requirement plants should consider. It does not mean impossible to other plants, as long as planted on appropriate location and not to interplant them with lower water requirements plants.

(5) Practical grass area. The grass is the best separator and watered separately from other plantation such trees, flowers, ground cover, shrubs, etc to decrease irrigation load. The grass area can use for the playground, sitting area, pathway, etc. It also can be replaced by ground cover low water consumption such as rubber material.

(6) Mulching. Mulches are needed to reduce rapid evaporation as well as the surface temperature of the soil. Also, the mulches can increase the aesthetic of the garden in proper materials and placement.

(7) Proper maintenance. The maintenance is not only related to mechanically or equipment and hardscape but also maintenance of the plant (soft scape). Regularly maintenance is needed to maximise implementation of these principles.

In general, the old towns in Arabia consist of narrow pedestrian streets and dense spaces inter-building. By exploiting the movement of the sun there will be shady areas around the building that can be used to create a public garden or private garden. The shady area also can reduce the surface temperature of the soil. Of course, water irrigation still needed to sustain the plants grow. It can use recycle grey water as apply in Bahrain.

IOP Conf. Series: Earth and Environmental Science **97** (2017) 012009 doi:10.1088/1755-1315/97/1/012009

2. Method

This study adopted the method that can categorise into two parts: (1) Literature studies regarding the sustainable garden in an arid climate. In this part was carried out to understand several kinds of literature regarding the issues of an arid climate garden, and (2) Investigation of some sites of the public parks and housing neighbourhood in Bahrain, and documentation.

3. Water Resources and Technology in Bahrain

Natural freshwater was utilised in Bahrain for agriculture and domestic purposes. The water was flew freely from the Northern area before 1925 (AQUASTAT website, 2016). On 1976, Bahrain's government commenced the desalination program of seawater which built desalination plant blended with ground water before supply to consumers (Al Noaimi, 2005). The annual growth of water use in Bahrain is about 8 % to 10% (Ahmed, 2010) because of population increase. For fulfil the demand of water, the Bahrain's Government built a re-use sewage water plant, Tubli Water Pollution Control Centre (Tubli WPCC) which total area is about 320,000 m^2 (Figure 2) to separate supply of water demand, domestic use and irrigation. This plant uses to recycle sewage water and to irrigate landscape and agriculture, started on 1985.

Figure 2. Tubli Water Pollution Control Centre

This Tubli plant consists of three types of treatment: primary treatment, secondary biological treatment and tertiary treatment, which divided into some sub treatment functions: the collection chamber, aeration ponds, settling tanks, sludge pumping station, sludge drying ponds, pipes and channels leading to the sea, ozonation facility (added on 1987), filtration complex, and chlorine additives complex for reuse of the treated wastewater for agriculture (Al Noaimi, 2005). Recently, this plant upgraded on the secondary treatment unit, using HYBACS (Hybrid Activated Sludge) technique which deploys SMARTTM reactor units that stimulate a special and specific bacteria. This technique has improved the quality of the treated water for agriculture and landscape irrigation as well as decrease the negative impact to ecology and Tubli environment. According to the Blue Water Bio company (the company that create HYBACS system), this system is 40 times faster clean the sludge compared to conventional sludge system. Also this system save time, cost effective, no interruption the existing plant processes during construction of this system, and increase the capacity or performance of treated waste water in the existing plant.

4. Gardens in Bahrain

Similar to another country in the arid climate zone, create and sustain the garden quality are a challenge. The concept of Bahrain's public garden mostly is for physical relaxation as well as a place for spiritual contemplation. This garden also provided some facilities such as kids playground, grass area for sitting, an open stage for events, sports track (jogging, cycling, walking), café, and visitor parking as shown on Figure 3. In some public garden, also provide specific facility such as family zone where family privacy is keep far away from public activities in the garden. A public garden (Arad Park) is proposed to support marine ecosystem conservation (Figure 4).

Figure 3. Khalifa Park, Hidd Area

Figure 4. Arab Park, Arad Area

Most of the public garden in this country use Xeriscape system which planned and designed garden base in the local context, improvement on soil quality, choosing the appropriate types of vegetation according to the local climate such as Pachypodium, Cactus, Dates, Thyme, Boojum, Acacia, Frangipani tree and Coconut. Applies mulches on the soil and surrounding of the vegetation to reduce excessive evaporation and keep temperature of soil and treated wastewater for irrigation with sprinkler lay on the soil (Figure 5).

The government of Bahrain through the Agriculture Affairs under Ministry of Works, Municipalities Affairs and Urban Planning has made a list of aromatic vegetation able to plant in the local garden (Table 1).

Table 1. List of several Aromatic Plants in Bahrain

English Name	Latin Name
Fennel	*Foeniculum Vulgare Miller*
Rosemary	*Rosmarinus Officinalis*
Turonj	*Citrus Medica L*
Lemon	*Citrul (L) Aurintifulia Swingle*
Blue/Red Gum	*Eucalyptus Camaldulensis L*
Jasmine	*Jasminum Grandflurum L*
Basil	*Ocimum Basillicum L*
Roses	*Rosa Sp-L*
Indian Jasmine	*Plumeria Acutifolia L*
Lemon Grass	*Cymbopogon Citratus Stapf*
English Name	**Latin Name**
Cummin	*Cumminum Cyminum L*
Parsley	*Petroselinum Sativum Hoffm*
Aniseed	*Pimpinella Anisum L*
Sage	*Savia Officinalis L*
Celery	*Apium Graveolens L*
Mary Gold	*Calendula Ofticinalis*
Mint	*Mentha Viridis L*
Majoram	*Majorana Hortensis*
Thyme	*Thymus Capitataus (L) Link*
Germander Mt.	*Teucrium Polium L*
Rose Geranum	*Pelargonium Odoratissmum*

The periodical maintenance also applies on this public garden. This maintenance is not only for big public garden, but street green area also do peridically. The main purpose of this maintenance is to maintain quality of public garden. Replacement of dead plants is also part of the maintenance work. Community participation also involve in public planning and implementing of the public garden to ensure the sustainable environment such as attend town meeting in creation and maintenance of public gardens as well as spreading awarness on people's roles in safeguarding their green spaces

Figure 5. Irrigation sprinkler lay on the soil

5. Conclusions and Suggestions

Climate has important influence in gardening activities as well as development to achieve sustainable goals which encourage to create appropriate system that can use in local garden development as well as improvement according to the existing situation. One of the appropriate garden system is Xeriscape that have been use in some arid climate regions and has possibility to use at outside of arid climate regions. However, the maintenance is one of the key succees to produce a sustainable and comfortable garden for humans. Water is important element on human life and also create human civilization since years ago. Fresh water is not only consume for human daily life, but also to help the plants growth. Through technology, the using of recycled water can replace fresh water for irrigation and it can save cost for long time. Also, the future research regarding recycle water technology should be encougared to produce appropriate and sustainable technologies in human life, especially in garden and landscape irrigation use as well as implementation to improve local garden quality.

6. References

[1] Crystal, Jill Ann & Smith, Charles Gordon. 2017. *Bahrain*, viewed 21 June 2017. http://www.britannica.com/place/Bahrain

[2] Ministry of Transportation and Telecommunications. 2014. *Bahrain Climate: 2014,* viewed 19 June 2017, http://www.bahrainweather.gov.bh/web/guest/climate

[3] Wescoat Jr, James L. 1996. A Geographical Perspective On Landscape Design in Arid Environment. *The Aga Khan Trust for Culture Symposium: Sustainable Landscape Design in Arid Climate*, pp.11-23

[4]] Lyle, John Tillman. 1996. Archetypes in The Arid Landscape. *The Aga Khan Trust for Culture Symposium: Sustainable Landscape Design in Arid Climate*, pp.25-33.

[5] FAO. 2016. *AQUASTAT website, Food and Agriculture of the United Nations, Bahrain*, viewed 21 June 2017. http://www.fao.org/nr/water/aquastat/countries_regions/Profile_segments/BHR-WR_eng.stm

[6] Al Noaimi, Mubarak A. 2005. *Water Use and Management in Bahrain: An Overview.* A Country Paper Presented at The Eleventh Regional Meeting of the Arab IHP National Committee. Damascus, September 2005, viewed 03 April 2017.

[7] Ahmed, Rehan. 2010. *Bahrain Water Consumption Soars. Trade Arabia*, 23 March 2010, viewed 20 June 2017, http://www.tradearabia.com/news/MISC_176867.html

[8] Ministry of Works, Municipalities Affairs and Urban Planning. *Tubli Wastewater Treatment Plant Upgrade Projec*, viewed 21 June 2017, https://www.works.gov.bh/English/Projects/sanitary-projects/Pages/TUBLI-WASTEWATER-TREATMENT-PLANT-UPGRADE-PROJECT.aspx

[9] Blue Water Bio. *HYBACS: Hybrid Activated Sludge*. brochure. Blue Water Bio: London

[10] Ministry of Works, Municipalities Affairs and Urban Planning. 2011. *Aromatic Plants in The Kingdom of Bahrain*, viewed 18 June 2017, websrv.municipality.gov.bh/north/pages/herbal_en.jsp

[11] Wisandanu, Rian. 2013. *Bahrain Great Parks for Picnic and Exercise. Indonesian Expatriate in Bahrain blog*, web log post, 24 August, viewed 23 June 2017, http://indonesianinbahrain.blogspot.com/2013/08/bahrain-great-parks-for-picnic-and.html

Determinants and impacts of public agricultural research in Japan: Product level evidence on agricultural *Kosetsushi*

Nobuya Fukugawa[1]

[1] *Graduate School of Engineering, Tohoku University*
6-6-11-804 Aramaki Aoba-ku Sendai 980-8579, Japan
fukugawa@tohoku.ac.jp

Abstract: The public sector is an important source of agricultural research as the agricultural sector in many countries consists of a number of individual farmers who have difficulty in bearing the cost of research and development. Public institutes for testing and research called *Kosetsushi* help agriculture and manufacturing improve labor productivity through technology transfer activities, whereby constituting an important component of regional innovation systems in Japan. This study establishes panel data of agricultural *Kosetsushi* and examines whether their research activities are responsive to local needs and which type of research effort is conducive to the promotion of agricultural product innovations. Estimation results reveal variations across plants in the impacts of agricultural clusters on research on the plant conducted by *Kosetsushi* located in the cluster. A positive impact is observed only for vegetable while negative or statistically insignificant relationships are found for rice, fruit, and flower. The impact of research on plant breeding on agricultural product innovations also varies across plants. Policy implications of the major findings are discussed.

Keywords: *Kosetsushi*; Japan; agricultural extension; new variety of a plant; regional innovation systems; product innovations; technology transfer

1. Introduction

The public sector is an important source of agricultural research for several reasons. First, agricultural research requires more time than research in physical-science-based sectors as it takes time to repeat experiments through growing a number of plants and selecting a few with appropriate characteristics. Second, price elasticity of demand to agricultural products, such as staple foods, tends to be small, which implies that a demand curve is steep. Furthermore, the elasticity of supply to price of agricultural products tends to be small as it is difficult for agricultural producers to increase production within a short period of time, which implies the steep supply curve. These demand- and supply-side conditions suggest that the impact of innovation tends to be absorbed by a great increase in consumer's surplus, leaving an increase in producer's surplus substantially small. Therefore, in agriculture it is more difficult for the private sector to appropriate the return to R&D than in other industries. The appropriation condition can be more difficult considering that the agricultural sector consists of a number of individual farmers who have no complementary assets. These supply-side and demand-side factors in innovation render the public sector a key source of agricultural research, and thus agricultural product innovations.

Public institutes for testing and research called Kosetsushi constitute an important component of regional innovation policies in Japan. Kosetsushi were initially established in the late 19th century in agriculture, textiles, and brewery (e.g., sake and soy sauce), and then gradually developed in manufacturing in general (Fukugawa and Goto 2016). An average prefecture has six or seven Kosetsushi headquarters (=318/47) in each prefecture (Appendix Table 1). Kosetsushi help small and medium-sized enterprises (SMEs) and individual farmers improve productivity through various technology transfer activities (Fukugawa and Goto 2016). They play three key roles in regional innovation systems. First, they diffuse technological knowledge through various routes, such as technical consultation and seminars for the introduction of new standards and technologies. Second, they conduct their own research, patent inventions, and license patents mainly to local SMEs. Agricultural Kosetsushi register new varieties of a plant in the same way as manufacturing Kosetsushi patent their inventions. Third, they act as a catalyst for local SMEs to develop innovative networks to external sources of knowledge, such as universities.

Regardless distinctive features in terms of regional embeddedness and research intensity, little has been known about structure and efficiency of research activities of agricultural Kosetsushi in regional innovation systems. Such tendencies contrast with an increasing research attention emerging mainly from developing countries to the design of effective agricultural extension services (Chi and Yamada 2002 ; Al-Sharafat et al. 2012 ; Cole and Fernando 2012 ; Hellin 2012 ; Issa and Issa 2013 ; Turner, et al. 2013 ; Kassem 2014 ; Agunga and Putra 2015 ; Asif et al. 2015 ; Elias et al. 2015 ; Carmen and Bautista 2016 ; Debnath et al. 2016). To fill this research gap, this study establishes panel data of research activities and new varieties of a plant registered by agricultural Kosetsushi and examines whether their research activities are responsive to local needs for research, research on plant breeding is efficiently transformed to agricultural product innovations, and whether research productivity varies across plants.

2. Method

This study estimates the two-way error components model described as follows.

$$Rplant_{ipt} = \alpha_1 + \beta_1 Clusterplant_{ipt} + u_{ip} + v_{it} + \varepsilon_{ipt} \quad (1);$$

$$\ln Y_{ipt} = \alpha_2 + \beta_2 \ln Rfield_{ipt-q} + u_{ip} + v_{it} + \varepsilon_{ipt} \quad (2).$$

Rplant denotes the number of researchers at agricultural Kosetsushi allocated to research on a specific plant i. Clusterplant denotes location quotients of a specific plant in a prefecture p where Kosetsushi are located. Y denotes the number of new varieties of a plant registered by Kosetsushi. Rfield denotes the number of researchers allocated to research on a specific field weighted by the proportion of Rplant to the number of researchers . Suffix p denotes a prefecture in which agricultural Kosetsushi are located. Suffix i denotes a type of a plant being developed (i.e., rice, flower, fruit, and vegetable). Suffix t denotes the year of registration of the new variety of a plant. Suffix q denotes a lag required for the research inputs to exert an effect on agricultural product innovations. One error component u denotes time-invariant individual effects. Another error component v denotes time effect which represents exogenous shocks to all Kosetsushi, such as a change in regulation. Last, ε denotes the error term. Information of Y is collected from the Database of Seed Registration and Application, compiled by the Ministry of Agriculture, Forestry and Fisheries. As shown in Table 1, the database provides information on ten plants. Only four of them (i.e., rice, flower, fruit, and vegetable) retain the sufficient number of observations for panel analysis. The balanced panel on which the present study builds is the time-series and cross-prefectural data. Concerning Equation 2 a dependent variable includes many zeros . Therefore, a random-effects Tobit regression model is used for estimation to analyze truncated data. Random effects model was selected as there is not a sufficient statistic allowing the fixed effects to be conditioned out of the likelihood.

The location quotient is defined as LQij=(Aij/Bj)/(Ai/B). Aij is the agricultural production of a plant i (i=1…10: the number of plants listed in Table 2), and in a prefecture j (j=1…47: the number of prefectures). Bj is the agricultural production in j. Ai is the agricultural production of i. B is the agricultural production in Japan. LQ is measured by plant and by prefecture. The more specialized the production in a specific plant, the greater the location quotient of that plant. For instance, if the agricultural production is concentrated more in rice than the population mean, LQij will be the highest in rice. Information of location quotient is collected from the Agricultural Income Statistics, complied by the Ministry of Agriculture, Forestry and Fisheries.

Information of research effort allocation is collected from the Ministry of Agriculture, Forestry and Fisheries “the Basic Survey of Agricultural Kosetsushi”. This survey provides two types of information on the allocation of research effort by plant and research field. The survey asked agricultural Kosetsushi to divide the number of researchers to fifteen research field, as shown in Footnote 4, and nineteen agricultural outputs, as shown in Footnote 3. In other words, the number of researchers R is described as where xi denotes the effort allocated to a research field i and yj denotes the effort allocated to a plant j. It should be noted that xi and yj can be decimal even though R is natural number.

Experts suggested in interviews I conducted in 2016 that q significantly varied across plants. To incorporate different lagged structures in the regression model, I have collected information of the “development lag” between commencement and completion of research projects associated with the breeding of a specific plant. Information is collected from the Annual Report of Agricultural Experiment and Research Institutes compiled by the Ministry of Agriculture, Forestry, and Fishery. This survey shows that average development lag is ten years for rice, 6.4 years for fruit, five years for flower, and 4.9 years for vegetable. An average lag between application and registration of new varieties of a plant is 2.8 years. Figure 1 shows that a lag between application and registration continued to decrease until 2011 and became longer after that. Taking an example of fruits, agricultural research inputs invested in 2000 exerted an effect on product innovations measured by registration of new varieties in 2009 (=2000+6.4+2.8), which was filed in 2006.

3. Result and Disccusion

Table 1 shows estimation results of Equation 1. Table 1 reveals the absence of a statistically significant relationship, except for fruit, between agricultural clusters of a specific plant and research effort by agricultural Kosetsushi on the plant. This may have to do with organizational characteristics of agricultural Kosetsushi. Unlike manufacturing Kosetsushi there is a clear division of labor between research and diffusion at agricultural Kosetsushi. Individual farmers are normally affiliated with the Japan Agricultural Cooperatives (JA). They let JA know about problems they face in everyday operation so that JA sends requests to the diffusion arm which is organized as a part of the prefecture. The diffusion arm is organized regionally as there are different agricultural clusters even within a prefecture. The subdivisions normally have a sufficient number of staff who can deal with problems local farmers encounter, such as diseases and pests. The diffusion arm transfers local research needs to agricultural Kosetsushi. The feedback from the diffusion arm constitutes one of the key factors for agricultural Kosetsushi to design a research plan . Based on local needs and requests from the local government and the major private buyers, agricultural Kosetsushi lays out the research plan and provides feedback to the diffusion arm which engages in the provision of solutions to individual farmers. Researchers at agricultural Kosetsushi do not directly engage in personal interactions with local clients (i.e., individual farmers), which creates the biggest difference from manufacturing Kosetsushi in terms of organizational structure. Concerning the relationship between organizational features and technology transfer activities, previous studies on technology transfer organizations (e.g., university-based technology licensing offices) show that the technology transfer productivity, measured by the number of patents licensed to firms and royalty revenue, is affected not only by physical factors, such as location, but also by organizational factors, such as incentive mechanisms designed for staff of technology transfer organizations (Lach and Schankerman

2008; Fukugawa 2009; Amico-Roxas et al. 2011; Adekunle, 2013 ; Hsu et al. 2015). In a similar way, the results suggest the possibility that a clear division of labor between research and diffusion makes research at agricultural Kosetsushi less responsive to local needs. It should be noted that there might have been different types of interactions between the diffusion and research arms according to plants they develop as Table 5 shows that potential impacts of organizational features are contingent on the type of a plant. Future study should collect information on the linkage between the diffusion and research arms by plant, such as vegetable and fruit which exhibit opposite results.

Table 2 shows estimation results of Equation 2. Table 2 shows that different types of research effort have different level of impacts on agricultural product innovations. Furthermore, there is a plant level variation in the way research effort exerts an effect on research outcome. In fact the development of new varieties of rice and vegetable requires agricultural Kosetsushi to arrange a broad set of research inputs (i.e., researchers) while no research effort in a specific field exerts an effect on product innovations in fruit and flower. Concerning the relationship between human resource development and technology transfer productivity, previous studies on technology transfer organizations (e.g., business incubators) show that the diversity and quality of human resources do affect the success of technology transfer organizations (e.g., the number of new firms created) but its impact is contingent on technological fields and life cycle stages of startups to which they aim to give support (Fukugawa 2013). In a similar way, the results suggest the possibility that the technology transfer productivity at agricultural Kosetsushi is contingent on plant level factors. The plant level factors would pertain not only to natural conditions, but also to economic conditions, such as appropriability. This implies that agricultural Kosetsushi play different roles in regional innovation systems according to these conditions which cannot be fully captured in the present study. Future study should incorporate detailed information to analyze the process of decision making on resource allocation at agricultural Kosetsushi.

4. Acknowlegement

This research was funded by the Japan Society for the Promotion of Science [15K03411], the Nomura Foundation, and the Murata Science Foundation

5. References

[1] Agunga, R. and Putra, R. 2015 Training *Needs of Indonesian Agricultural Extension Workers for the 21st Century: A Recommendation Based on a Field Study, Asian Journal of Agriculture and Development*, Vol. 12, No. 2, 45-57.

[2] Al-Sharafat, A., M. Altarawneh, and E. Altahat 2012 *Effectiveness of Agricultural Extension Activities*, American Journal of Agricultural and Biological Sciences 7 (2): 194-200.

[3] Asif, M., Khan, N., Khan, F. and Zakria, M. 2015 *Effectiveness of agricultural extension services in dissemination of information to the farming community of District Bhimber*, Azad Jammu and Kashmir, Pakistan Journal of Agricultural Research, Vol. 28 No.2, 210-216.

[4] Baig, M. and F. Aldosari 2013 Agricultural extension in Asia: *Constraints and options for improvement*, The Journal of Animal & Plant Sciences, 23(2): Page: 619-632.

[5] Carmen, D. and Bautista, O. 2016 *Nature and Extent of Extension Delivery on Postharvest Handling of Horticultural Perishables in the Philippines*, Asian Journal of Agriculture and Development, 13, 1.

[6] Debnath, A., R Saravanan, J Datta 2016 *Effectiveness of public agricultural extension services in Tripura state of North-East India, Economic Affairs*, Year : 2016, Volume : 61, Issue : 1 pp 153-158.

[7] Elias, A., M. Nohmi, K. Yasunobu, and A. Ishida (2015) Farmers' *Satisfaction with Agricultural Extension Service and Its Influencing Factors: A Case Study in North West Ethiopia*, Journal of Agriculture, Science and Technology, Vol. 18: 39-53.

[8] Fukugawa, N. *2013 Heterogeneity among science parks with incubators as intermediaries of research collaborations between startups and universities in Japan*, International Journal of Technology Transfer and Commercialisation, Volume 12, Issue 4, pp 231-262.

[9] Fukugawa, N. and Goto, A. 2016 *Problem solving and intermediation by local public technology centers in regional innovation systems: The first report on a branch-level survey on technical consultation*, RIETI Discussion Paper 16-E-062.

[10] Fukugawa, N. 2016 *Knowledge creation and dissemination by Kosetsushi in sectoral innovation systems: Insights from patent data, Scientometrics*, Volume 109, Issue 3, pp 2303–2327.

[11] Fukugawa, N. 2017 *University spillover before the national innovation system reform in Japan*, International Journal of Technology Management. forthcoming.

[12] Hsu, D., Shen, Y., Yuan, B., and Chou, C. 2015 *Toward successful commercialization of university technology: Performance drivers of university technology transfer in Taiwan*, Technological Forecasting and Social Change, Volume 92, 25–39.

[13] Issa, F. and Issa, M. 2013 *Building the capacity of agricultural extension personnel for effective implementation of agricultural transformation agenda in Nigeria*, Journal of Agricultural Extension, Vol. 17, No. 1, pp
78-88.

[14] Kassem, H. 2014 *Effectiveness of Different Agricultural Extension Methods in Providing Knowledge and Skills in Disease Prevention: A Case of Smallholder Poultry Production Systems in Dakhalia Governorate of Egypt*, Asian Journal of Agricultural Extension, Economics & Sociology 3(2): 91-107.

[15] Ofuoku, A. and Agbamu, J. 2013 *Leadership Effectiveness among Farmers' Self-Help Groups in Central Agricultural Zone of Delta State, Nigeria and Its Implication for Improved Agricultural Extension Service*, Asian Journal of Agricultural Extension, Economics & Sociology 2(1):
41-53.

[16] Takemura, K., Uchida, Y., and Fujino, M. 2014 *Extension officers as social coordinators: Comparisons between agricultural and fishing communities in Japan*, Psychologia, 2014.

[17] Turner, J.; Rijswijk, K.; Williams, T.; Barnard, T. and Klerkx, L. 2013

[18] *Challenges to effective interaction in the New Zealand agricultural research and extension system: An innovation systems analysis*, Extension Farming Systems Journal, Vol.

Pattern of Cleanliness with Technology Intervention for Innovation Life

Sitti Hutari Mulyani[1], Billy Hendrik[2,] Rio Andhika Putra[3] and Mardhiah Masril[4]

[1,2] Universiti Kebangsaan Malaysia, Bangi, 43600, Malaysia
Phone: +62811661013
E-mail: sittihutarimulyani@upiyptk.ac.id
E-mail: billy_hendrik@upiyptk.ac.id
[3,4] Universitas Putra Indonesia YPTK, Padang, 25221, Indonesia
Tel/Fax: +62751 776666
E-mail: rio.upi29@gmail.com
E-mail: mardhiah_m@upiyptk.ac.id

Abstract: Environmental solutions around the world today are making people dependent on technological developments. However, the technology cannot be separated from the community either in the form or its influence, therefore the pattern of behavior of society must also get arrangements for technology to occur as an acceleration of life done properly. This study aims to obtain patterns of community behavior on non-organic waste by using technology intervention. Gap exploration is essential for theoretical and experimental analysis of humans who dispose of unorganic and organic waste out of place. But the field of behavior analysis is uniquely tailored to contribute to this body of work. Sustainable development depends on changing technology to achieve its goals. We report on data collected form an on-line survey, which possible solutions for trash problems. In this paper we present an integrated waste management system with IT that we called I-BSC (Indonesia *Bersih* (Clean), Sehat (Healty) and *Cerdas* (Smart)). This I-BSC is not only for deposit bottles and systems but the system aims also to create awareness of waste production and management, which serves as an educational platform in urban environments for further life innovation.

Keywords: Technology Intervention, cleanliness behavior, trash management, environmental psychology

1. Introduction

Indonesia is one country that has a very wide area in the world, the total area of Indonesia is 5.193.250 km² which includes land and sea [12,14]. It put Indonesia as the 7th largest country in the world after 6 other countries, namely Russia, Canada, USA, China, Brazil and Australia. Compared to the wide range of countries in Asia, Indonesia is ranked 2nd. Meanwhile, when compared with countries in Southeast Asia, Indonesia is the largest country in Southeast Asia.

Meanwhile, based on the report from the United Nations Population Division on the prospect of the world population that estimates the population of the world by medium fertility method for July 1, 2015, Indonesia itself has a population of 255,708,785 people or 3.49% of the world population and occupies the fourth position in the world.

But for the pattern of people's behavior on waste, Indonesia is ranked second in the world of plastic waste to the sea which reached 187.2 million tons after China reached 262.9 million tons. Based on Jambeck data (2015) [13]

The data proves that the hygiene behavior of Indonesian people, especially on garbage is still not good. It can also be seen from the activities of citizens in everyday life, where although the government has tried to provide training as one strategy to provide information to all levels of society about the implementation of Reduce, Reuse, Recycle and Composting (R3C) in waste management or Recycle waste into products that have economic value [8] [9] [15] [16] [17]. Then the government has also appealed to households, shops and business establishments not to mix materials that can be recycled (unorganic) with organic decomposition by placing it in separate trash bins by providing trash cans Two containers. However, these efforts are still experiencing difficulties because there are still few people who are able to apply the R3C pattern in daily life and still lack of public awareness to separate organic and unorganic waste directly when disposing garbage.

The hygiene behavior that people do to waste, including environmentally responsible behaviors. Hines, Hungerford and Tomera (1986) conducted a meta-analysis of studies pertaining to environmentally responsible behavior, obtaining a number of variables associated with the intended behavior, knowledge of issues, knowledge of action strategies, locus of control, the verbal commitment and sense of responsibility one has. [3]

In this study the authors examine how the pattern of public hygiene behavior if influenced by technological intervention in everyday life. The technological intervention to be provided is a system that supports the government's efforts to create a clean and healthy environment that is "I-BSC System (*Indonesia Bersih Sehat dan Cerdas*)". The I-BSC system is an integrated system with technology and community empowerment and several other variables to manage plastic bottle waste, where every plastic bottle waste disposed of by the community on the I-BSC system will be rewarded with the value of dollars per each bottle so that people can enjoy Cash out system, in addition to the value of rupiah owned can be used to perform electronic transactions and of course will increase public awareness to sort out organic and unorganic waste directly.

2. Methodology

In accordance with the purpose of this study is to see how the pattern of public hygiene by applying technology to waste management [10] [11] [18] [19]. In this research, the form of waste management technology used the I-BSC system, then the approach used in this study is quantitative method, this method is called as a positivistic method because it is based on the philosophy of positivism. This method as a scientific or scientific method because it has met the scientific norms are concrete or empirical, objective, measurable, rational, and systematic. This method is also called discovery method, because with this method can be found and developed various new science and technology. This method is called quantitative method because the research data in the form of numbers and analysis using statistics. [4]

The location of the research was conducted at Putra Indonesia University "YPTK" Padang located in West Sumatera province, site selection was based on the consideration that: (1) This location is where the innovation of the I-BSC system originated; (2) there are variation of education level, occupation type and number of respondent's income which enables variation of community hygiene level with existence of I-

BSC system. The subjects of this study were students of Universitas Putra Indonesia "YPTK" Padang and the community residing in Putra Indonesia University "YPTK" Padang, where sampling technique used according to Krejcie and Morgan (1970) [21] with error rate 5% then the population is 500000 people with The number of samples was 384 respondents. Where the characteristics of students and community residing in the University of Putra Indonesia "YPTK" Padang from samples tested based on:

- Gender can be seen in figure 1
- The last education can be seen in figure 2
- Job type can be seen in Figure 3
- Total revenue can be seen in figure 4

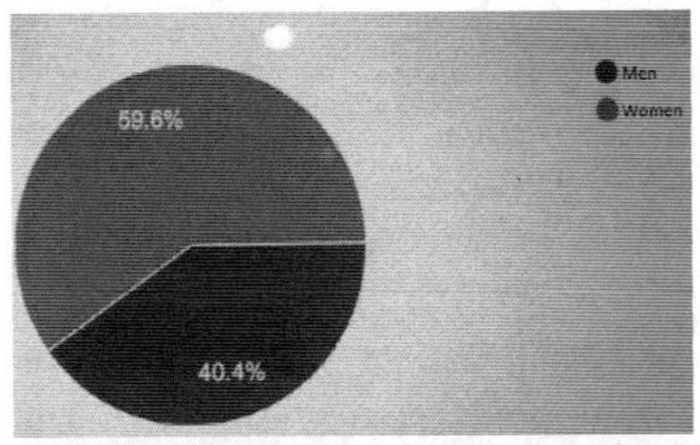

Figure 1. Percentage of respondents by sex

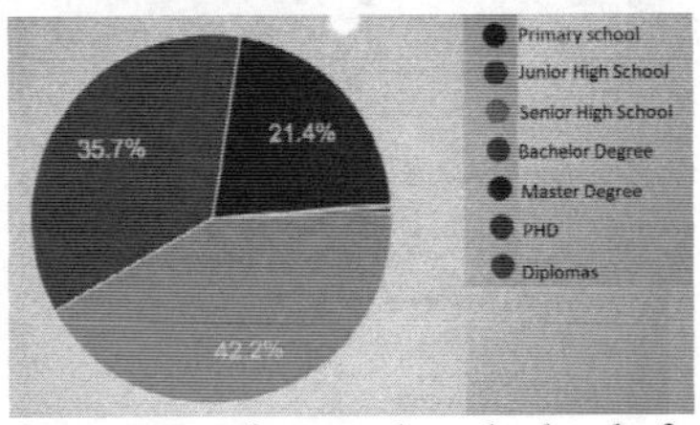

Figure 2. Percentage of respondents by level of education

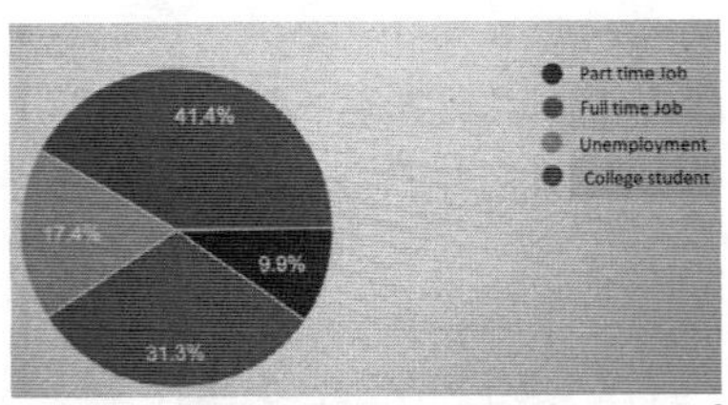

Figure 3. Percentage of respondents by type of work

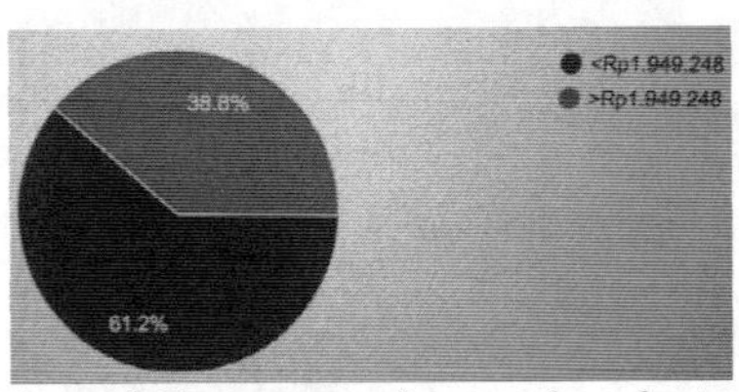

Figure 4. Percentage of respondents by total revenue

The samples were collected by distributing questionnaires online, to find out how the behavior of people in the management of unorganic waste during this time.

In general, community behavior in the management of unorganic waste can be measured from the important components in showing a person's behavior in environmental management (Albert, 2009) [5] [7] [20]. Where the components underlying a person's behavior are related to the understanding of knowledge possessed about waste and waste management, how the habits or behaviors of unorganic waste generated from daily life, the impulse of the social environment and how the society's acceptance of technology.

- Knowledge of waste
- The level of public understanding about the waste based on data obtained is 78.41%.
- Community behavior on waste in daily life

Table 1. The results of respondents about community behavior on waste in daily life.

No	Behavior	Yes	No	Sometimes
1	Do you throw rubbish on public roads ?	0.7%	72.7%	26.6%
2	Do you destroy garbage by burning it	27.3%	31.8%	40.9%
3	Do you separate organic and unorganic waste bins ?	29.4%	39.6%	31%
4	Do you bring your own baskets while shopping at markets to reduce plastic waste?	18.8%	45.6%	35.7%
5	Do you throw out plastic bottles immediately?	41.9%	22.9%	35.2%
6	Do you bring your own bottle to reduce plastic waste?	46.4%	20.6%	32.8%
7	Do you reuse or recycle plastic bottle products?	18.2%	41.9%	39.8%
8	Do you collect waste plastic bottle for resale?	21.1%	50.3%	28.6%

1) The impulse of the social environment

Table 2. The result of respondent about influence of the social environment

No	Behavior	Yes	No	Sometimes
1	Have you ever flooded around your neighborhood ?	35.9%	45.8%	18.2%
2	Whether floods happen every year in your residence?	17.4%	60.2%	22.4%
3	Whether in the PKK forum or your community have an independent waste management program or waste bank ?	22.7%	65.6%	11.7%
4	Have you ever been invited by other peoples in your residence to process plastic waste?	28.4%	56.6%	15.1%
5	Have you ever been invited by other peoples in your residence to recycle plastic waste?	30.2%	57.8%	12%
6	Has anyone offered a reward in the form of money or other things if your recycle your trash?	9.4%	83.3%	7.3%

2) The society's acceptance of technology.

After the implementation of the I-BSC system in daily life, the data collecting through the online questionnaire then analyzed to find out how big the influence of technology to the hygiene behavior of the students and the community around Universitas Putra Indonesia "YPTK" Padang. So obtained the results of acceptance of technology with research instruments that have Likert measurement scale with the measurement scale used is 1 (very unattractive) 2 (not interesting) 3 (enough) 4 (interesting) 5 (very interesting). Measurement scale is an agreement used as a reference to determine the short length of intervals in the instrument (instrument) so that the measuring tool when used in the measurement will produce quantitative data [6]. The data analysis was done by using statistic by survey, based on the analysis, then the results obtained as shown in Figure 5 to Figure 13.

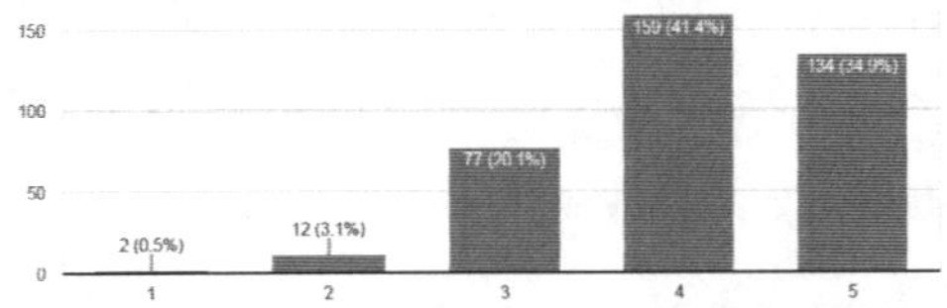

Figure 5. The level of respondent's satisfaction with the menu and facilities offered by the I-BSC system

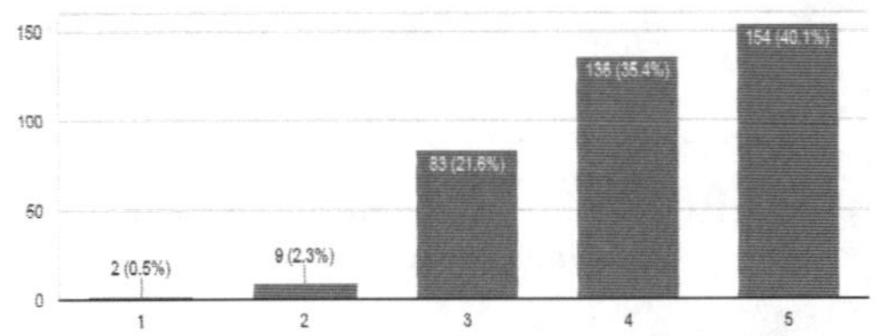

Figure 6. Desire to support zero waste movement through I-BSC system

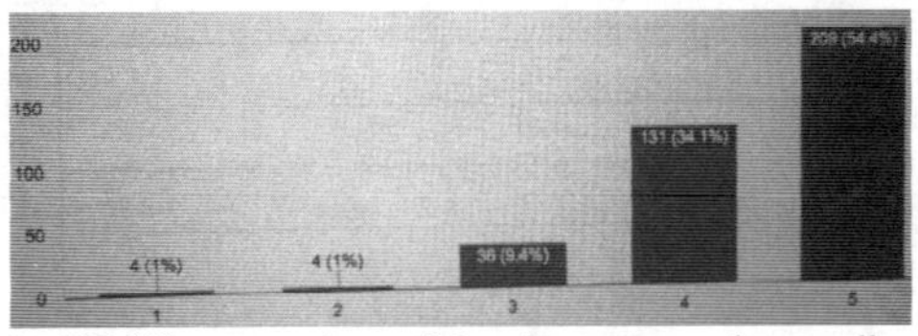

Figure 7. The existence of the I-BSC system is useful to improve understanding of waste, especially plastic bottle waste

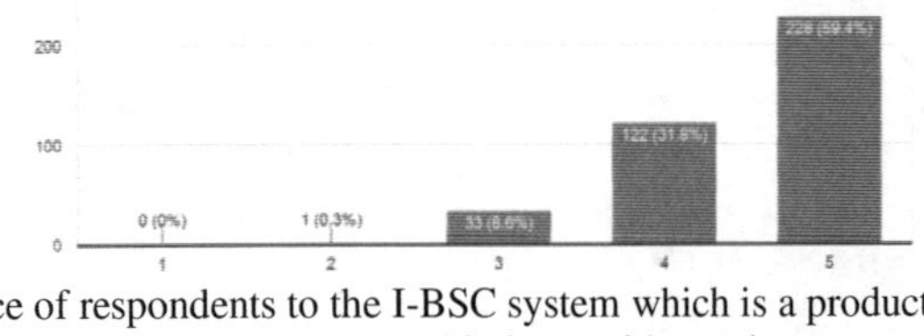

Figure 8. The confidence of respondents to the I-BSC system which is a product of the domestic children who can compete with the outside products.

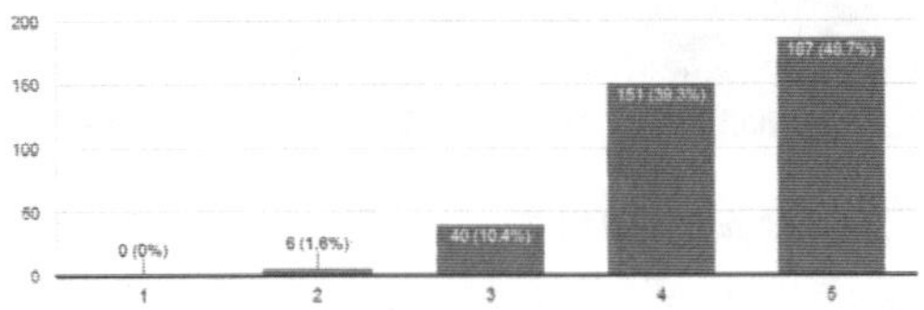

Figure 9. The I-BSC system facilitates the respondents in studying plastic waste waste

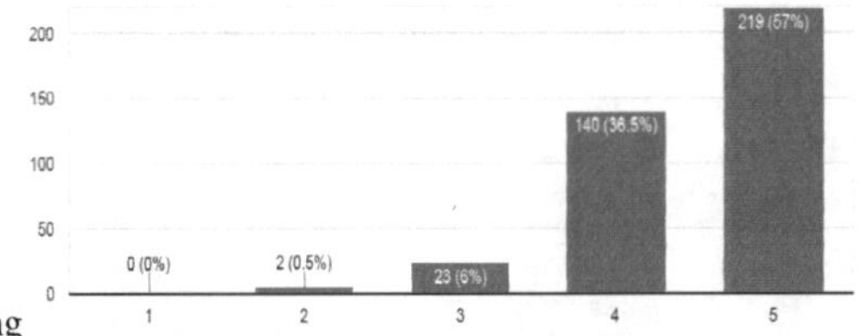

processing

Figure 10. I-BSC Machine will be useful for waste management, especially plastic trash bottles

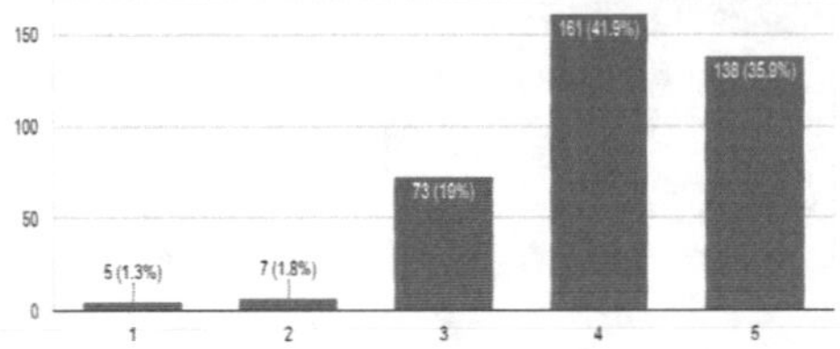

Figure 11. Respondents accessing I-BSC to find information on plastic bottle receptacle and information about buying and selling media bottles that have been processed

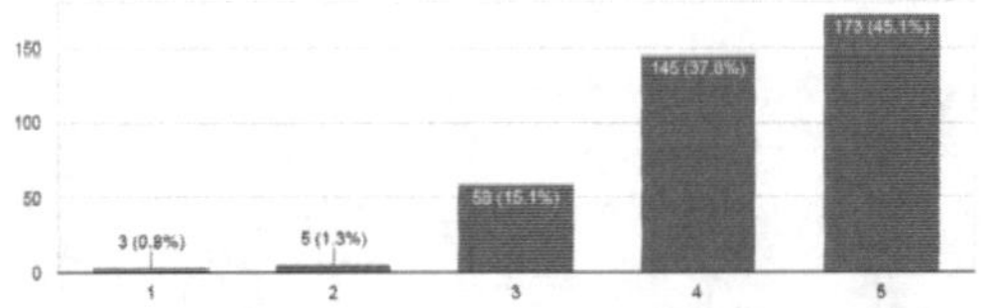

Figure 12. Respondents access the I-BSC system to learn waste plastic bottle management into creative goods.

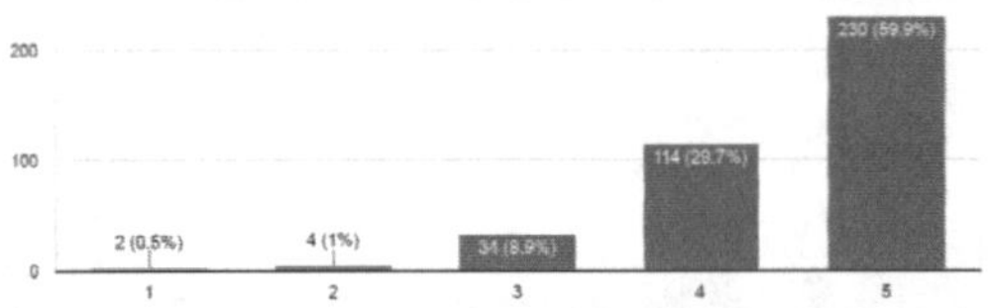

Figure 13. The level of support of respondents to the I-BSC system as an innovative tool for the community in grind unorganic waste.

3. Conclusion and Recommendation

Based on the above analysis, it can be concluded as follows:

Table 3. Results of respondents to the I-BSC system

I-BSC System	Scale				
	1	2	3	4	5
The level of menu satisfaction and facilities	0.5 %	3.1 %	20.1%	41,4 %	34,9 %
Support zero waste movement	0.5 %	2.3 %	21.6 %	35.4 %	40.1 %
Increase understanding of bottled waste	1.0 %	1.0 %	9.4 %	34.1 %	54.4 %
Products that can compete	0 %	0.3 %	8.6 %	31.8 %	59.4 %

Makes it easy to learn plastic waste bottle processing	0 %	1.6 %	10.4 %	39.3 %	48.7 %
Useful for waste management	0 %	0.5%	6 %	36.5%	57%
Provide the media buying and selling of waste processing	1,3 %	1,8 %	19 %	41.9 %	35,9 %
Learning waste management to be creative	0.8 %	1.3 %	15.1 %	37.8 %	45.1 %
Innovative means	0.5 %	1.0 %	8.9 %	29.7 %	59.9 %

Based on Table 3, it can be concluded that respondents said accepting the presence of I-BSC systems, it can be seen from 48.37% of the respondents stated I-BSC system is a very interesting technology, where the system is able to give positive impact to the hygiene behavior of the community where the community is getting used to separating organic and unorganic waste directly, then people are also interested to utilize Facilities - facilities in the system I-BSC both cash out system, electronic transactions and media sales of plastic waste processing products as well as educational information presented by the I-BSC system.

4. References

[1] B. M. Dwiyanto. 2011. *Model Peningkatan Partisipasi Masyarakat Dan Penguatan Sinergi Dalam Pengelolaan Sampah Perkotaan. J. Ekon. Pembang.*, vol. 12, no. 2, pp. 239–256.

[2] G. J. Amador and D. L. Hu. 2015 . *Cleanliness is next to godliness : mechanisms for staying clean.* pp. 3164–3174.

[3] I. Wibowo. 2009. *Pola prilaku kebersihan : studi psikologi lingkungan tentang penanggulangan sampah perkotaan. Makara Sosial Humaniora*, vol. 13, no.1, pp.37-47, 2009.

[4] N. Hayati. *Pemilihan metode yang tepat dalam penelitian (metode kuantitatif dan metode kualitatif). Tarbiyah al-awlad*, vol. IV, no. 1, pp. 345-357.

[5] E. S. Albert, S.Y.Hadi and R. G. Sri. 2011. *Perilaku Masyarakat Dalam Pengelolaan Sampah Anorganik di Kecamatan Abepura Kota Jayapura. Majalah Geografi Indonesia*, vol. 25, no. 1, pp. 1-14.

[6] Sugiyono. 2008. *Metode penelitian kuantitatif, kualtitatif dan R&D.* Bandung: Alfabeta

[7] Wardhana. A.W. 2004. *Dampak Pencermaran Lingkungan.* Yogyakarta: Pustaka Belajar.

[8] Mulyono. H. 2010. *Model Pengelolaan Tempat Pemrosesan Akhir (TPA) Sampah Secara Berkelanjutan di TPA Cipayung Kota Depok. Jawa Barat. Institute Pertanian Bogor.*

[9] S. Isti, H. Akhmad dan S. Ade. 2009. *Model Dinamis Pengelolaan Sampah Untuk Mengurangi Beban Penumpukan, Teknik Industri*, Vol. 11, No.2, pp.134-147.

[10] P.O. Tawiah, P.Y. Andoh, A. A. Agyei, F. Nyarko. 2016. *Characterization of Recycled Plastic for Structural Applications, International Journal of Science and Technology*, Vol.5, No.6.

[11] J. Siddiqui, G. Pandey. 2013. *A review of Plastic Waste Management Strategies, International Research Journal of Environment Sciences*, Vol.2, No.12.

[12] R. Lasabuda. 2013. *Pembangunan Wilayah Pesisir dan Lautan Dalam Perspektif Negara Kepulauan Republik Indonesia, Ilmiah Platax,* Vol. I-2, pp.92-101.

[13] Jenna R. JAmbeck. 2015. *Plastic Waste Input from Land into the Ocean, Science*, Vol. 347.

[14] M. Zulkarnain, P. Purwanti, E. Indrayani. 2013. *Analisis Pengaruh Nilai Produksi Perikanan Budidaya Terhadap Produk Domestik Bruto Sektor Perikanan Di Indonesia*, *ECSOFiM*, Vol.1, No.1.

[15] A. H. Chowdhury, N. Mohammad, R. U. Haque, T. Hossain. 2014. *Developing 3Rs (Reduce, Reuse and Recycle) Strategy for Waste Management in the Urban Areas of Bangladesh: Socioeconomic and Climate Adoption Mitigation Option*, *IOSR Journal of Environmental Science, Toxicology and Food Technology*, Vol.8, pp.09-18.

[16] D. Asteria, H. Heruman. 2016. *Bank Sampah Sebagai Alternatif Strategi Pengelolaan Sampah Berbasis Masyarakat Di Tasikmalaya*, *J. Manusia dan Lingkungan,* Vol.23, No.1.

[17] Y. Puspita, M. Rahdriawan. 2012. *Kajian Pengelolaan Sampah Berbasis Masyarakat Dengan Konsep 3R (Reduce, Reuse, Recycle) di Kelurahan Larangan Kota Cirebon*, *Pembangunan*

Wilayah dan Kota, Vol.8, No.4.

[18] Y.P. Deshmukh, A. B. Borade. 2015. *Plastic Waste Management – Present Practice and Future Possibilities, International Journal on Recent and Innovition Trends in Computing and Communication*, Vol.3, 2015

[19] T.Z. Quazi, M.D. Jagtap. 2015. *Sustainable Plastic Waste Management – A Case Study of Thane Municipal Corporation, International Journal of Innovations in Engineering Research and Technology*, Vol.2.

[20] Nazaruddin, 2014. *Analisis Prilaku Dalam Upaya Menciptakan Kebersihan Lingkungan Di Kota Pekanbaru (Studi Kasus Kelurahan Simpang Baru), Jom FISIP*, Vol.1, No.2.

[21] A. M. Zahran, I. M. Shah, A. Jusoh. 2016. *Consideration and Methodological Approaches in Studying Transformational Leadership Impact on Work Performance Behaviors, International Journal of Advances Research*, Vol.4, 2016

Allocative Efficiency Analysys Production Function Of Cassava Farming (Manihot, Sp.) In Tenayan Raya Sub-District Of Pekanbaru City

Rini Nizar[1], Niken Nurwati[2], Amalia[3]
Universitas Lancang Kuning, Pekanbaru, 28265, Indonesia
Telp/Fax: +6276152439
E-mail: rininizar@unilak.ac.id
E-mail: nikennurwati@yahoo.co.id
E-mail: amaliamasjkur@unilak.ac.id

Abstract: Cassava (Manihot sp) has long been known and cultivated by Indonesian farmers. The economic and social potential of cassava aside from foodstuffs can also be used as raw materials for industrial use and animal feed. In Riau Province, Cassva has the potential to be developed considering Cassva is a plant that can easily grow on low altitude to high altitude lands. Cassava does not need a complex maintenance. Conventionally, this plant can be planted and left alone by itself. Cassava roots can be developed to be a processed products that society needs as main foodstuffs ingredients. This research is done in three months and the purpose is to know the influence of input use (pesticide, seeds, fertilizers and labor) on cassava farming to cassava farming by the model of cobb-douglas. Other than that is also the effect on economical efficiency. The method used in this research is a quantitative research by using Cobb-Douglas Function Model. This research was done in the Tenayan Raya sub-district with 55 farmer samples. This research shows Cobb-Douglas Production Function can be used as the predictor for Cassava production function in Tenayan Raya Sub-district of Pekanbaru City. Altogether the production factor used by farmers influence production. Partially only usage of organic fertilizer that does not affect production, while other production factor such as, seeds, pesticides, an-organic fertilizer (urea) and labor affect production by quite a bit. Usage of production factor seeds, urea and pesticides is not yet efficient while usage of organic fertilizer is not efficient and usage of labor on cassava agriculture by respondent farmers is relatively efficient
Keyword: Production Function, economical efficiency

1.Introduction

The value of economic and social potential of Cassava (Manihot sp) other than foodstuffs also function as materials to other industry and animal fodders. In Riau Province, cassava has the potential to be developed on the other hand cassava can be also developed as main product needed by society. Because the carbohydrate content is high enough and it can also be processed into other culinary foodstuffs such as kemojo cake which is a typical Riau food, tapioca flour, mocaf flour, and snacks. Derivative products from cassava such as chips are dried raw cassava pieces that can be reprocessed into cassava flour, cattle feed,bioethanol and dried cassava. all of them are export commodities. Cassava as one of the food commodities experienced significant demand spike in 2016 (Detik Finance, Monday 17/7/2016). on the other hand, eventhough the productivity of cassava in Riau Province is still higher compared to the national average productivity, but the productivity growth rate from 2015 to 2014 has decreased.

Problems in the research is low productivity allegedly due to the cultivation technology used by farmers is still not maximal so it affects the production and income of farmers who are also low, from Cassava farming.This study aims to determine the effect of inputs (seeds, fertilizer, pest, labor and fertilizer) production through the cassava production model on cassava farmers in Tenayan Raya City Pekanbaru and the determination of economic efficiency.

2.Research Method

This research was conducted in Tenayan Raya District Pekanbaru City. Data type used in this research is primary data and secondary data. Samples were taken by using the Skaran table with a Margin of Error of 5%, as many as 55 people. The analysis method used in this research is Cobb-Douglas Function. . Mathematically, the Cobb-Douglas function can be written as follows:

$Y = \propto X_1^{b1} X_2^{b2} \ldots\ldots X_n^{b}\ e$, To facilitate the estimation of the equation above, the equation is converted to multiple linear form by Alogarythm with equation as follows:

$$\text{Log } Y = \text{loga} + b_1 l\ {}_1 + b_2 l\ {}_2 + b_3 l\ {}_3 + b_4 l\ {}_4 + b_5 l\ {}_5 + e$$

To see Optimal Use of Production Factor using the formula (b.Y.Py) / (X.Px) = 1

3.Result and Discussion

3.1 Characteristic of Respondent

The average land area cultivated by the farmers of respondents is 1.45 ha, decreased when compared to the research conducted Fatmiwati (2012) that is 2.5 Ha.

Seeds used by farmers of respondents are sanjai which is a local superior seed from West Sumatara. The average use of farmer seedlings is 3,525 stem cuttings. While the yield is 18 tons. This result is still lower than the productivity of cassava Riau Province (2015) which is 28.96 tons per ha. Seeds used by farmers in cassava farming is obtained from cassava stems that have been harvested before. In general, farmers have used about 3 times the original seedlings of origin. This condition also affects the resulting yields decrease productivity.

The average use of inputs is urea 303,64 kg, ponska 241,09 kg and organic Fertilizer is 120,55 ku for each farmer farm. Other production factors used by farmers of respondents are pesticides and labor. The average use of production and production factors and prices can be seen in table 1 below:

Table 1

The average use of factors of production and factor prices of production by respondent farmers

No	Description	Amount	Price Input per Unit
1	Production (ton)	18	1.500
2	Seeds (sticks)	3.525	500
3	Pesticides (liter)	18,38	48.500
4	Urea (Kg)	303,64	5.000
5	Fertilizer (ku)	120,55	20.000
6	Labor (hok)	50,67	80.000

3. 2 Model Production Cobb-Douglas on Respondent Farmers

In terms of statistics and econometrics, the cassava production function model is good enough as an estimator. Based on the estimation result of production function obtained from Cobb-Douglas model is as follows:
$\text{Ln } Y = 6.27 - 0.606 \text{ Ln } X1 - 0.4770\ X2 + 0.0953 \text{ Ln } X3 + 0.0707 \text{ Ln } X4 + 0.1785 \text{ Ln } X5 + u$
Then the Cobb-Douglas equation can be written as follows:

$$Y = 528.48\ X_1^{-0.606}\ X_2^{-0.477}\ X_3^{0.0953}\ X_4^{0.0707}\ X_5^{0.1785}$$

Desc : Prod. = Y; Seeds = X_1; Pest = X_2; Urea = X_3; organic Fertilizer = X_4; Labor = X_5

From the test results known the value of termination coefficient (R2) of this study amounted to 0.8326, this value indicates simultaneously (simultaneously) production of cassava farmers affected by seeds, pest, urea, Fertilizer and labor of 83.26% Other factors. The value of interception coefficient in this research is 528.48, it states total productivity for seed, pest, urea, Fertilizer and labor is equal to index of production efficiency that is equal to 528,48. The higher the production efficiency index means the process of transforming inputs into ouput becomes more efficient. Based on the results of t test for seeds obtained value t - hit - 3.06 at 95% confidence level with the value of elasticity of - 0.606 this shows there is a significant or significant influence between seeds on cassava production. The value of regression coefficient for seedlings is -0.606. This means that if the seeds increased by 1% then cassava production will be reduced by 0.606%. This can happen because it is caused by the seeds used by the farmer is already a third derivative of the parent seed, and has not paid attention to the requirements of good seeds to be planted.

Result of t test for pesticide that have been done obtained result that is t-count value -8.65 at 95% confidence level, there is real influence between pesticide with cassava production. The value of regression coefficient for pesticide is -0.477. This means that if the pesticide increases by 1% then the production of cassava is reduced by 0.477%. This condition is made possible by the use and manner of giving crops not in accordance with the correct instructions or recommendations. Based on the t test for urea that has been done obtained the result that is t-count value 2.42 at 95% confidence level, there is a real influence between urea with cassava production. The regression coefficient value for urea fertilizer is 0.0953. This means that if urea increased by 1% then cassava production will increase by 0.0953%. Based on t test that has been done for fertilizer obtained result that is t value count 1,61 at 95% confidence level, there is influence which is not real between fertilizer with cassava production. The regression coefficient value for Fertilizer is 0.0707. This means that if fertilizer is increased by 1% then cassava production will increase by 0.0707%. Effect of Labor Usage on Increasing Cassava Production. Based on the result of t-test calculation for labor usage, the result is t-count 2.32 at 95% confidence level, that labor has significant effect to cassava production. The coefficient value for labor is 0.1785. This means that if the labor increased by 1% then cassava production will increase or increase by 0.1785%.

Alocative Efficiency Analysis on Production Factor

The calculation results for the allocative efficiency of each production factor used in cassava farming can be seen in table 2 below:

Table 2.
Value of Alocative Efficiency on Production Factor

No	Description	Value	Explanation
1	Seeds	10.06	Not yet Efficient
2	Pesticide	14,45	Not yet Efficient
3	Urea	1,69	Not yet Efficient
4	Organic Fertilizer	0,079	Not efficient
5	Labor	0,95	Relatively Efficient

Table 2 above shows that the use of production factors of cassava farming by respondent farmers. Seedlings, Pesticides and urea are inefficient, while organic Fertilizer is inefficient and labor is relatively efficient. Judging from the analysis of allocative efficiency indicates the use of seeds is not efficient, meaning that the use of seeds should be added. Field conditions indicate that farmers use superior local cassava seedlings from West Sumatra (sanjai) obtained from propagation of third derivative crops not from F1 plant propagation. According to the recommendation of Plant Nutrition Research Institute and Tuber, the material used in the propagation of cassava cuttings for seedlings is F1 plant. Propagation of the F1 plant is used as a source seed with due regard to the origin, purity and quality of the cuttings should be kept in mind. The recommended number of cassava cuttings / seedlings used with the normal cultivation distance (100 x 80) cm2 was 12,500 stems while the sample farmers used 3,525 stem cuttings. This condition explains that the use of cassava seeds should be superior seeds and in accordance with the correct way of planting.

Pesticides are negatively related to production. This is due to: (a). The use of pesticides does not fit the rules. Farmers often ignore the rules of pesticide use correctly, (b). High prices of pesticides. The high price of pesticides resulted in the giving of pesticides to cassava plants are not as recommended. From the analysis of the efficiency of the allocation of pesticides has not been efficient, meaning its use should be added. The weakness in this research is pesticide data used is the sum of the use of round up and other pesticides by farmers, so less can be analyzed the influence of each type of pesticide on production. Pesticide. However, pesticides have a significant effect on production. From result of analysis of efficiency of allocation of urea fertilizer by farmer not yet efficient this mean the use of urea fertilizer must be added by farmer. Urea fertilizer has a significant effect on cassava production. Analysis of the efficiency of the allocation of Fertilizer is not efiisen means that the use of Fertilizer needs to be reduced. Fertilizer has no significant effect on cassava production. The use of labor is a factor that must be met for the sustainability of cassava business activities. The use of labor must be careful and really - taken into account. The use of excessive labor will surely increase the production cost so that the income earned will be reduced and even have the potential to bring losses. Based on the analysis of the efficiency of the allocation of labor by the farmers in the cultivation of cassava relatife efficient and have a significant effect on the production of cassava.

4.Conclusion

It can be concluded that firstly, the Cobb-Douglass production function model can be used as a predictor of the cassava production function. Together the factors of production used to influence cassava production. Partially production factors that have real effect on the production are seeds, pesticides, urea fertilizer and labor. Secondly, the use of seed, pesticide, and urea production factors has not been efficient while the use of organic Fertilizer is inefficient and labor is relatively efficient in cassava farming by farmers in Kecamatan Kulim Tenayan Raya District Pekanbaru City.

5.Suggestion

The use of seeds by farmers should use superior seeds or should pay attention to the requirements of good cassava seedlings to be planted. The use of doses of fertilizers and pesticides should take into consideration the recommended recommendations. Need to do research to determine the appropriate dosage of fertilizers and pesticides at cassava farming in Kulim Village, Tenayan Raya District, Pekanbaru City.

6.Reference

[1] Badan Pusat Statistik. 2013. *Riau Dalam Angka.* Badan Pusat Statistik Propinsi Riau. Pekanbaru

[2] Balai Penelitian Tanaman Kacang-Kacangan dan Umbi- Umbian. 2016. Penyiapan Bibit Ubi Kayu. http://pangan.litbang.pertanian.go.id. Diakses tanggal 1 Mei 2017

[3] Dinas Pertanian. 2012. Profil, *Investasi Biofuel Ubikayu.* Pphp.deptan.go.id. Posting 9-3-2012

[4] Dinas Tanaman Pangan. 2006. *Sistem Budi Daya Ubi Kayu.* Provinsi Riau

[5] Gayatri Indah Cahyani. 2008. *Analisis Faktor Sosial Ekonomi Keluarga Terhadap Keragaman Konsumsi Pangan Berbasis Agribisnis di Kabupaten Banyumas.* Thesis. Jakarta State University.

[6] Gaspersz, Vincent 2004. *Production Planing and Inventory Control. Gramedia.* Jakarta

[7] Nasir.S, Rahayuningsih, dan Muchlis. A. 2012. *Peningkatan Produksi dan Kualitas Umbi-umbian.* www.opi.lipi.go.id. Diakses 23-4-2013

[8] Paidi, hurip Suripto, Sunarty.S. 2010. *Analisis Pendapatan Usahatani Ubi kayu dan Kesempatan Keja di Sekitar Agroindustri Tapioka di Kecamatan Tulang bawang Tengah.* www.dostoc.com. Posting 26-3-2010

[9] Puteri, G.A. 2009. *Analisis Respond an Proyeksi Pemasaran Ubikayu di Indonesia. Skripsi. Fakultas Ekonomi dan manajemen. Institut Pertanian Bogor.* Bogor

[10] Risbianini Fardaniah. 2013. *Pemupukan Berimbang Bisa Menghasilkan 120 ton Singkong/Hektare.* Http://www.antaranews.com. Upload: 7 Juni 2013

[11] Soekartawi. 1993. *Agribisnis, Teori dan Aplikasinya. PT. Raja Grafindo Persada.* Jakarta.

IOP Conf. Series: Earth and Environmental Science **97** (2017) 012013 doi:10.1088/1755-1315/97/1/012013

Land Husbandry : Biochar application to reduce land degradation and erosion on cassava production

E.D. Yuniwati
[1] Wissnuwardhana University, Malang, East Java
Telp: +0341-713604, Fax: +0341-719297
E-mail: nieyuniwati @wisnuwardhana.ac.id

Abstract: This field experiment was carried out to examine the effect of increasing crop yield on land degradation and erosion in cassava-based cropping systems. The experiment was also aimed at showing that with proper crop management, the planting of cassava does not result in land degradation, and therefore, a sustainable production system can be obtained.
The experiment was done in a farmer's fields in Batu, about 15 km south east of Malang, East Java, Indonesia. The soils are Alfisols with a surface slope of about 8%. There were 8 experimental treatments with two replications. The experiment results show that biochar applications reduce of soil erosion rate of the cassava field were not necessarily higher than those of maize in terms of crop yield and crop management. At low-to-medium yield, also observed the nutrient uptake of cassava was lower than that of maize. At high yield, only the K uptake of cassava was higher than that of maize, whereas the N and P uptake was more or less similar. Soil erosion on the cassava field was significantly higher than that on the maize field; however, this only occurred when there was no suitable crop management. Simple crop managements, such as ridging, biochar application, or manure application could significantly reduce soil erosion. The results also revealed that proper management could prevent land degradation and increase crop yield. In turn, the increase in crop yield could decrease soil erosion and plant nutrient depletion.
Keywords: Land husbandry, biochar, application, land degradation, soil erosion, cassava production

1.Introduction

Cassava can be considered a super-crop. It can be utilized for many purposes, and it can grow well in very marginal environmental conditions. In Indonesia, although cassava is an important crop and is planted in a very large area, it is still considered a minor crop by both the government and private sectors so it only receives little attention (2). Recently, with the food crisis due to the difficulty in increasing cereal crop production and decreasing oil fuel reserves, cassava has attracted more attention, especially from businessmen. However, the development of cassava still faces many constraints; one of which is the assumption that planting cassava will accelerate soil and land degradation. In addition, some people believe that land degradation accelerates with the increase of cassava yield.

The suggestion is usually based on the assumption that cassava removes a lot of plant nutrients from the soil, due to both high plant nutrient uptake and soil erosion (4). This opinion is considered to be true by most people due to the fact that land planted with cassava is usually in a very poor condition with a very low productivity, or even already in a degraded condition.

Notwithstanding the fact that land planted with cassava is usually in a degraded or nearly degraded condition, the hypothesis that growing cassava and/or increasing cassava yield causes soil and land degradation is still questionable. A better interpretation of the fact would probably be that only cassava crops can grow and produce a reasonable yield in such a poor soil condition. The common assumption that the nutrient uptake by cassava is higher than other crops is also not entirely correct. Howeler (2) calculated that with a harvest of 35.7 fresh tubers/ha (equal to 13.53t/ha dry) cassava removes 55kg/ha of N, 13.2kg/ha of P, and 112kg/ha of K. As a comparison, Heckman *et al.* (3) show that nitrogen, phosphorus and potassium removed by 10.3t/ha maize grain yield was 141 kg N, 39 kg P, and 49 kg K. In more recent studies conducted in Thailand, Puttacaharoen *et al.* (4) showed that the amounts of N and P removed in the harvested plant areas were also much lower than those removed by other crops, while the amount of K removed by cassava was similar to other crops but much lower than that of pineapple or cassava grown for forage. In term of concentration of plant nutrient, a good comparison of nutrient removal by cassava and some other crops had been summarized by Howeler (5).Indeed, nutrient removal with harvesting will be influenced by crop management, and hence crop yields. Amanulah *et al.* (6) showed that application of organic manure in cassava growing increased both tuber yields and nutrient uptake. Application of composted poultry manure had a positive soil nitrogen balance, while farm yard manure application had a negative soil nitrogen balance.

High soil erosion in cassava field is believed to be another factor that accelerates soil and land degradation. This suggestion arises because cassava stake is usually planted with wide spacing, and cassava stake has a slow initial canopy development. Study by Puttacaharoen *et al.* (4) showed that soil losses due to erosion were highest in cassava grown for roots, followed by cassava for forage, sugarcane, mungbean, sorghum, peanut, maize and pineapple).. Soil erosion in cassava crops mainly occurs at the early phase of growth in which poor land coverage exists. Ardjasa *et al.* (7) showed that in wet areas like those in Sumatra with an annual rainfall of about 2,500 mm/year, the amount of eroded soil during the first 4-month period in the rainy season was about 90% of the annual amount .Therefore, if there is any practice that can speed up and improve coverage of the land surface, it will help to reduce soil erosion. A common management to reduce soil erosion for this purpose is mulching (11; 12), and intercropping the cassava with other short maturity crops such as maize, upland rice, peanut, and soybean. The benefits of the cassava intercropping system in reducing soil erosion has been shown elsewhere (13, 14)

The aim of this study is to demonstrate that planting cassava does not necessarily accelerate soil and land degradation. The study is also aimed at showing that increasing the crop yield does not necessarily speed up land degradation, and at determining a planting management system which could simultaneously increase crop yield and decrease land degradation, or even maintain and increase land productivity.

2. Materials and Methods

The experimental treatments were set up according to the farmer's suggestions and the experiment was carried out on the farmer's field at Batu Village, 10 km south west of Malang, East Java, Indonesia. The soil was Alfisol, developed from volcanic materials with the top soil (up to 25cm depth) properties as given in Table 1. The experiment was carried out from September 2004 through to August 2008.

The experimental treatments were:

1. Cm: Control - cassava was planted on flat land (without ridging) in a mono-cropping system with no fertilizers applied.
2. CmR: Cassava was planted on ridges in a mono cropping system with no fertilized applied. The ridges were constructed across the slope with a distance of 1.0m between ridges.
3. CmRF: Cassava was planted on ridges in a mono cropping system. The crops were given 300kg/ha urea, 150kg/ha SP36, and 100kg/ha KCl.
4. MmRF: Maize was planted on ridges in a mono cropping system. The crops were given 300kg/ha urea, 150kg/ha SP36, and 100kg/ha KCl.
5. CmRF+Mnr: Cassava was planted on ridges in a mono cropping system. In addition, it was fertilized with the same treatments as in 3 (CmRF), 5t/ha of farm yard manure was also added.
6. (C+M) RF: Cassava was intercropped with maize and planted on ridges. The crops were given 400kg/ha urea, 200kg/ha SP36, and 150kg/ha KCl.
7. (C+M)RF+B: Cassava was intercropped with maize and planted on ridges. The crops were given 400kg/ha urea, 200kg/ha SP36, and 150kg/ha KCl and planted in an alley cropping system with Biochar
8. (C+M) R+Eg: Cassava was intercropped with maize and planted on ridges. The crops were given 400kg urea, 200kg SP36, and 150kg KCl and were planted in an alley cropping system with Elephant grass as the hedgerow with spacing of 6m between hedgerow and 0.3m as the width of the hedgerow.

These 8 treatments were arranged in Randomized Block Design with three replications. The crops were planted in field plots with the size of each 12m x 5m on a slope of about 8%.

Cassava cutting of about 25cm length was planted at a distance of 1.0 x 1.0m. For the treatment of maize intercropping, two maize seeds were planted with a 30cm interval on the sides of the cassava row (±25cm from the cassava row). After 2 weeks, the maize was thinned to one plant per hill. For the mono crop maize, maize was planted at a distance of 0.8 x 0.25m. Cassava, maize, and the hedgerow crops were planted at the same time. For the mono crop maize, maize was planted twice a year.

Farm yard manure (the properties are given in Table 1) was applied during land preparation. All SP36 (36% P_2O_5), KCl (50% K_2O) and a third of the urea (45% N) fertilizers were given at the planting date, and then a third of the urea fertilizer was given at 60 days after planting. The remaining urea fertilizer was given after harvesting the maize intercrop (105 days after planting). Observations were made for runoff water, eroded soil, total harvested biomass of cassava and maize, cassava tuber yields, grain maize yield, gliricide stem and leaves, and elephant grass biomass. Soil properties before the experiment and after the first and forth year harvests were also observed.

Laboratory analysis done for the soil includes: soil pH (in H_2O), soil organic matter content by wet digestion method Walkley and Black, total N (Kjeldhal), available P (Bray II), exchangeable K (NH_4OAc) as well as soil physical properties which include soil bulk density, aggregate stability (wet sieving, Yoder, 1928) and water content at ψm of -33kPa and -15MPa. Plant analysis was done (wet destruction) for total N (Kjeldhal), total P (spectrophotometer), and total K (Flame photometer).

3. Results and Discussion

3.1 *Nutrient Uptake*

Nutrient uptake in the harvested tuber yield of cassava and grain maize is given in Table 1. The amount of nutrients removed by cassava tuber harvesting is not necessarily larger than that by maize grain harvesting. In the Cm treatment, for example, with a yield of 9.12 t/ha, the nutrients removed by tuber harvesting only stood at 14.04 kg N/ha, 2.50 kg P/ha and 24.02 kg K/ha. For maize, with a yield of 6.21 t/ha, the nutrients removed in the grain are 87.08 kg N/ha, 20.53 kg P/ha and 29.86 kg K/ha. While in the CmRF treatment, only the absorption of potassium at the high yield of cassava was higher than that of maize. Nitrogen and phosphorus absorption at this yield level is still lower by cassava compared to that by maize.

Table 1. Nutrient uptake in the harvested tuber yield of cassava and grain maize

Crop Mgmt	**Csv Yield (t/ha)**	**Nutrient uptake**			**Maize Yield (t/ha)**	**Nutrient uptake**		
		N (kg/ha)	**P (kg/ha)**	**K (kg/ha)**		**N (kg/ha)**	**P (kg/ha)**	**K (kg/ha)**
CM	**9.12**	**14.04**	**2.50**	**24.02**	**0**	**0**	**0**	**0**
CmR	**9.83**	**14.49**	**2.58**	**24.79**	**0**	**0**	**0**	**0**
CmRF	**27.33**	**54.42**	**9.07**	**75.28**	**0**	**0**	**0**	**0**
MmRF	**0**	**0**	**0**	**0**	**6.21**	**87.08**	**20.53**	**29.86**
CmRF +Mnr	**31.70**	**61.32**	**10.22**	**84.83**	**0**	**0**	**0**	**0**
(C+M)RF	**23.72**	**48.72**	**8.12**	**67.40**	**3.54**	**38.22**	**9.01**	**13.10**
(C+M)RF+B	**26.73**	**49.68**	**8.28**	**68.72**	**2.78**	**34.58**	**8.15**	**11.86**
(C+M)RF+Eg	**20.67**	**37.44**	**6.24**	**51.79**	**2.09**	**25.90**	**6.11**	**8.88**

The macro nutrient uptake per unit of yield as given in Table 2 is obtained by dividing the total plant nutrient in the tuber or grain yield and that in the harvested biomass (Table 4) with the respective yields. Again, except potassium, the removal of macro nutrients of cassava is lower than that of maize. Howeler *et al.* (1) have summarized the previous research results and found that N and P removal in the harvested part of the cassava plant was actually lower, and K removal was similar to that of other crops tested.

Table 2. Nutrient uptake per unit of yield of cassava or maize with different crop managements

Crop Mgmt	Csv Dry tuber (t/ha)	Nutrient uptake			Maize Dry tuber (t/ha)	Nutrient uptake		
		N (kg/t)of tuber	P (kg/ha) of tuber	K (kg/t) of tuber		N (kg/t) of tuber	P (kg/t) of tuber	K (kg/t) of tuber
CM	3.48	4.15	0.73	7.1	0	0	0	0
CmR	3.82	3.79	0.67	6.48	0	0	0	0
CmRF	10.96	4.96	0.82	6.86	0	0	0	0
MmRF	0	0	0	0	5.46	15.9	3.76	5.46
CmRF +Mnr	11.32	5.41	0.9	7.49	0	0	0	0
(C+M)RF	9.89	5.02	0.83	6.94	3.11	12.28	2.89	4.21
(C+M)RF+B	8.78	4.26	0.71	5.89	2.52	13.72	3.23	4.7
(C+M)RF+Eg	6.34	5.91	0.98	8.17	1.88	13.77	3.25	4.72

3.2 Runoff, soil erosion and nutrient lost

The measurement results given in Table 3 and Figure 1 show that runoff and soil erosion from cassava planted on surface land without fertilizers throughout the time were higher compared to those from the other crop management systems. Simple management by planting cassava on ridges has significantly reduced both runoff and soil erosion. However, care should be taken with making ridges because it can increase the erosion rate. Odemerho and Auwunudiogba (13) showed that erosion from flat surface land is lower than that from ridging, and erosion from contour ridging is lower than that from ridging following the contour. The reason for these phenomena is that ridges can accumulate surface runoff water, and if the ridges fail to function, the erosive energy of this concentrated surface runoff is much higher.

Table 3. Nutrient removal by erosion from different crop managements

Crop Mgmt	The removal of plant nutrient					
	2015/2016			2016/2017		
	N (kg/ha)	P (kg/ha)	K (kg/ha)	N (kg/ha)	P (kg/ha)	K (kg/ha)
Cm	65.7	6.7	34.6	49.6	5.6	23.2
CmR	45.6	5.6	23.3	42.8	4.2	20.7
CmRF	75.5	16.7	38.5	45.5	13.8	24.6
MmRF	74.7	17.6	36.6	48.2	19.2	26.3
CmRF +Mnr	70.5	16.4	31.5	34.6	11.4	22.9
(C+M)RF	72.7	14.7	28.7	39.7	11.2	18.4
(C+M)RF+B	75.8	14.2	32.4	46.4	12.6	22.3
(C+M)RF+Eg	70.6	16.3	29.4	47.3	13.4	20.6

Figure 1: Effects of crop managements on soil loss and runoff

Surface runoff water and soil loss can be decreased further by improving crop management such as by fertilizer application, intercropping cassava with maize, applying farm yard manure, or practicing alley cropping system. Fertilizer application will improve crop development, and hence speed up land coverage. Similarly, in intercropping systems, the addition of other crops will speed up the coverage of the land surface. This will reduce both the erosive energy of rain fall and the surface runoff. A decrease of the erosion rate on a cassava plot with intercropping of upland rice and maize has also been shown by Ardjasa *et al.* (7).

Compared to maize (treatment CmRF and MmRF), it can be concluded that the amount of surface runoff and soil loss from a cassava plot is not larger than that from a maize plot. Indeed, in the early growth phase, cassava has a lower speed of covering the land surface. However, with two times of planting maize, there are two times of land preparation, in which the conditions are very susceptible to erosion. Thus, it is not surprising that the amount of surface runoff and soil loss from maize plots is larger than that from cassava plots.

The removal of the macro nutrients by erosion is given in Table 3. It is interesting to notice that a high level of surface runoff and soil loss are not followed by a high nutrient removal. In the Cm treatment, for example, with a surface runoff of more than 400m^3/ha and a soil loss of more than 45t/ha per year, the removal of nutrients is less than 65,7 kg N/ha, 6,7 kg P/ha, 34,6 kg K/ha in the 2015/16 planting season, and 49.6 kg N/ha, 5.6 kg P/ha and 23.2 kg K/ha in the 2016/2017 planting season. For a lower level of surface runoff and soil loss, such as in the treatment of CmRF, the amount of removal of nutrients in both planting seasons was much lower. This phenomenon is merely due to the higher concentration of nutrients in the eroded soil of the fertilized plot. The nutrient contents in eroded soil of CmRF, for example, were 1.88% N, 0.57% P, and 1.01% K, whereas the concentrations of N, P, and K in the eroded soil of CmR treatment were 1.20%, 0.12%, and 0.59% respectively.

4. Conclusion

Looking at the experimental results discussed above, it can be concluded that planting cassava does not necessarily speed up land degradation, either by plant nutrient uptake or soil erosion. Except for potassium, the nutrient uptake by cassava is not higher than that by maize. In terms of nutrient utilization for biomass production, cassava is more efficient than maize. To produce one ton of dry tubers in treatment (CmRF), for example, only requires 5.94 kg N, 0.99 kg P, and 8.21 kg K. On the other hand, the production of one ton of grain maize requires 15.90 kg N, 3.76 kg P, and 5.46 kg K.

Another important result to point out is that increasing the crop yield does not necessarily increase the rate of land degradation. It is true, however, that increasing the yield will increase the nutrient uptake. A proper crop management system such as the addition of farm yard manure or the use of alley cropping systems can minimize this negative effect, notwithstanding the fact that it increases the level of plant nutrients in the soil (see Table 3 for treatment of alley cropping with Biochar). In addition, increasing the yield is usually associated with the improvement of crop growth; hence, there will be a better land management, addition organic of matter from biochar, and improvement of soil properties, leading to the decrease of soil erosion. In terms of production sustainability, the addition of either organic materials such as farm yard manure (treatment CmRF+Mnr) or fresh biomass (treatment (C+M) RF+B) has proven able to maintain soil productivity. If it is only added with organic fertilizers, the cassava yield decreases by 30% (see Figure 1). With the addition of 5t/ha of manure, the yield is relatively constant at about 30t/ha.

5.Acknowledgments

Thanks to the Directorate of Research and Dedication to the community, the research ministry, and higher education who have funded the research of Superior University (PUPT)on DIPA Kopertis Nomor 042.06.1.401516/2017, tanggal 07 Desember 2016.

6.References

[1] Howeler, R.H., Oates, C.G., and Allen, A.C., 2001, *Strategic Environmental Assessment: An Assessment of the Impact of Cassava Production and Processing on the Environment and Biodiversivity FAO, Rome.*

[2] Howeler, R.H. 1991. *Long Term Effect of Cassava Cultivation on Soil Productivity.* Field Crops Research 26, pp. 1-18.

[3] Heckman, J.R., Sims, J.T., Beegle, D.B., Coale, F.J., Herbert, S.J., Bruulsema, T.W., and Bamka, W.J. 2003. *.Nutrient Removal by Corn Grain Harvest.* Agron. J. 95, pp. 587-591

[4] Putthacharoen, S., Howeler, R.H., Jantawat, S., and Vichukit, V., 1998. Field Crops Research 57, pp. 113–126

[5] Howeler, R.H., 2002. :"Cassava Mineral Nutrition and Fertilization." In Hilloks, R.J., Thresh, J.M., and Belloti, A.C. (Eds.). *Cassava: Biology, Production and Utilization* CAB International. New York. pp 115-147

[6] Amanullah, M.M., Vaiyapuri, K., Sathyamuoorthi, K., Pashanivelan, S., and Alagesan, A., 2007, "*Nutrient Uptake, Tuber Yield of Cassava (Manihot Esculenta Cranzt.) and Soil Fertility as Influenced by Organic Manures,*" *Journal of Agronomy 6,* pp. 183-187.

[7] Ardjasa, W .S., Ando, H., and Kimura. M., 2001, "*Yield and Soil Erosion among Cassava-Based Cropping Patterns in South Sumatra,*" Soil Sci. Plant Nutr. 47, pp. 101 -112

[8] Lal, R., 1984, "Soil Erosion from Tropical Arable Lands and its Control," *Advances in Agronomy 37,* pp. 183–248.

[9] Bhatt, R., and Khera, K.L., 2006, "Effect of Tillage and Mode of Straw Mulch Application on Soil Erosion in the Sub-mountainous Tract of Punjab, India," *Soil & Tillage Research* 88, pp. 107–115.

[10] Howeler, R.H., Tongglum, A., Jantawat, S., and Utomo, WH., 1998, "The Use of Forage for Soil Fertility Maintenance and Erosion Control in Cassava in Asia," In Sturr EW. (Ed). *Proc. Third Regional Meeting of the Forages for Smallholder Project (FSP),* held in Samarinda, Indonesia. March 23-26, 1998. CIAT Working Document No. 188. pp. 196-211

[11] Iijima, M., Izumi, Y., Yuliadi, E., Sunyoto, Arjasa, S.B., 2004, "Cassava Based Cropping System in Sumatra in Indonesia: Productivity, Soil Erosion and Rooting Zone," *Plant Prod. Sci.* 7, pp. 347 – 355.

[12] Odemerho, F.O., and Avwunudiogba, A., 1993, "The Effect of the Changing of Management Practices on Soil Loss: A Nigeria Example," *The Geographical Journal* 159, pp. 63 – 68.

[13] Yuniwati, E.D., N. Basuki, E.W. Wisnubroto, and W.H. Utomo. 2012. "*Combating land degradation in cassava field by crop yield improvement:. Basic. Appl. Sci.* Res., 2: 4975-4982,

[14] Yuniwati E.D., W.H. Utomo, and R.H. Howeler. 2015. Farmers' Based Technology Development for Sustainable Cassava Production System. *International Journal Agriculture Research.* ISSN 1816-4897 Vol 10 (No 02): http://scialert.net/archivedetails.php?issn=1816-4897&issueno=84

Projection of Big Cities Waste Management and Cost Based on Economic and Demographic Factors in Indonesia

Gita Prajati[1], Tri Padmi[2], dan Benno Rahardyan[3]
[1] Universal University, Batam, 29433, Indonesia
Telp : +62778-473399 Fax : +62778-473593
Email : prajati@uvers.ac.id
[2,3] Bandung Institute of Technology, Bandung, 40132, Indonesia
Telp : +62-22-2502647 Fax : +62-22-2530704
Email : benno.rahardyan@gmail.com

Abstract : Nowadays, solid waste management continues to be a major challenge in urban areas, especially in developing country. It is triggered by population growth, economic growth, industrialization and urbanization. Indonesia itselfs categorized into developing country. Indonesia's government has many program in order to increase the economic growth. One of them is MP3EI (Masterplan Percepatan dan Perluasan Pembangunan Ekonomi Indonesia. This program should be suppported by right waste management system. If Indonesia's waste management system can't afford the economic growth, it will trigger health and environmental problems. This study's purpose is to develop the socio-economic-environment model that can be used as a basis planning for the facility and cost of waste management systems. In this paper we used the development of Khajuria model test method. This method used six variables, which are GDP, population, population density, illiteracy, school's period and economic growth. The result showed that development of Khajuria test could explained the influence of economic and demographic factors to waste generation, 65.6%. The projection of waste generation shows that Pangkalpinang, Pekanbaru and Serang are the cities with the highest waste generation for the next five years. The number of dump truck and TPS in DKI Jakarata is the highest within another city, which is 39.37%. For the next five years, the waste management system in our study areas cost maximum 0.8% from GDP (Gross Domestic Products).

Keywords: waste generation, test model, projection, GDP.

1. Introduction

Population growth, economic growth, development of industrialization, and urbanization are caused the increasing of waste generated in the big cities significantly[7]. Wang and Wang[11] state that economic and population growth in Beijing, caused the increasing of waste generated, 2,960,000 tons in 2000 to 6.20 million tons in 2007, and it fluctuates in to 6.35 million tons in 2010. Solid waste generated globally in 1997 was around 0.49 billion tonnes with an estimated annual growth rate of 3.2-4.5% in developed countries and 2-3% in developing countries[6]. The waste management system will face critical phase because its facility unavailable to manage waste in larger quantities. Waste management system need the right infrastructure, maintenance and enhancement to run its activity. And it will make the cost more expensive and complex[10].

Indonesia categorized into developing country with middle low GDP. Indonesia's government has been done various efforts to encourage our economy growth. One of them is MP3EI [8]. If Indonesia's waste management system can't afford the economic growth, it will trigger health and environmental problems. Furthermore, we need the right strategy in waste management system. Sosio-economic-environment model can project waste generated. This projection is based on economy, demography and policy. The model system can be used as base in waste management system. Adlina[1] had done research about sosio-economy-environment model in West Java. That research used theree models, which are Khajuria, Weng and Daskalopoulus. The result showed that the accuracy of this three models compare to the real datas are 90.32% for Daskalopoulus, 67.76% for Khajuria and 68.89 % for Weng.

Khajuria et al[9] state that there are three factors affect waste generated, which are GDP, urbanization (population) and illiteracy. Beside this three factors, there are another factors, economic growth. Population density can be used as a variable. People who lives in area with high population density, usually disposal their waste into undue place, like river because the lack of waste dump[4]. Furthermore, this research used the development of Khajuria test model. Variables that used in development Khajuria model are GDP, population, population density, illiteracy, school's period and economic growth

Therefore, this study did reserach to identify the corelation between economic, demographic and social factors to waste generated in the capital city of Java and Sumatera provinces. Whereas, the aim of this study to develop socio-economy-environment model that can be used as base for planing and cost of waste management system. This model using economic, demographic and social factors to waste generated in each capital city of Java and Sumatera provinces. This study used the development of Khajuria Model.

This paper is structured as follows, in Section 2 we describe Waste Management, Influences of Economic and Demographic Factors to Waste Generated and Khajuria Test Model. In Section 3 we describe Methodology, the development of Khajuria model test to projected waste generated as basis of waste management and cost in the future. In Section 4 which is the result of an analysis of the findings of this paper. Section 5 is part of the Conclusion and Suggestion.

2. Theoretical Framework

2.1 Waste Management

Waste management is an effort to handle waste generated from human activities. This waste management is classified in to six elements, which are controling, storage, collecting, processing and disposal. The six elements interdependent with each other, forming a waste management system[10].
Local government need partners to do the integrated waste management. The society becomes the important stakeholder for waste treatment. They have to involved actively for achieving the better waste management by changing their behaviour[1].

2.2 Influences of Economic and Demographic Factors to Waste Generated

Waste generated affected by economic growth, industrialization, lifestyle and local climate. Generally, the higher economy growth and urbanization, the greater waste generated become. Level income and urbanisation have the close relevance. As income and living standards increase, consumption of goods and services will also increase. This situation will make waste generated increase. In addition , by increasing the purchasing power of the community for various types of essential commodities and the outcome of technology also made a significant contribution to the quantity and quality of waste generated[5].

Theoretically, demand for energy can improve economic conditions of a country and is expected to produce technology efficient and friendly to the environment. Unfortunately, economic growth in the developing and less developed countries are caused the increasing of CO_2 globally[2].

2.3 Khajuria Test Model

Khajuria et al.[9] stated there are three factors that can influence waste generated. The three factors are :

- Urbanisation (population)
  The last decades, waste generated increase significantly in developing countries. It is caused by urbanisation that triggered the population growth in urban areas.
- GDP
  Economic growth caused the changing of society's life style. That life style increase the consumption of society and it caused the increasing of waste generated.
- Illiteracy
  Illiteracy can influence waste generated because the ignorance of society about waste processing made the increasing of waste generated.

3. Research Methodology

The study areas are Banda Aceh, Medan, Pekanbaru, Padang, Palembang, Bengkulu, Jambi, Pangkalpinang, Tanjungpinang, Bandar Lampung, Serang, DKI Jakarta, Bandung, Semarang, Yogyakarta and Surabaya. There are six independent variables which are population, population density, GDP, economic growth, literacy rate and school's period. Whereas the dependen variable is waste generated.

Secondary datas collected through the relevant department. There are five departments, Central Bureau of Statistics, Regional Office of Sanitary Agency, Ministry of Public Works, Environmental Agency and Ministry of Environment.

Method used is linear model of development Khajuria test that is assumed the linear realtion between waste generated and factors that influences it.

$$\mathbf{WG = a + b_1x_1 + b_2x_2 + b_3x_3 + b_4x_4 + b_5x_5 + b_6x_6} \quad \mathbf{(1)}$$

WG is waste generated, x_1 is population, x_2 is GDP, x_3 is literacy rate, x_4 is school's period, x_5 is economic growth dan x_6 is population density. Whereas a is constant and b is coeficient from each factors.

4. Result and Discussion

4.1The Development of Khajuria Test Model

First the development of Khajuria tested to all of the capital cities simultaneously. **Table 1.** shows the result of development Khajuria Test Model for all of the capital cities. The analysis result of SPSS shows that this test model has *R square* 0.656. It means that independent variables has capability to explain the variance of dependent variable 65.6%.

Next, this model tested separately for each of that capital cities. It is meant to know the differences between this two treatment. The result of development Khajuria model for each cities show the better result. Population, GDP, population density, economic growth, litiracy rate, and school's period capable to explain waste generated above 50%. It happened because the characteristic of these cities are different. So that, the development of Khajuria model can't be used in the capital cities of Java and Sumatera Provinces generally.

Table 1. Coeficient of variable from development khajuria test model

Variable	Coefficient
Constanta	-44.877
Population	-1.819×10^{-7}
GDP	-5.07×10^{-9}
School's period	-2.518
Litiracy rate	.0.733310
Population density	0.000104
Economic growth	0.065944

4.2The Projection of Waste Generated

The projection of waste generated is done by using **Equation 2**. This equation is formed by the development Khajuria test model's result from the first treatment. Population variable not included because this variable has significant value greater than the other variables.

$$\mathbf{WG = -44{,}877 - 5{,}076 \times 10^{-9} GDP + 0{,}40\ Literacy\ Rate - 2{,}518\ School's\ period + 0{,}000104\ Population\ Density + 0{,}066\ Economic\ Growth} \quad (2)$$

The projection of waste generated in Pekanbaru and Semarang were done using **Equation 3** and **Equation 4**. Those equatians were formed by the development Khajuria test model's result from the second treatment. This is because significant values of that two cities from the second treatment better than the first treatment.

$$\mathbf{WG = -490{,}2 - 4 \times 10^{-6}\ Population - 1{,}69 \times 10^{-7}\ GDP + 4{,}87\ Literacy\ Rate + 0{,}80\ School's\ Period - 0{,}035\ Economic\ Growth} \quad (3)$$

$$\mathbf{WG = -61{,}29 + 8{,}47 \times 10\text{-}8\ GDP - 0{,}49\ Literacy\ Rate - 5{,}05\ School's\ Period - 0{,}00054\ Population\ Density - 0{,}20\ Economic\ Growth} \quad (4)$$

Table 2. shows the increasing of population and waste generated for each cities per five years. This table proves that the increasing of waste generated linear with the increasing of population. Population in Pangkalpinang, Serang and Pekanbaru have increased two times in 2030, as well as the waste generated.

The increasing of waste generated per five years for each cities shown in **Figure 1**. Pangkalpinang, Pekanbaru and Serang have the highest waste generated per five years. It stated before that waste generated increase as the population increase. The population of Pangkal Pinang, Pekanbaru and Serang shown in **Table 2** have increase about two times.

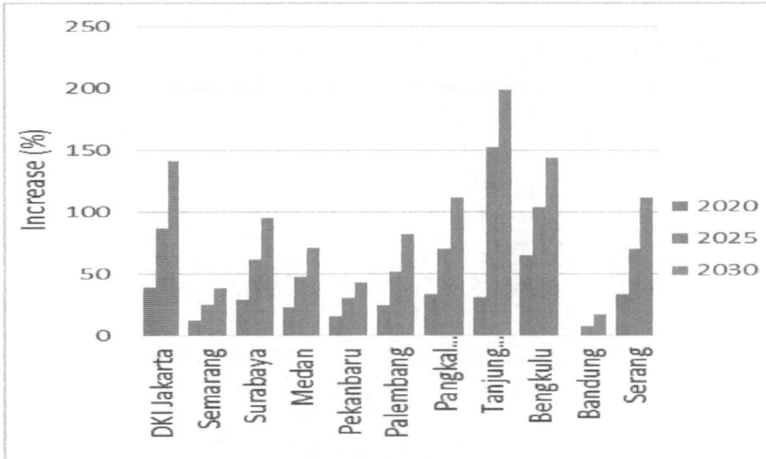

Figure 1. Increasing of waste generated per next five years

Table 2. Projection of population and waste generated per the next five years

City	Population (x 10^6 people)			Waste Generated (x 10^3 m^3/h)		
	2020	2025	2030	2020	2025	2030
DKI Jakarta	12,3	13,6	14,9	45.8	61.5	79.4
Serang	0,9	1,1	1,2	2.9	3.8	4.7
Semarang	1,8	1,9	2,1	6.1	6.8	7.5
Surabaya	2,9	2,9	3,0	8.1	10.1	12.2
Medan	0,3	0,3	0,4	1.7	2.0	2.4
Pekanbaru	1,3	1,6	1,8	4.1	4.6	5.0
Palembang	1,7	1,8	1,9	4.2	5.1	6.1
Bengkulu	0,4	0,5	0,6	0.5	0.8	1.1
Pangkal pinang	0,3	0,4	0,5	2.0	2.5	3.0
Tanjung pinang	0,3	0,3	0,3	0.6	1.1	1.7
Bandung	2,6	2,7	2,8	1.5	1.6	1.8

4.3 The Projection of Dump Truck and Temporary Waste Dump (TPS)

The higher waste generated, the higher the need of dump truck and TPS. **Figure 2**. shows that DKI Jakarta needs more dump truck and TPS than other cities. It is because the increasing of waste generated in DKI Jakarta is the highets among the other cities.

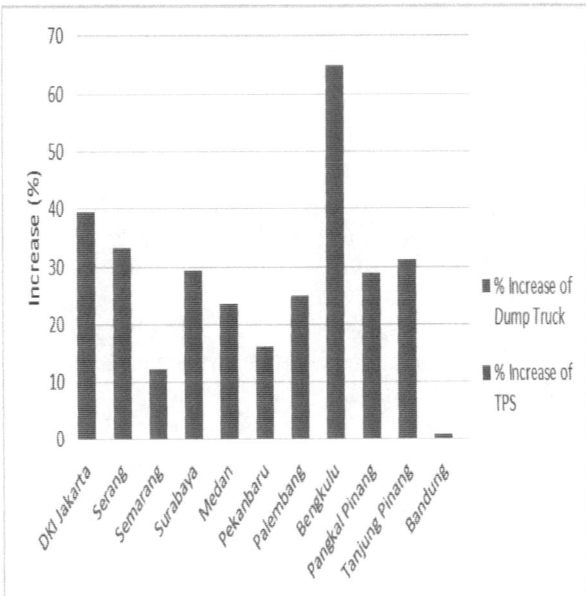

Figure 2. Increase of dump truck and TPS per the next five years

4.4 Ratio of Waste Management Cost to GDP

Brunner and Fellner[3] states that the developed countries spend around 0.4% of the gross domestic product (GDP) for the cost of waste management. As for the developing countries about 0.2-0.4%. In this study the calculation of waste management cost based on KSNP-SPP 2006 rules. KSNP-SPP 2006 rules divides the cost into several stages. Table 3. shows the stages required in waste management along with the fees charged per ton of waste generated. First, the cost in 2015 is conversed because there's changes of inflation value from 2006 to 2015.

Table 4. shows the cost of waste management based on Table 3. From Table 4. we can state that even the waste generated and waste management's cost have increased, the ratio of waste management's cost to GDP have decreased. Economic growth triggers the increasing of GDP. The higher GDP can cover the cost of waste management. So that, the waste management cost should not be an issue in the future.

Table 3. Componenets of waste management cost

No	Componenets	Cost per ton in 2006	Cost per ton in 2015
1	Transporting	119,268.00	208,215.13
2	Composting	103,973.00	181,513.50
3	Processing	141,684.00	247,348.43

Table 4. Projection of waste management cost based on KSNP-SPP 2006 rules

City	Year	Total Cost of Waste	GDP (x 10^6	Ratio of

		Management (Billions/year)	Billions/ year)	waste management cost to GDP
DKI Jakarta	2015	902.85	5.2	0.0012
	2020	1258.35	0,78	0.0016
Serang	2015	61.04	0.000284	0.2144
	2020	81.39	0.000442	0.1840
Semarang	2015	149.45	0.11	0.0127
	2020	167.82	0.17	0.0097
Pangkal pinang	2015	42.93	0.0000041	1.0268
	2020	55.35	0.0000065	0.8448
Tanjung pinang	2015	11.61	0.000084	0.1374
	2020	15.23	0.00012	0.1301

5. Conclussion

The development of Khajuria Model can explains waste generated and can be used to project waste generated in the future. Waste generated and waste management cost have increased but the ratio of waste mangement cost and GDP have decreased. Data base development is required to made better analysis as a basis of development waste management.

6. References

[1] Adlina, A. 2013. *Identification of Socioeconomic Factors and Population's Effects to Waste Generation In West Java*, Tesis Program Magister, ITB.

[2] Alsheyab, M. and Kusch, S. 2013. *Decoupling resources use from economic growth chances and challenges of recycling electronic communication devices*, *Journal of Economy, Business and Financing*, 1, 1, 2013, *1615-1619.*

[3] Brunner, P. H. and Fellner, J. 2007. *Setting priorities for waste management strategies in developing countries*, *Waste Management & Research,* 25, 3, *234-240.*

[4] Bhavannarayana, C., Prakash, K.S., and Saritha, V. 2011. *Estimation of municipal solid waste generation a case study*, *International Journal of Research and Reviews in Pharmacy and Applied science*, **2, 3,** *473-481.*

[5] Jaelani, A., H. I. Purwanti and M. R. Aziz. 2011. *Utilization Of Simple Composter As An Alternative Solution To Overcome The Waste In Podosugih Housing Pekalongan City*, Tesis Program Magister, Fakultas Pertanian Universitas Pekalongan.

[6] Johari, A., Ahmed, S. I., Hashim, H., Alkali, H., and Ramli, M. 2012. *Economic and environmental benefits of landfill gas from municipal solid waste in Malaysia*, *Renewable and Sustainable Energy Reviews,* 16, *2907– 2912.*

[7] Kaushal, R. J., Varghese, G. K., and Chabukdhara, M. 2012. Municipal solid waste management in india-current state and future challenges : a review, *International Journal of Engineering Science and Technology (IJEST*), **4, 04**, *1473.*

[8] Kementerian Keuangan Republik Indonesia. 2011: *Masterplan Percepatan Dan Perluasan Pembangunan Ekonomi Indonesia 2011-2025*, Lampiran Peraturan Presiden Republik Indonesia Nomor 32 Tahun 2011.

[9] Khajuria, A., Yamamoto, Y., and Morioka, T. 2010. *Estimation of municipal solid waste generation and landfill area in asian developing countries*, *Journal of Environmental Biology*, **31, 5**, *649-654.*

[10] Saraswati, E. 2007. *Institutional Development Model of Urban Waste Management Based on Community Participation (Case Study : Bandung)*, Sekolah Pasca Sarjana Institut Pertanian Bogor.

[11] Sharholy, M., Ahmad, K., Vaishya, R.C., and Gupto, R.D. 2007. *Municipal solid waste characteristics and management in allahabad*, india, *Journal of Waste Management,* **27, 4**, *490-496.*

Control of Solar Power Plants Connected Grid with Simple Calculation Method on Residential Homes

Kiki Kananda[1] and Refdinal Nazir[2]
[1] Institut Teknologi Sumatera, Lampung Selatan, 35365, Indonesia
E-mail: kiki.kananda@itera.ac.id
[2]Universitas Andalas, Padang, 25163, Indonesia
E-mail: refdinalnazir@yahoo.co.id

Abstract: One of the most compatible renewable energy in all regions to apply is solar energy. Solar power plants can be built connected to existing or stand-alone power grids. In assisting the residential electricity in which there is a power grid, then a small scale solar energy power plants is very appropriate. However, the general constraint of solar energy power plants is still low in terms of efficiency. Therefore, this study will explain how to control the power of solar power plants more optimally, which is expected to reactive power to zero to raise efficiency. This is a continuation of previous research using Newton Rapshon control method. In this study we introduce a simple method by using ordinary mathematical calculations of solar-related equations. In this model, 10 PV modules type of ND T060M1 with a 60 Wp capacity are used. The calculations performed using MATLAB Simulink provide excellent value. For PCC voltage values obtained a stable quantity of approximately 220 V. At a maximum irradiation condition of 1000 W / m2, the reactive power value of Q solar generating system maximum 20.48 Var and maximum active power of 417.5 W. In the condition of lower irradiation, value of reactive power Q almost close to zero 0.77Var. This simple mathematical method can provide excellent quality control power values.

Keywords: solar, control, power

1. Introduction

Now, electrical energy needs is something to be fulfilled. The need for electric energy has moved closer to the primary needs of some people or groups, as well as in national issues. Various efforts doing by Indonesia in meeting the needs of various energy sources, the improvement of all aspects for each source of energy. Renewable energy sources is one of the energy sources that currently occupy the first position as an alternative energy. So the development of power plants with renewable energy is highly recommended in the acceleration of national development. The power plants on a medium or small scale, scattered or centered following the wisdom of local resources.

One source of renewable energy that is very flexible to be built anywhere is solar energy. Indonesia's geographical condition is very supportive for the construction of solar energy power plants on a large scale or even small. The common obstacle faced is the efficiency of solar panels that are still low. Efforts to improve efficiency have been largely undertaken by previous researchers [1] [2]. Research has also been done by author using the Newton Raphson method in optimizing the flow of active power from solar energy generation to PLN network or load [3]. Design system built on the residential scale and connected to the grid (PLN network). The aim is to maximize the power flow in the previous review of the buy-back system [4]. This research will continue the previous research but by introducing a more simple and easy to understand control method. By utilizing simple mathematical calculations to produce a system control that still gives maximum results.

The design of solar energy generation system connected to the grid will be simulated with MATLAB.

1.1. System Description

Photovoltaic (PV) systems are devices that convert solar energy into electrical energy. PV consists of several solar cells, which each cell is connected to another in series or parallel to form a series of PVs commonly called PV modules [5]. The maximum operating efficiency of a solar cell depends on the maximum power point (MPP) of the PV system. In its operation, PV performance is strongly

Influenced by local climatological conditions (environmental temperature and solar radiation) and electrical parameters (short circuit current, opern circuit voltage, PV cell temperature, MPP, etc.) [5] [6] [7]. The characteristics of PV systems are not very linear because they are influenced by external factors (environment). Solar irradiation, temperature and wind velocity are the major environmental factors affecting PV [6].

To build a solar power plant (PLTS) system are used a single phase inverter. Single-phase inverters are used as DC voltage converters into AC for load fulfillment and connection to the grid by using SPWM technology. Control of SPWM to set output voltage to have frequency according to fundamental frequency of load. The output voltage waveform can be seen from Figure 1 above.

For PLTS system connected grid where PLTS is connected to grid through inverter without battery. These systems can be classified as small systems, such as residential roof systems or connected to large grid systems. In the case of PLTS inverter connection with the grid, then the voltage and frequency should be in sync. Interactive inverters with the network not only the power condition of the PLTS output but are assured that the PLTS output is synchronized with the PLN network.

1.2. Optimization of PLTS Systems Connected Grid

The PLTS inverter is connected through a transformer with an equivalent impedance L at the PCC point. The phase and voltage magnitude of the inverter is controlled to regulate the active and reactive power of the PLTS inverter supplied to the grid. The real power of inverter the PLTS is equal to the output power of the array PV minus the inverter losses due to the energy balance between the DC input and the inverter PV AC output [10]. And the average output power of the PV inverter to the grid is described by equation (2) below. From the equation the Inverter PLTS output voltage can be controlled. With the output of the inverter is affected by the modulation index according to equation (1). This control system can be analog system or with microprocessor. This control system produces waveforms and amplitude and phase waveform regulation to control the power flow between the PLTS and the grid. The grid interfaced with a PV inverter, controlled voltage or controlled current, has a bidirectional power flow potential. Not only supplying local loads, but also can export active and reactive power overload to the grid. A control system approach is needed to reduce some error in power transfer due to synchronization error, which can make the inverter overload [9]. The simple interface of a grid connected inverter with a first order filter can be shown in Figure 3 below.

PLTS systems in grid-connected residential homes are designed using circuits such as Fig. 4 Controls are performed so that the accumulated energy absorbed from the grid within a given time range is zero. By knowing the characteristic pattern of the output of PLTS system and the house load, it can be calculated controlling so that the total power of PLTS can meet the load in certain period, so there is only temporary borrowing / storage of electrical energy in the grid system. The amount of energy supplied from the grid for the residential load and the amount of power supplied by the PLTS system to the grid is controlled for the same total value in a given period.

2. Methode

In this system used 60W peak solar module as much as 10 pieces and planned to supply a residential house with 450VA installed power.

Table 1. Electro Optical Characteristic SHARP ND T060M1

Model	ND-T060M1				
Parameters	Symbol.	Min	Typ.	Unit	Condition
Open circuit voltage	Voc	—	22.0	V	Irradiance: 1000 W/m^2
Maximum power voltage	Vpm	—	17.4	V	
Short circuit current	Isc	—	3.90	A	
Maximum power current	Ipm	—	3.45	A	Module temperatur: 25°C
Maximum power	P_m	57	60	W	
Module efficiency	ήm	—	11.36	%	

The modeling of solar panels in the MATLAB simulation is modeled based on a series of solar cell equivalents. By modeling the component of current source and diode as a controlled DC current source. The control system used for the PLTS system based on the power system flow equations connected to the PLTS grid.

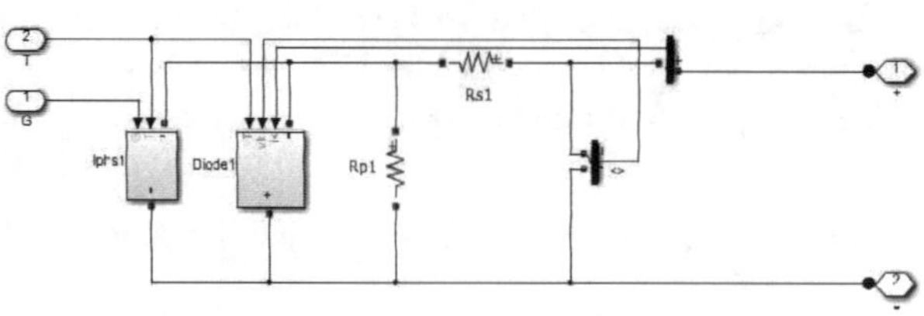

Figure 1. Solar Panel Series Simulink Model

For the system control system of PLTS using equations (2) and (3) we obtain the values of α and m_a by determining that the constant Vpcc value of 220 and Xl is the reactance of the system channel connecting the inverter terminal with the grid and load terminals (L and transformer). By wanting the reactive power of the PLTS system to close to zero, the value of Q in equation (3) is made zero. The equations P and Q are the equations of the functions of α and m_a which have the roots of the equations at certain α and m_a values. At the root values, P and Q are optimum as desired. Calculation of the root value by means of completion of substitution. From equation (3), the value of P is set as the desired maximum P value, according to the PV array output and the output of the PLTS system must meet the synchronization requirement that is the voltage and frequency value equal to the grid network [9]. The α and m_a control equations are described below.

$$P_{max} = \frac{VV_{pwm}}{X} \sin \delta \qquad (4)$$

With a is the ratio of the transformer used 60/230. Then the equations (6) and (7) are used as equations to control the sinusoidal signal SPWM gate inverter. The value of XL constant used is = 10.04407446 ohm. This value is calculated using equations (5) and (7) by inputting manual-controlled initial simulation data for some variations of load and irradiation. Here is a complete series of simulation models of PLTS systems connected to a single phase grid at home (Figure 6).

PV System Connected Single Phase Grid at Home

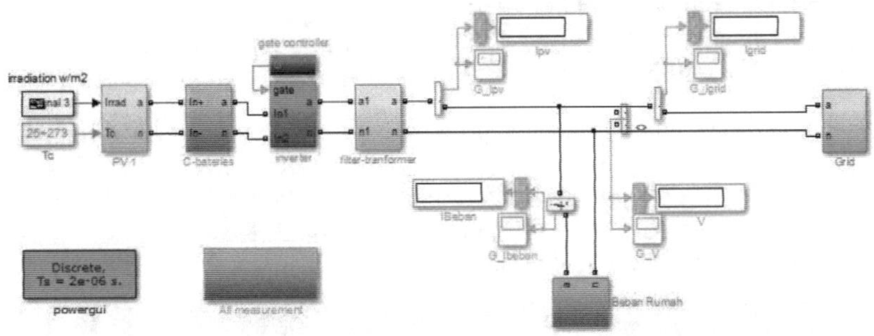

Seen that the absorbs, supplies and also neutral conditions on the grid. When the output power of the PLTS system exceeds the load requirement, the grid will absorb it, while when the output of the PLTS system approaches the same as the load requirement, the grid also supplies or absorbs with a value close to zero. This means that at the time the house load is filled by the PLTS system without add with the grid. But when the PLTS system conditions are changed to insufficient load demand, the grid returns as a supplier.

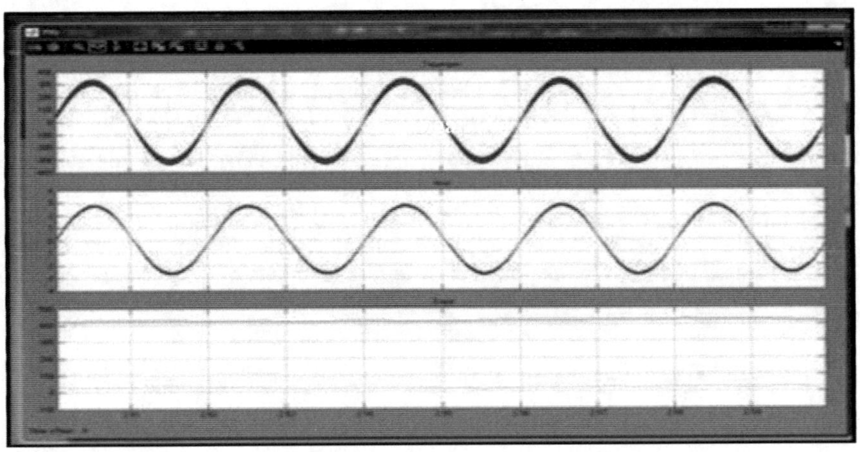

Figure 2. Simulink Circuits of PLTS Connected Grid System

3. Result and Discusion

The purpose of solar panel modeling is to achieve the IV curve similar to the graph given on the solar panel manufacturer's data sheet. The more precise the value, the better the model. From the simulation results obtained curve characteristics of solar panel models are almost the same and in accordance with the desired results. The IP-V curve corresponds to the theory of solar panel characteristics. Solar panel modeling is said to be successful by looking at the output characteristics of the modeling. Such as testing performed by previous studies [1] [5].

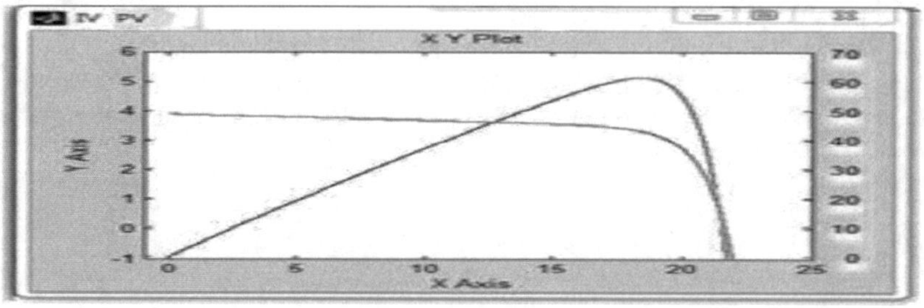

Figure 3. Curve of IP-V Characteristics of Solar Panels

The voltage value of Vm and Im current at module maximum power is almost the same as indicated by SHARP NDT060M1 solar panel manufacturer. Where the maximum power Pmax 63.87 W and the maximum current Im 3.65A at a voltage of 17.51 Volts. The characteristics of IP-V simulation results shown by Figure 7. Next on the test, the system is fed irradiated inputs as well as residential loads that change. In Figure 9. The simulation test results, the power output of the PLTS system is proportional to the irradiation input value. It is

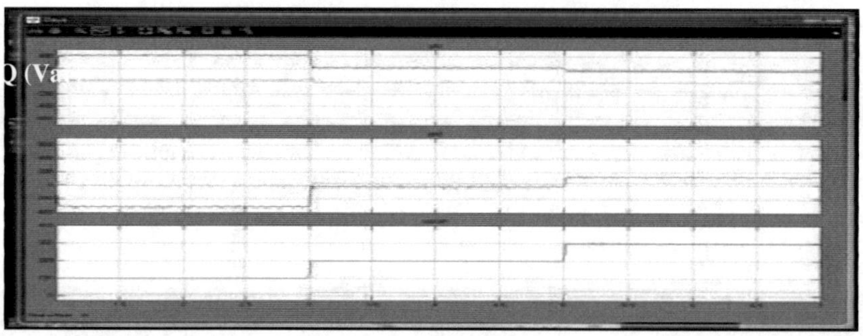

Figure 4. The Power Flow Curve of Iradiation and Load Systems Changes

The output waveform of the PLTS system for each irradiation variation can be seen in the figure below. From the current waveform and voltage can be the work area of each system. Where for the side of the PLTS system, the current and voltage wave are in phase, as well as the home side load waveform. As for the grid side, voltage and current waveforms can be in phase and different phase 180o according to whether the grid conditions absorb or supply power in the system. In this experiment it is clear for each working of PLTS connected grid system. The conditions under which the grid becomes a supplier and absorber for a certain time. The flow of power that occurs is as in Table 2 below.

Table 2. Power Flow of PLTS System Connected Grid with Irradiated and load Changed

No	Irradiation W/m2	Home Load	Vpcc (Volt)	Active Power P (W)			Reaktif Power Q (var)		
				PLTS	Grid	Home	PLTS	Grid	Home
1	1000	100W 10Var	220.6	417.5	-317.2	100.3	20.48	-10.46	10.02
2	600	200W 20Var	220	226.1	-26.32	199.8	-0.7744	20.75	19.98
3	500	300W 30Var	219.7	170.2	128.9	299.1	-7.904	37.8	29.9

In this system, PLTS system is designed using topology without DC-DC converter and without battery storage. So the system control is applied to the inverter. The selection of a simple topology and with simple system control to be applied to a residential house with a PLTS grid connected system. Cost savings and components by reducing battery and DC-DC converter. While the control by using MPPT can cause rise in PCC point voltage, so it will damage in the inverter and also the grid network. Then the designed control system must be able to keep the PCC voltage constant but still deliver maximum power to the system. This designed control system keeps the inverter working area always close to being an inverter unity, so the possibility of inverter being loaded or acting as a bidirectional converter is very small. Using equations (5) and (7) has been tested for some experiment mode conditions by the calculation of the equation roots substituted. From the test results can be said that the control system works quite well. The voltage at the PCC point approaches a stable value of 220Volt. For the reactive power of the PLTS system as shown in Table 2, the maximum value of the PLTS system Q on the test is 24.64Var. The values are still tolerable. While the maximum value of P active power of PLTS system at irradiation 1000 W / m2 is 419.2W. Although small of the maximum value given on the topheet is 60Wx10 = 600 W. This decrease in the maximum value of power is due to losses in the inverter to balance the input DC and AC output. By looking at the reactive power and phase phases and the voltage of the PLTS system, the active power supplied is the maximum power of the existing system after the reduction of the interface circuit of the PLTS system.

4. Conclusion

Simulation modeling for controlling solar power systems connected to the grid at home by using simple mathematical methods can work well. Point voltage PCC systems can be kept constant ≈220 Volts. In a maximum irradiation condition of 1000 W / m2, the reactive power value of Q solar generating system is maximum 20.48 Var and maximum active power of 417.5 W. In the lower irradiation conditions the reactive power Q value is almost close to 0.77Var. This control method is simple so it can be applied easily. This method is not only at home but is also suitable for other small-scale generators as it provides excellent quality power control values.

IOP Conf. Series: Earth and Environmental Science **97** (2017) 012015 doi:10.1088/1755-1315/97/1/012015

5. References

[1] Jaleel, Dr. J. Abdul., et.al. 2012. *Simulation on Maximum Power Point Tracking of the Photovoltaic Module using LabVIEW*. International Journal of Advanced Research in Electrical, Electronics and Instrumentation Engineering Vol. 1, Issue 3, September

[2] Ki, Byung., et.al. 2012. *Implementation of Monitoring and Control Systems for 50KW PV Systems Using the Wire-Wireless Network*. International Journal of Grid and Distributed Computing Vol. 5, No. 1, March,

[3] Nazir, Refdinal., et al. 2014."*Optimization Active and Reactive Power Flow for PV Connected to Grid System Using Newton Rapshon Method*" *ICSEEA Energy Procedia Vol 68*, pp.77-86

[4] Nazir, Refdinal., et al. 2014."*Optimization Active and Reactive Power Flow for PV Connected to Grid System Using* *Kananda*, Kiki., et.al. 2013. "Concept of Power Flow Arrangements for PLTS Connected to Grid System at Home" *JNTE Vol 2 No 2,* pp.65-71.

[5] Hernanz, J.A. Ramos-., et.al. 2012. *Two Photovoltaic CellSimulationModelsInMatlab/Simulink*. International Journal on "Technical and Physical Problems of Engineering. March 2012 Issue 10 Volume 4 Number 1 Pages 45-51.

[6] Mahmud, M. A., et.al. 2012. *Dynamic Stability of Three-Phase Grid-Connected Photovoltaic System Using Zero Dynamic Design Approach*. IEEE Journal Of Photovoltaics, Vol. 2, No. 4, October 2012.

[7] Fu Wu, Tsai-., et.al. 2011. *Power Loss Comparison of Single- and Two-Stage Grid-Connected Photovoltaic Systems. IEEE Transactions On Energy Conversion*, Vol. 26, No. 2, June 2011.

[8] Azzakir, R. 2009. *Simulation of One-phase Inverter Vehicles with Electrical Networks. The undergraduate's final assignment*. Electrical Engineering Faculty of Andalas University Engineering

[9] M. H. Rashid. 1999. *Power Electronics: Circuit, Device and Its Application*, Volume 1, Ary Prihatmanto MS, PT Prenhallindo.

[10] Li, Huijuan, et.al.(_)*Real and Reactive Power Control of a Three-Phase Single-Stage PV System and PV Voltage Stability*. UT-Battelle, LLC, under Contract No. DE-AC05-00OR22725 with the U.S. Department of Energy.

Planning Of Drainage Channel Dimension In The Core Zone Of Muara Takus Temple

Alfian Saleh
Universitas Lancang Kuning, Pekanbaru, 28265, Indonesia
Jl,Yos Sudarso Km.8 Rumbai-Pekanbaru
E-mail:alfian.saleh@gmail.com/alfian.saleh@unilak.ac.id

Abstract: Preservation of Cultural Heritage is a dynamic effort to maintain the existence of cultural heritage by protecting, developing, and utilizing the cultural heritage in the contemporary context. To protect the cultural heritage in term of conservation called protection of which the effort to prevent and overcome from damage, it needs to do destruction or obliteration through rescue, security, zoning, maintenance, and restoration of cultural heritage. The most fundamental issue is the hydrological impact of the existence of Hydroelectric Power Koto Panjang located around Muara Takus temple that could threaten the sustainability of the region. In this case, hydroelectric dam frequently causes Kampar Kanan River overflowed thus potentially floods, especially in the rainy season that could eventually submerges Muara Takus area. The total area of the region Muara Takus enshrinement is ± 94.5 hectares that are divided into two main parts. Those are the terrestrial land of ± 56.44 m^2, and PLTA Koto Panjang lake of ± 38.06 m^2. Consequently, it is necessary for drainage planning of economical dimension in the core zone of Muara Takus temple. Furthermore, from the data of the maximum rainfall of 101 mm/ day obtained a discharge of rainfall of 0.38 m^3/ second so that this discharge of rainfall can be designed drainage channel dimension to accommodate the discharge of rainfall. From the analysis of dimension designed drainage is the size of 30 cm x 45 cm. this dimension can accommodate the discharge rainfall that is equal to 0.43 m3 / second. Regarding the finding, it can be concluded that the discharge of rainfall that occurred less than discharge calculation of dimensional analysis of drainage channel so that the size of this dimension can accommodate discharge rainfall occurs.

Keywords *: Muara Takus Temple, Discharge, Channel Dimension*

1. Introduction

Muara Takus Enshrinement of cultural heritage viewed from the geographical location, has a very strategic layout, i.e. in the middle of the cruise lines trade between India and China. Therefore, it has become a liability to the people of Indonesia as his successor to attempt to preserve it. However, the activity of preservation is not an easy thing. Various issues must first be identified so that the concept of preservation can be designed accordingly. The observation in the field showed that the existence of two things that need to be taken into account in the planning of preservation, that is associated with the presence of cultural heritage Muara Takus Enshrinement and social issues. The economy and culture are estimated to be a challenge or obstacle to doing preservation.

The most fundamental issues in the area of Muara Takus temple complex is about the hydrological impact of the existence of Koto Panjang Hydroelectric Power Plant (PLTA) located around the Muara Takus temple that can threaten the sustainability of the area. PLTA dams often cause Kampar Kanan River overflowed thus potentially flooding, especially in the rainy season which could eventually sink the area of Muara Takus Enshrinement. The total area of Muara Takus Enshrinement area is ± 94,5 hectares. Land use in the Muara Takus Temple area is divided into two main parts, ie terrestrial land of ± 56.44 m², and PLTA Koto Panjang lake of ± 38.06 m².

The existence of a pool of reservoirs which almost surrounds the entire complex, only about 16% of the circumferences of the complex are not inundated in water reservoirs face elevation 85 m above sea level. n addition, the drainage conditions around the temple complex that are still no existence. Therefore, it is necessary to have a suitable drainage design applied in the temple complex and the appropriate drainage dimension.

2. Literature Review

2.1. The concept of Site Preservation

Conservation is a dynamic effort to preserve the existence of the Heritage and its value by protecting, developing and utilizing it. Protection is an effort to prevent and cope with damage, destruction, or obliteration through rescue, security, zoning, maintenance, and restoration of cultural heritage. The terms of protection discuss in the following.

1) Rescue is an effort to prevent and/or to cope with cultural heritage from damage, destruction, or ruin.
2) Security is an effort to maintain keeping cultural heritage and to prevent from threats and/or disruption.
3) Zoning is the determination of the boundaries of the spatial site of cultural heritage and the heritage area according in accordance with needs.
4) Maintenance is the effort of keeping and caring for the physical condition of cultural heritage in order to remain sustainable.
5) Restoration is the attempt returns the physical condition of objects of cultural heritage, Heritage Buildings and structures damaged cultural heritage in accordance with the authenticity of the material, shape, layout and/or machining techniques to prolong its age.

The site of Muara Takus Enshrinement seemed prepared also as a well-planned settlement. As the boundaries of settlements made land dikes that serve as a fortress, therefore the area of this site was once known as *Koto* which means the *Benteng*. The land surface elevation is a reference elevation of this site is 86.5 m above sea level, higher than 1.5 m from the stakes made by Koto Panjang Hydroelectric ie at 85 m above sea level.

In the site there are buildings that are used for religious purposes as well as non-religious interests. The largest religious building is called Muara Takus Temple. The building that is not for religious interest can be known from the foundation ruins, such as the excavation from 1985, the building of the excavation by Balai Pelestarian Cagar Budaya Batusangkar , and new findings of the excavation of Balai Medan. It is estimated that there is archaeological data now in the Koto Panjang hydropower lake that cannot be traced back.

2.2 Drainage

The word *drainase* comes from the word *drainage* which means to dry or drain. Drainage is a system designed to deal with the problem of excess water both excess water above the surface ground and water under the surface ground. Drainage has several types. They are:

1) Based on the way of formation of drainage

   Types of drainage on the basis of how the formation, can be grouped into:

   a. Natural drainage; Natural drainage is formed through a long-lasting natural process. Drainage channels are formed by water scouring according to the soil contour. This natural drainage is formed on soil conditions sufficient slope, so the water will flow by itself, into the rivers. On the ground are quite porous, water on the surface of the soil will soak into the ground (infiltration). The water that seeped turned into a stream between (sub-surface flow) flows to the river, and can also flow into the ground (perlokasi) up to groundwater which then together with the groundwater flow as a stream of ground water (ground water flow) to the river. Generally this natural drainage of a River along with its tributaries that form a network of river channels.

b. Artificial Drainage; Artificial drainage is a system made with a specific purpose and is the result of engineering based on the results of the calculations performed for efforts to perfect or complete the lack of natural drainage system. At require artificial drainage system costs both in planning and in making.

2). Drainage based on the layout; This type of drainage is reviewed based on its layout, can be grouped into:

a. Surface Drainage; Surface drainage is drainage channel system is above the soil surface water drainage occurs because of the height difference of the surface of the channel (slope).
b. Subsurface drainage; Subsurface drainage is a drainage system that flowed underground (grown) usually for artistic system or in an area that does not allow it to drain the water above the ground as on the sports field, airports, parks and more.

3). Drainage based on its function; Type of drainage based on its function, can be grouped into:

a. Single Purpose Drainage; Single purpose drainage is a drainage channel that functions to drain one type of water such as rain water or waste water or other.
b. Multi-Purpose Drainage; Multi-purpose drainage is a drainage channel that functions to drain more than one waste water either mixed or alternately e.g. mixture of rainwater and waste water.

4). Drainage based on its construction; The type of drainage reviewed based on its construction can be categorized as:

a. Open Channels Drainage; Open drainage is a drainage system whose surface water is affected by the outside air (atmosphere). Open drainage usually has a sufficient area and is used to drain rain water or wastewater that does not endanger the health of the environment and does not interfere with the beauty.
b. Closed Channels Drainage; Drainage of closed channels is a system of channels whose surface water is not affected by the outside air (atmosphere). Closed drainage channels are often used to drain wastewater or dirty water that interferes with environmental health and disrupts beauty.

3. Research Method

This research employed field survey method, namely conducting the research through observing the geographical condition of complex area of Muara takus temple directly. Muara Takus Temple Site is administratively located in Muara Takus Village, District XIII Koto Kampar, Kampar District, Riau Province. In astronomical site is located at position 0°20'9,7 "LU and 100°38'31,3" BT, and topographically situated in a hilly area with a weak wave height of ±86.5 meters. To design this drainage dimension required some data that must be obtained such as:

3.1 Flow capacity due to rain

The rain that occurs causes the rainwater that possibility, largely stagnant and flowing on the ground surface (run off) and a small sink in (infiltration) into the subsoil. If the surface of the soil occurs greater inundation of infiltration, then to stream water drainage used advance land. Maximum flow (discharge) capacity is analyzed by rational method:**Error! Bookmark not defined.**

$$Q = \alpha \cdot \beta \cdot It \cdot A \quad \text{.....................................(1)}$$

where:

Q = flow debit (m^3/second)
α = Coeficient of run off
β = Coefficient of rain dispersion
It = Rain intensity
A = area

3.2 Coeficient of run off (α)

It is categorized becomes: The run off coefficient is the comparison between the rainy part that ran on the earth with the total rainfall. The run off coefficients are mostly intermediate, but preferably for analysis, the largest or maximum value is used. Where the run off value is as follows in accordance with the existing land use.

Type Of Land	Coeficient of run

	off
The steep mountains Bumpy land and Forest Cultivated plains Waterproof roof Asphalt Pavement and concrete Soil solid, hard to be infused The soil is rather easily impregnated Park / open field Garden Housing is not so tightly (20 houses/hectares) Medium density housing (21 - 60 houses/hectares) High density housing (61 - 160 houses/hectares) Recreational Area Industrial Area Commercial Area	0,75 – 0,90 0,50 – 0,75 0,45 – 0,60 0,75 – 0,90 0,80 – 0,90 0,40 – 0,55 0,05 – 0,35 0,05 – 0,25 – 0,20 0,25 – 0,40 0,40 – 0,70 0,70 – 0,80 0,20 – 0,30 0,80 – 0,90 0,90 – 0,95

3.3 Coefficient of Rain Dispersion (β)

The rain-dispersion coefficient depends on the area of the rain-fed region according to the study area. The coefficient of rain dispersion can be seen in the following table:

Land Area (km^2)	Koefisien β
≤ 4	1
5	0,995
10	0,980
15	0,955
20	0,920
25	0,875
30	0,820
50	0,500

3.4 Rain intensity (It)

Rain intensity was analyzed based on rainfall data and rainfall time concentration data. The formula used is the *Mononobe Formula* as follows:

$$It = (R/24)(24/tc)^{2/3} \qquad (2)$$

where:

It = Rain intensity
R = Duration, rainfall
Tc = Time concentration

3.5 Channel Dimensions

The flow capacity due to rain should be drained through the drainage channel to the point of downstream planning. Rainfall discharges are analyzed to become volumetric flow to make the channel dimension.

$$Q_{Rain} = Q_{Channel} = Fs \cdot v \qquad (3)$$

where:

Fs = Channel Design
v = Speed of water flow in the channel

to analyze the speed of water flow determined based on *Manning* atau *Chezy* formula:

Manning Formula:

$$v = 1/n \, . \, Rs^{2/3} \, . \, I^{1/2} \quad \text{....................................(4)}$$

where:

v = Speed of water flow in the channel (m/second)
n = Coefficient of wall roughness
Rs = Hydraulic radius = Fs/Ps (wide channel /Channel wet circumference)
I = Slope channel plan

Chezy Formula:

$$v = C\sqrt{Rs.I} \quad \text{....................................(5)}$$

Coeficient of *Chezy* :

$$C = (100\sqrt{Rs}) \, / \, (0{,}35 + \sqrt{Rs})$$

$$v = \frac{100.Rs.I^{\wedge 1}/2}{0{,}35 + Rs^{\wedge 1}/2} \quad \text{.......................(6)}$$

where:

v = Speed of water flow in the channel (m/second)
Rs = Hydraulic radius = Fs/Ps (wide channel /Channel wet circumference)

4. Findings and Discussion

Based on the results of field surveys Muara Takus temple complex area directly adjacent to Kampar River, Thus, flow of water around Muara Takus temple complex was to be large and easily flooded. So, it is necessary an action of countermeasures the large water discharge so as not to inundate Muara Takus temple complex that is by designing the drainage channel in the temple complex.

4.1 Analysis of Drainage Channel Dimension

Data obtained from that maximum rainfall in the region of the Temple of muara takus i.e. R = 101 mm/day. With an area of 94.5 Ha = 0.945 km^2
to find the discharge by using this formula:

$$Q = \alpha \, . \, \beta \, . \, It \, . \, A$$

So, from the result of field survey obtained score:
α Data from the area around the temple was a plantation area α = 1 obtained from land area < 4 km^2 where the data was obtained from all of land area of temple which was 94,5 Ha = 0,945 km^2

$$It = (R/24)\,(24/tc)^{2/3}$$

Where:

R = 101 mm/day = 0,0042083 m/hours
tc = Seen from channel slope table

Channel Slope I (%)	Average Speed v (m/second)
< 1	0,40
1 - <2	0,60
2 - <4	0,90
3 - <6	1,20
4 - <10	1,50
5 - <15	2,40

So, seen from the topographic map, it can be said that the level of slope from complex area in candi muara takus temple is <1% so the value of v = 0.40 m/ second so that the value of tc = 20 seconds

$$It = (R/24)(24/tc)^{2/3}$$
$$= (101/24)(24/20)^{2/3}$$
$$= 2{,}02$$

$$Q = \alpha \,.\, \beta \,.\, It \,.\, A$$
$$= 0{,}2 \,.\, 1 \,.\, 2{,}02 \,.\, 0{,}945$$
$$= 0{,}382\ m^3/s$$

So that the dimension of cross-sectional surface drainage to around the pedestrian path is made rectangular dimension by using formula:

$$Q = Fs \,.\, V$$

$$v = \frac{1}{n} R_s^{2/3} I^{1/2} \quad \text{(m/s) manning}$$

Rs = Hydraulic radius (Fs/Ps)
Fs = Wet cross-sectional area
Ps = Circumference of wet cross section
I = Channel slope

Where the slope of channel made 2%

$$v = (1/0{,}018)\ (Rs)^{2/3} \,.\, (0{,}02)^{1/2}$$
$$= (1/0{,}018)\ (Fs/Ps)^{2/3} \,.\, (0.02)^{1/2}$$
$$= (1/0{,}018) \,.\, (0{,}135/1{,}05)^{2/3} \,.\, (0{,}02)^{1/2}$$
$$= 0{,}313$$

By using a trial error then obtained the drainage dimension is 30 cm x 45 cm

$$Q = Fs \,.\, v$$
$$= 1{,}350 \,.\, 0{,}313$$
$$= 0{,}43$$

So Q rainfall < Q dimension calculation

0,38 < 0,43

So the drainage dimension used 30 cm x 45 cm can accommodate existing rainfall that is 0.38 m3 /s

5. Conclusion and Suggestion

From the survey results and data analysis obtained the size of the drainage dimension that can accommodate the discharge of rainfall is the size of 30 x 45 cm rectangular shaped under it is given an absorption hole.

6. References

[1] Pemerintah Daerah Kabupaten Kampar. *Rencana Program Investasi Jangka Menengah (RPIJM) Wisata Terpadu Candi Muara Takus Kawasan Agropolitan Kecamatan XIII Koto Kampar*. Pemerintah Daerah Kabupaten Kampar. 2010.

[2] Balai Pelestarian Cagar Budaya Provinsi Sumatera Barat, Riau dan Kepulauan Riau. Laporan Focus Group Discussion (FGD) Pelestarian Candi Muara Takus Kabupaten Kampar, Provinsi Riau. Pekanbaru: Kementerian Pendidikan dan Kebudayaan. 1 Juni 2015.

[3] Simoen, Soenarso. "*Dampak Hidrologis Pembangunan Waduk Kotapanjang terhadap Kompleks Candi Muara Takus di Riau*" dalam Majalah Geografi Indonesia Vol 14, No 1 (2000) Fakultas Geografi Indonesia dan Ikatan Geografi Indonesia, 2014.

[4] Pemerintah Daerah Kabupaten Kampar dan P4W. *Masterplan Kawasan Agropolitan Kecamatan XIII Koto Kampar*. Pemerintah Daerah Kabupaten Kampar. 2009.

IOP Conf. Series: Earth and Environmental Science **97** (2017) 012016 doi:10.1088/1755-1315/97/1/012016

[5] Jalil, Abdul. "*Konflik Sosial Dalam Pengembangan Objek Wisata Candi Muara Takus Di Desa Muara Takus Kecamatan XIII Koto Kampar Kabupaten Kampar*" dalam *Jom FISIP Volume 2 No. 1 – Februari 2015*. FISIP Universitas Riau. 2015.

The Variation Test and Extraction Equipment to Optimum Asphalt by Using Gasoline Solvent

Fitridawati Soehardi
Universitas Lancang Kuning, Pekanbaru, 28265, Indonesia
Tel/Fax: +62761 52324
E-mail:fitridawati@unilak.ac.id

Abstract: Based on the Binamarga Specification 2010 at third revision, the extraction test should be carried out using the specimen from the loose asphalt mixture extracted from the back of the finisher bitumen machine. The purpose of this research is to find out the result of pretest and posttest extraction asphalt content. The Extraction test using two equipment, they are Soklet and Centrifuge. The specimens was used AMP, Asphalt Finisher and Core, which involved gasoline solvent. Based on the asphalt level extraction test results, the appropriate equipment was used centrifuge with the level accuracy as requirement of Binamarga Specification 2010 at third revision and the level of ease used as equipment in the field study. The asphalt content obtained for AMP 5,51%, Asphalt Finisher5,46% and Core 5.34%. As for the socket asphalt content obtained is AMP 5.55%, Asphalt Finisher 5.50% and Core core 5. 41%. The extract test value of asphalt content decreased, so it can be formulated KA JMF <KA AMP <KA Asphalt Finisher <KA Core. The value of Job mix used was 5.56% with the tolerance given according to Binamarga Specification 2010 at third revision is ± 0.30%. In accordance with the results obtained then the results of a centrifuge tool that matches and meets the requirements of time, accuracy of results and economic value.

Keywords: Extraction, centrifuge, socket, Asphalt Level, Gasoline

1. Introduction

Based on the 2010 Revised Specification of Binamarga 2010, the extraction test be carried out using the specimen from the loose asphalt mixture extracted from the back of the finisher bitumen machine. Implementation of the field found asphalt content of extracted test results are not appropriate and reduced from the value of bitumen levels of *Job Mix Formula*.

Asphalt content is one of the most important parameters in the overall quality of concrete asphalt pavement in addition to its gradation and aggregate properties. If the bitumen content is less than required, there will be surface defect in the form of hole, release of grain while if the asphalt content is bigger than required, there will be crack in the form of crack shrink, curly, obesity, etc. Gasoline is a mixture of various hydrocarbons obtained by distillation or refining of crude oil. The quality of gasoline is expressed by the octane number of 80. The reason gasoline is used in this study that gasoline is easy found in the field as well as facilitates the extraction testing.

The purpose of this research is to find out the result of bitumen content in pretest and posttest extraction. Extracting test using two tools that is Soklet and Centrifuge. This study uses the test specimens which was taken from the combination of asphalt production, gasoline was used during slap and compactions field. Several related researches, first Hadijah, conducted that extraction tests of appropriate bitumen content that used in this research was using Centrifuge with gasoline solvent with the level of accuracy to meet the requirements and level of ease of tools to be brought to the field (project) and solvent easily obtained. the second researcher is Shamier who was conducting asphalt test using socklet to core drill result but asphalt content of average value of asphalt STA 1 + 175 = 6,020% and STA 3 + 875 = 6,015%, but the result did not match with job specification. The third research is Anggraini who used Pertamax plus as solvent in extraction of content dissolves bitumen than gasoline solvent. Where the deviation is obtained in AMP -0,03%, from finisher -0,01%, and core -0,02%. The next researcher is Putri she conducted extracting test on mixed AC-WC Gradation Mixture. The Extracting test on mixed AC-WC Gradation mixture The extraction asphalt level for the mixture behind the asphalt finisher has a -0.05% deviation to mixed asphalt content in AMP.

The asphalt content of the extraction from the core had a deviation of -0.08% against the mixed asphalt level behind the asphalt finisher. Soehardi conducted extraction analysis of asphalt content of AC-WC gradation mixture with coarse gasoline. The value of pore content after extraction decreased from JMF pore value of AMP, finisher, and coredrill test object was 1.062%, 0.823%, 0.878%, with the mean value of deviation of 0.273%. With the average value of pore content for AMP, finisher, and coredrill test object is 0.673%, 0.667%, 0.602% with the average value of deviation of 0.273%.

2. Literature Review

Asphalt or bitumen is a material that is visco-elastic blackish brown color that will soften and melt when it gets warm up and vice versa. This viscoelastic property makes the asphalt able to envelop and hold the aggregate firmly in place during the production process and the service life. Generally, asphalt is produced from petroleum refining, so called as hard asphalt. The degree of control performed at the stage of the distillation process will produce asphalt with special properties suitable for specific applications. The asphalt content is the percentage of the asphalt weight to the mixture that ranges from 4.5 to 7.5 percent. Appropriate asphalt content should be determined on the basis of Marsall's way of testing to obtain a mixture that meets the requirements.

The main function of asphalt for pavement forming process is mixing process before it is spread out and after it is spread differently. In the process of pre asphalt mixed with aggregate would be wrapped or enveloped totally grains, fill the pores between the grains, and seep into the pores of each grain. In the post-asphalt process fill the pores of the total layer.

2.1 Paving Alloy Extraction.

Extraction according to RSNI M-05- (2004) is the process of separating the mixture of two or more ingredients by adding a solvent which can dissolve one of the ingredients provide in the mixture. The extraction of asphalt mixture is necessary to recover the material composition according to the plan in order to maintain the quality of the asphalt mixture. Tests of asphalt content in asphalt mixtures should be made by companies producing at least twice a day mixture of asphalt morning and day or any production of 200 tons of asphalt mixture. There are three methods of testing the asphalt content of solvent extraction (extraction with solvent), nuclear asphalt content gauge and ignition furnace.

This research used two extraction methods in determining the bitumen content in the mixture that is the method of extracting test with Centrifuge device according to SNI 03-6894-2002. and Method of extraction testing with Soklet tool in accordance with SNI 03-3640-1994.

The formula for determining the extraction bitumen content is as follows:

$$H = (A - (E + D) / A \times 100 \% \quad (1)$$

Information :

H = bitumen content of sample (%)
A = Weight Sample before extraction (gram)
D = The weight of the paper filter (gram)
E = Weight sample after extraction (gram).

2.2 Gasoline.

Gasoline is also called Petrol. Gasoline is a mixture of various hydrocarbons obtained by the distillation / refining process of crude oil. The quality of gasoline is expressed by the octane number. Premium is a clear yellowish color distillate oil. The yellow color is due to the presence of an additional dye. The octane number is a number that indicates how much pressure cann be give before the gasoline burns spontaneously. The name octane comes from octane (C8), because of all the molecules of the gasoline compound, octane which has the best compression properties. Octana can be compressed to a small volume without spontaneous combustion, unlike the spontaneous combustion. The heptane function is as solvent in the extraction. By having a high octane the fuel will have a complete combustion.

3. Methodology

The study was conducted using laboratory testing. Te aim of this research is to determine the bitumen content using two methods that refer to extraction testing with Centrifuge tools using SNI 03-6894-2002 and extraction testing with Soklet tool using SNI 03-3640-1994. Extracting test of bitumen content using gasoline solvent. In each specimen of AMP, asphalt finisher, and core. Sampling on Marpoyan-Lipat Kain road segment. The results were compared opposite the parameters reviewed ie the method of testing and the retrieval of specimens.

4. Results and Discussion

The recapitulation of asphalt extraction test results can be seen in a follow.

Table 1. Asphalt Levels In Mix After Extraction Using Centrifuge Method.

No	Sample	Asphalt Level (%)		Specification Tolerance (%)
		Extraction Results (%)*	JMF (%)	
1	*AMP*	5,51	5,56	± 0,3
2	*Aspal*	5,46	5,56	± 0,3
3	Core	5,34	5,56	± 0,3

* Fitri analysis results (2014)

From the table it can be seen that the bitumen content of extracted using centrifuge method with gasoline solvent on the three test specimens decreased at AMP 0.13% deviation. Finisher 0.1% deviation, and core deviation 0.11%. Tests of asphalt content were carried out by using Gasoline solvent as much as ± 7 liters with the weight of specimen for each sample was ± 1200 kg.

The recapitulation of asphalt extraction test results which can be seen in Table 2 is the result of extraction test Using Soklet Method.

Table 2. Asphalt Levels In Mix After Extraction Using Soklet Method

No	Sampel	Asphalt Level (%)		Specification Tolerance (%)
		Extraction Results (%)	JMF (%)	
1	*AMP*	5,55	5,56	± 0,3
2	*Aspal*	5,50	5,56	± 0,3
3	Core	5,41	5,56	± 0,3

From Table 2 n be anazed tahtit c that asphalt level after extraction with Soklet method with gasoline solvent at AMP deviation 0,01%., In Asphalt Finisher 0,06% deviation, at Core 0,15% deviation. Tests of bitumen content were conducted by using Gasol solvent as much as ± 1 liter with the weight of specimen for each sample is ± 150 kg.

The recapitulation of asphalt extraction test results which can be seen in Table 3 is the result of extraction test Using Centrifuge Method.

Table 3. Asphalt Test Result Result Extraction from AMP, asphalt finisher, and core

	Sample	Asphalt Level (%)	Specificatio

No		Centrifuge*	Soklet	n Tolerance (%)
1	*AMP*	0,05	0,01	± 0,3
2	*Finisher*	0,10	0,06	± 0,3
3	Core	0,22	0,15	± 0,3

* Fitri analysis results (2014)

From Table 3 it can be seen that the bitumen content of extracted by using two extraction methods has decreased from JMF value but still in tolerance of Binamarga Specification 2010 revision 3. Deviation value of decrease of bitumen content Soklet Method for AMP is 0,04%. The finisher is 0.4%, and the core is 0.7% lower than using the centrifuge method. Extraction testing method used Soklet tool with gasoline solvent which more precisely used in extraction testing, it can be seen from the test result has obtained asphalt level at AMP is 5,55% close to asphalt level value at JMF. This proves that asphalt content is more extracted. However, this test can only be carried out in laboratories with a sample weight of ± 150 gr. So it is difficult to determine the aggregate filter analysis. Extraction testing method using Centrifuge tool with gasoline solvent is more appropriate to be used in field extraction testing because it is practical in use, and the extract test data obtained is aggregate filter analysis and bitumen content on AMP is 5,51% smaller than using soklet method . This test can be used in the field because of its practical use with a sample weight of ± 1200 gr. So it can be easier in determining aggregate filter analysis.

5. Conclusions.

Extraction test results of asphalt content decreased, so it can be formulated KA JMF <KA AMP <KA Asphalt Finisher <KA Core. The value of Job mix used is 5.56% with the tolerance given by Bina marga Specification 2010 the third revision is ± 0.30%, the AMP deviation value is 0.05%. Asphalt Finisher is 0.10%, and the core is 0.22%. Extraction testing method using Centrifuge tools with solvents Gasoline is more appropriate to be used in extraction testing in the field because it is practical in the use meet the requirements of time, accuracy of results and economic value.

References

[1] Hadijah, Ida. 2011. "*Evaluasi Variasi Bahan Pelarut Untuk Penentuan Kadar Aspal Optimum*", *Jurnal Tapak,*Vol 1 No.1.pp1-11.

[2] F.Soehardi, S.Wiyono, dan A.Wanim. 2014. "*Kajian Perbandingan Kadar Aspal Hasil Ekstraksi Campuran AC-WC Gradasi Kasar Dengan Cairan Ekstraksi Menggunakan Bensin*", Program Magister Teknik Sipil Universitas Islam Riau, Pekanbaru.

[3] Hadijah, Ida. 2011. "*Evaluasi Variasi Bahan Pelarut Untuk Penentuan Kadar Aspal Optimum*", *Jurnal Tapak,*Vol 1 No.1.pp1-11.

[4] Shamier, Mochamad. " *Evaluasi Karakteristik Campuran Laston Asphalt Concrete Wearing Cou*rse", *Tugas Akhir*, Fakultas Teknik Sipil Universitas Kristen Maranatha.

[5] M.Anggraini, S.Wiyono, dan A.Wanim.2015."*Kajian Kadar Aspal Hasil Ekstraksi Penghamparan Campuran AC-WC Gradasi Kasar Dengan Job Mix Formula*", *Prosiding Seminar Teknik Sipil Annual Civil Engineering Seminar*, ISBN 978-979-792-636-6, pp 96-103.

[6] L.D.Putri, S.Wiyono, dan A.Puri.2015." *Kajian Kadar Aspal Hasil Ekstraksi Penghamparan dan Mix Design Pada Campuran Asphalt Concrete Wearing Course (AC-WC) Gradasi Halus*", *Prosiding Seminar Teknik Sipil Annual Civil Engineering Seminar*, ISBN 978-979-792-636-6, pp 117-123.

[7] F.Soehardi, S.Wiyono, dan A.Wanim. 2015. "*Kajian Perbandingan Kadar Aspal Hasil Ekstraksi Campuran AC-WC Gradasi Kasar Dengan Cairan Ekstraksi Menggunakan Bensin*", *Prosiding Seminar Teknik Sipil Annual Civil Engineering Seminar*, ISBN 978-979-792-636-6, pp 137-144.

[8] Surat. 2015. "Perbandingan Metode *Ektraksi* Kadar Aspal Alat *Centrifuge Extractor* dengan *Reflux*", *Jurnal INTEKNA, Tahun XIV No.1,PP 1-9.*

[9] Badan Standardisasi Nasional (2004). *Cara Uji Ektraksi kadar Aspal dari campuran beraspal menggunakan refluks Gelas.* RSNI M-05-2004.

[10] Departemen Pekerjaan Umum (2002). Standar Nasional Indonesia Cara pengujian Kadar Aspla dari Campuran Beraspal Dengan Cara sentrifus. SNI 03-6894-2002.

[11] Departemen Pekerjaan Umum (1994). *Standar Nasional Indonesia Metode Pengujian Kadar Beraspal Dengan Cara Ektraksi Menggunakan Alat Soklet.* SNI 03-3640-1994.

[12] M.Anggraini, 2014, "Kajian Kadar Aspal Hasil Ekstraksi Penghamparan Campuran AC-WC Gradasi Kasar Dengan Job Mix Formula", *Tesis*, Program Magister Teknik Sipil Universitas Islam Riau, Pekanbaru.

Batako Quality Optimization with Addition of Palm Oil Stem Fiber from Kampar District and Dumai City

Zainuri[1], Gusneli Yanti[2] and Shanti Wahyuni Megasari[3]
Universitas Lancang Kuning, Pekanbaru, 28265, Indonesia
E-mail: zainuri@unilak.ac.id
E-mail: gusneli@unilak.ac.id
E-mail: shanti@unilak.ac.id

Abstract: The waste of dry palm oil produced by 148 trees per hectare is 3,108 ton/month or 37,296 ton/year as calculated. Riau province has oil palm plantations covering an area of 2.399.172 hectares (BPS Riau Province, 2014). It can be estimated the amount of waste generated. Palm stem waste can be utilized, one of which is the utilization of midrib fiber as an added material in the manufacture of batako. Batako-fiber that is made still must be examined feasibility as building materials. The purpose of this study was to determine the optimization of the quality of batako works by the addition of palm stem fiber originated from the districts of Kampar and Dumai. This research used experimental method with laboratory research. Batako-fiber with the addition of palm fiber stem 1% of the weight of cement can increase the value of compressive strength above the normal batako and a batako with first quality according to SNI 03-0349-1989 standard. The use of palm stem fiber originating from the Kampar district resulted in better batakos with higher average compressive strength values than the dumai-derived fibers, especially in the addition of 1% fiber by weight of cement. The finding of this research is that the batakos originating from Kampar district are better than those from Dumai city. The most optimal addition of palm fiber burrs to batako-fiber products is 1% of the weight of cement.

Keywords: batako, fiber, compressive strength.

1. Introduction

Innovation of the product is always done for refinement. Innovation is done through a series of trials with testing of products. Innovating by utilizing waste is quite often done with hope that waste can still be utilized with a new look as an innovation. Batako-fiber is one of innovation products by utilizing waste of fiber that has not been utilized optimally. In this case the used fiber is palm stem fiber. The waste of dry palm oil that produced by 148 trees per hectare is 3,108 ton/month or 37,296 ton/year as calculated. Riau Province has oil palm plantations covering an area of 2.399.172 hectares (BPS Riau Province, 2014), it can be estimated by how much palm oil waste is produced continuously. The oil palm stems until now have not been optimally utilized and left to accumulate in oil palm plantations, so if it stays there, it will potentially to be pollutants such as smoke due to people's habit of burning garbage.

The palm oil stem has a straight fiber and more strong than the fiber from the coconut husk. Coco fiber has been used as raw material for some products. It may also be imposed on palm fiber. With a series of trials of palm stem fiber will be as efficient as any other organic fiber. The problem raised in this research is how to optimize the quality of the batako work with the addition of palm stem fiber coming from Kampar district and Dumai city. With these problems, the palm oil stem fiber is tested as an added material in the manufacture of batako which is used as building material/construction. The purpose of this study was to determine the optimization of the quality of batako works by the addition of palm stem fiber originated from the districts of Kampar and Dumai. The addition of palm stem fiber to different origin or place of growth. With a certain portion, palm fiber can be selected which have better quality to be used as an added ingredient in the manufacture of batako. The origin of palm-oil stem fiber that is studied is palm-oil stem fiber originating from Kampar district (26 - 100 m asl) and Dumai city (average 3 m asl).

2. Research Method

This research was conducted by experimental method with laboratory research. The first step in the preparation of oil palm stem fiber is to prepare the fiber to be cleaned from the bonding network. The palm oil stem is first processed to obtain the desired palm-oil stem fiber. Separation of fiber is done using chemical method by soaking the midrib into NaOH solution for 4 - 5 days. The palm oil heap has been gently brushed with an iron brush to separate the fibers from the bonding tissue, the fibers are washed and dried. In the dry conditions of palm stem fiber ready to be used as an ingredient added to the batako maker. Batako made with size 20 cm x 10 cm x 6 cm, job mix design batako with ratio 1 : 5 from weight of material which then modified to 1 cement : 3 sand cast : 2 sand pile. Comparison of materials based on the weight of the material is used with the consideration that the product to be made in small size and calculated instead of basket based on cultivation as prevalent for concrete calculations. The number of batakos required for each group of at least 6 pieces with the details of 5 pieces as a sample of test specimens and 1 piece for the prototype of the product being made. To print 6 pieces of batako-normal made job mix design as shown in table 1 below.

Table 1. *Job mix design*

Material	Amount of Material	Information
Cement	2,7168 kg	Semen Padang
Concrete sand	8,1510 kg	District Kampar
Sand pile	5,4342 kg	District Kampar
Water	0,1000 lt	Pekanbaru
Fiber :		
1%	0,0272 kg	Kampar and Dumai with dried condition
3%	0,0815 kg	
5%	0,1358 kg	

(Source: The design of the batako, 2017)

Batako made can be grouped on solid concrete batako material in accordance with the requirements of SNI. Physical requirement of solid concrete batako according to SNI 03-0349-1989 is seen in table 2 below.

Table 2. The physical requirements of solid concrete batako

Quality	Gross Press Strength Minimum Average* (Kg/cm^2)	Average water absorption (%)
I	100	25
II	70	35
III	40	-
IV	25	-

(Source: SNI 03-0349-1989)

*) By area the real size of the batako includes the width of the hole as well as the edge basin.

3. Results and Discussion

The quality of a product is usually determined by the quality of the ingredients used. The cement used is portland cement produced by PT. Semen Padang. Concrete sand and sand pile originated from Kampar district. Water used from clean ground water without pollutants.

The oil palm stem fiber used comes from two different places namely Kampar district (26 - 100 m asl) and Dumai city (average 3 m asl). To know the quality of palm fiber is done by tensile and elongation test to calculate fiber strength and elasticity. The test results are listed in table 3 below.

Table 3. physical properties of palm-oil palm fiber

Origin of Fiber	Test Result		
	Tensile Strength (gr/hl)	Elongation (%)	Diameter (mm)
Kab. Kampar	4400	4,23	0,81
Kota Dumai	3890	6,85	0,74

(Source: Bandung textile testing laboratory, 2017)

Tensile strength and elongation are tested according to SNI 7650: 2010 standard. The value of tensile strength of palm fiber from Kampar district is better than fiber originating from Dumai city with value of 4400 gr/strand and 3890 gr/strand respectively. This figure shows that palm fiber from Kampar district is stronger than fiber originating from Dumai city.

The value of elongation obtained places the city of Dumai at a higher value than fiber originating from Kampar district. The palm-oil palm fiber from Dumai can flourish up to 6,85%, while the fiber from Kampar district has a 4,23% grade. These values indicate that palm fiber from Dumai city is more elastic than fiber originating from Kampar district.

The oil palm fiber used is separated according to their respective portions and cut into lengths of ± 2 cm. All the ingredients are stirred and mixed evenly and then put into the mold and pressed. The batakos are made up of normal batako and fiber. Batako is dried at room temperature, after 28 days or more of compressive strength test. The results of the compressive strength test of the batako as sample are shown in the following table.

Table 4. Mean compressive strength value

Origin of Fiber	Percentage of Fiber (%)	Tensile Strength Mean (kg/cm^2)	Quality Standard SNI
Normal	0	104,24	I
Kampar District	1	111,34	I
	3	70,61	II
	5	68,93	III
Dumai City	1	110,80	I
	3	79,30	II
	5	57,57	III

(Source: Concrete laboratory PT. Riau Mas Bersaudara, 2017)

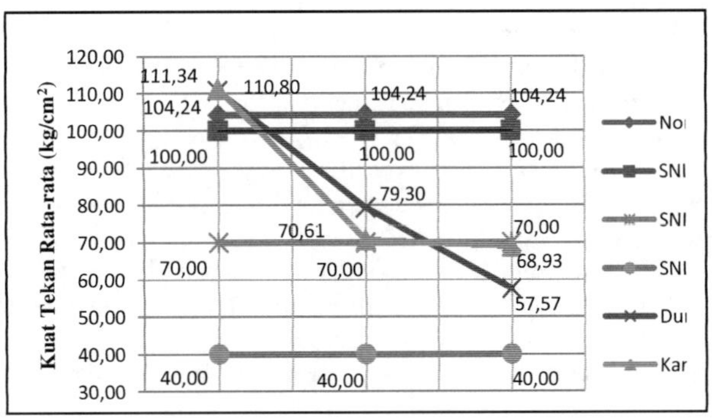

Figure 1. Comparison of compressive strength of batako

Batako-normal made to achieve quality I SNI 03-0349-1989 for solid concrete batako with tensile strength value 104,24 kg/cm^2. The addition of palm oil fiber as much as 1% of the weight of cement has a positive impact with the increase of the compressive strength value of the batako-normal. The 1% strength of the 1% fiber-reinforced concrete batakos originated from Kampar district was higher than the 1% dampai batako-fiber compressive strength value of Dumai city with an average value of 111,34 kg/cm^2 and 110,80 kg/cm^2 respectively. The addition of palm-oil stem fiber around 3% of the weight of cement causes the value of batako-fiber press down and lower than the batako-normal and is in the range of quality II according to SNI. Likewise with the addition of fiber 5%, causing the value of batako-fiber suppressive strength continues to decline to the level of quality III of SNI standards. Although it still meets the SNI quality standard, but the decline in the value of the compressive strength of the batako-fiber as the increased portion of fiber used as an additional material for building the batako-fiber must be considered and adjusted to the needs in the field.
In addition to the compressive strength test, to determine the quality of batako-fiber made then also test water absorption. The results of water absorption test are listed in table 5 below.

Table 5. The average water absorption test

Origin of Fiber	Percentage of Fiber (%)	Water Absorption (%)	Information
Normal	0	4,50	Maximum water absorption according to SNI quality I = 25%
Kampar District	1	3,50	
	3	4,80	
	5	5,70	
Dumai City	1	3,00	
	3	3,60	
	5	4,10	

(Source: Concrete laboratory of Civil Engineering Unilak, 2017)

The percentage of water-absorption of batako-fiber that is sampled is all below the maximum allowed value of SNI. The value obtained ranges from 3,0%-5,7% less than 25% of the maximum value of water absorption test of SNI of quality I. Several tests conducted on batako-fiber with 1%, 3% and 5% additional fibers from two different sources of oil palm fiber origin, namely Kampar and Dumai show that the resulting fiber batakos are still in accordance with the standard Quality set by SNI 03-0349-1989. The more palm fiber stem added, the resulting compressive strength showed a downward trend. The addition of midrib fiber up to 5% of the weight of cement is still acceptable because it entered the quality of III SNI. Some conducted experiments on materials and finished products batako-fiber are made to show some of the things that become product superiority. In addition to the use of palm oil stem fiber to contribute to the maintenance of the environment, the resulting product was proven to be more powerful when compared with conventional batako without fiber. Lower water absorption will have an impact on product durability. The strength and durability of the products are the consideration of consumers in choosing the materials to be used in construction projects.

The compressive strength test of the batako samples made show another advantage of the batako-fiber product. In the presence of fibers spreading inside the batako, when the value of compressive strength has been maximum and the test object is broken, the broken batako-fiber is not immediately destroyed, but still hanging on each other tied to the fiber contained in the product. This suggests that batako-fiber have a higher level of security when compared to non-fiber batako. Can be imagined if an earthquake with a scale that can undermine the building. The use of non-fiber batako will immediately collapse with a strong vibration and can override all the objects around it. With the same shock force, the walls that use the batako-fiber will be more durable and will not collapse immediately because the fibers are still tied to other elements. These advantages can be highlighted and to be more convincing can be done a series of trials on the resistance of batako-fiber products to earthquakes. In this study it has not been studied in depth, but at other times and other studies, it can be done.
The advantages of batako-fiber found in a series of studies are strength, durability and safety. These advantages can attract consumers to choose the product. With proper socialization, the batako-fiber can be commercialized so that the bricklayers will be interested to produce it. Thus, palm oil stem fiber that were not valuable to sale and even potentially become waste, with batako-fiber produced that has several advantages, the palm oil stem

fiber into a commodity that is worth selling. Furthermore, it is possible to estimate coconut palm oil stem fiber separation centers even very likely later on there will be machines that can separate the fiber as well as the coco fiber separator machine.

Continuous innovation done as a step of product improvement is a plus of a research and its application. Product innovation which is the finding of this research is that the batako-fiber originating from Kampar district are better than those from Dumai city. Where the most optimal addition of palm fiber burrs to batako-fiber products is 1% of the weight of cement.

4. Conclusions

The palm-oil palm fiber that is represented by the sample of the test specimen is a batako size of 20 cm x 10 cm x 6 cm with the design job mix defined and after going through several tests, it can be concluded as follows:

1) The fibers with the addition of 1% of palm oil stem fiber increase the value of compressive strength above the batako-normal (104,24 kg/cm^2) and is a bracket with quality I according to SNI 03-0349-1989 standard with strong average value Press obtained 111,34 kg/cm^2 batako-fiber originating from Kampar district and 110,80 kg/cm^2 of batako-fiber originating from Dumai.
2) The use of palm oil fiber from Kampar district resulted in better batako-fiber with higher average compressive strength values than the Dumai-derived batako-fiber, with optimal addition of 1% of the weight of the cement.

5.References

[1] Adibroto, F. 2014. *Pengaruh Penambahan Berbagai Jenis Serat Pada Kuat Tekan Paving Block.* Jurnal Rekayasa Sipil. 10 (1): 1-11.
[2] Andriyani, Y. dan Nursyamsi. 2014. *Pemanfaatan Serbuk Kaca Sebagai Bahan Tambah Dalam Pembuatan Batako.* Medan: Jurnal Teknik Sipil USU. 3 (2).
[3] Badan Penelitian dan Pengembangan Industri Laboratorium Pengujian Balai Besar Tekstil. 2017. *Laporan Uji Serat Pelepah Kelapa Sawit.* Bandung.
[4] Badan Pusat Statistik Provinsi Riau. 2014. *Luas Areal Perkebunan Menurut Jenis tanaman.* Pekanbaru.
[5] Fauzi, Y. 2006. *Kelapa Sawit : Budidaya, Pemanfaatan Hasil dan Limbah, Analisis Usaha dan Pemasaran.* Jakarta: Penebar Swadaya.
[6] Hadi, M. M. 2004. *Teknik Berkebun Kelapa Sawit.* ed. I, cet. I. Yogyakarta: Adicita Karya Nusa.
[7] Hermanto, D., Supardi dan Edy Purwanto. 2014. *Kuat Tekan Batako Dengan Variasi Bahan Tambah Serat Ijuk.* e-Jurnal Matriks Teknik Sipil. 491-497.
[8] Intara, Y. I. dan Banun Diah P. 2012. *Studi Sifat Fisik dan Mekanik Parenkhim Pelepah Daun Kelapa Sawit Untuk Pemanfaatan Sebagai Bahan Anyaman.* Jurnal Agrointek. 6 (1): 36-44.
[9] Kristiawan, A. dan Putri Anggi Permata Suwandi. 2015. *Pengaruh Penambahan Kapur dan Sabut Kelapa Terhadap Bobot dan Daya Serap Air Batako.* Jurnal Ilmiah Teknosains. 1 (1): 29-35.
[10] Laboratorium Beton PT. Riau Mas Bersaudara. 2017. *Laporan Uji Kuat Tekan.* Pekanbaru.
[11] Laboratorium Beton Universitas Lancang Kuning. 2017. *Laporan Uji Penyerapan Air.* Pekanbaru.
[12] Multazzam, K. A. dan Priyanto Saelan. 2014. *Studi Mengenai Perancangan Komposisi Bahan Dalam Campuran Mortar Untuk Pembuatan Bata Beton (Paving Block).* Jurnal Online Teknik Sipil Itenas. xx (x): 1-12.
[13] Mulyono, T. 2005. *Teknologi Beton.* ed. II. Andi. Yogyakarta.
[14] Standar Nasional Indonesia (SNI 03-0349-1989). 1989. *Bata Beton Untuk Pasangan Dinding.* Jakarta: Badan Standarisasi Nasional (BSN).
[15] Standar Nasional Indonesia (SNI 7650:2010). 2010. *Tekstil-Benang dari Gulungan-Cara Uji Kekuatan Tarik dan Mulur Per Helai.* Jakarta: Badan Standarisasi Nasional (BSN).

[16]Suhardiman, M. 2011. *Kajian Pengaruh Penambahan serat Bambu Ori Terhadap Kuat Tekan dan Kuat Tarik Beton.* Jurnal teknik. 1 (2): 88-95.
[17] T. Sofyan, B. 2011. *Pengantar Material Teknik.* Jakarta: Salemba Teknika.
[18] Widiastuti, R. dan Dana Kurnia Syabana. 2015. *Serat Pelepah Kelapa Sawit (Sepawit) Untuk Bahan Baku Produk Kerajinan.* Prosiding Seminar Nasional 4th UNS SME's Summit & Awards. Hal. 7-14.
[19] Zainuri, Gusneli Yanti dan Shanti Wahyuni Megasari. 2015. *Analisis Beton Ringan Tanpa Agregat Kasar Dengan Penambahan Polymer Concrete.* Jurnal Sainstek. 3 (1): 1-9.

IOP Conf. Series: Earth and Environmental Science **97** (2017) 012019 doi:10.1088/1755-1315/97/1/012019

A Review of Vehicles Speed on School Safety Zone Areas in Pekanbaru City

Lusi Dwi Putri [1], Fitridawati Soehardi [2], and Alfian Saleh [3]
[1] Universitas Lancang Kuning, Pekanbaru, 28265, Indonesia
E-mail: lusidwiputri @ unilak.ac.id
[2] Universitas Lancang Kuning, Pekanbaru, 28265, Indonesia
E-mail: fitridawati@unilak.ac.id
[3] Universitas Lancang Kuning, Pekanbaru, 28265, Indonesia
E-mail: alfian.saleh@unilak.ac.id

Abstract: School Safety Zone is a location or region on particular roads that are time-based speed zone to set the speed of the vehicle in the school environment. The maximum speed limit permits entering a School Safety Zone, especially in Pekanbaru City is 25 km / h and an outline of the speed limit permit vehicles that pass through the School Safety Zone in Indonesia is generally 20-30 km / h. However, the vehicles speeds that pass School Safety Zone are higher than permit speeds.To ensure the level of vehicle offense across the territory of the School Safety Zone so it is necessary a primary data which is taken randomly based on field survey for 3 days at schools that has that facility ie SDN 3 Jalan Kesehatan Pekanbaru City, SDN 68 Jalan Balam Ujung Kota Pekanbaru and SDN 143 Jalan Taskurun Kota Pekanbaru. Furthermore, the data were taken in good condition that is at 6:30 to 7:30 am and at 12:00 to 13:00 pm. In addition, the data obtained is mileage and travel time of the vehicle. Both of these data can generate good speed value that passes through the area of School Safety Zone. Based on the research findings, the vehicle speed passing through the area of School Safety Zone is incompatible with speed permit at 35 km / h with a maximum average percentage of the rate of offense in the area of the school zone is 91.7%. This indicates that the vehicle passes School Safety Zone not following the rules of the maximum limit area and can be potentially harmful to elementary school students.
keywords: Mileage, Speed, School Safety Zone, Travel Time

1. Introduction

Many schools in big cities are located on the edge of the highway, where many vehicles are driving at high speed. One of the busiest traffic roads is the West Cross Road. West Cross Road is a road that connects Pekanbaru City with Kampar regency, even some roads in Pekanbaru City called 'Trajectory' like Jalan HR Soebrantas because of the many trucks that pass.

The conditions mentioned above will be very dangerous for school students when going across the street. School students, especially elementary school students (elementary school) are still vulnerable in traffic such as when crossing the road. According to kalantas data from Polresta Pekanbaru City, Pelalawan Regency and Kampar regency of 2016 is not very significant from the previous year and some have decreased, but the risk we need to anticipate before the victims occurred considering the object of this study is the elementary and junior high school in fact inhabited by children, children who do not understand the ethics of traffic, especially during school hours and when returning from school. Not all schools have ZoSS facilities even though the visual conditions should be eligible for ZoSS. Even for some schools that have facilities ZoSS current conditions such as unkempt for example street color of the school (usually red) is no longer clear color.

The implementation of Zoss is one form of traffic management in order to fulfill the sense of security in crossing the road for pedestrians. According to the Department of Transportation, the Director General of Land Transport (2006) "School Safe Zone (ZoSS)" is a location on a particular road segment which is a time-based velocity zone to regulate the pace in the school environment ". This is also explained by the Ministry of Transportation of the Directorate General of Land Transport (2014) "Zoss is controlling traffic activities through speed regulation by placing markers and signs on roads in schools that aim to prevent accidents in order to ensure safety in schools".

2.Literature Review

The School Safe Zone (ZoSS) is a zone for certain road segments in a school environment with a time-based speed. Through traffic engineering, this zone is equipped with supporting facilities that can adjust the speed of the vehicle. At Zona Saving School (ZoSS) expected safe, convenient, easy and economical traffic. The ZoSS facility is installed at schools located on arterial roads and collectors. The installation of ZoSS on a national road which is the primary arterial or collector 1 (KP 1) road is designated specifically for schools that have been built on the national side of the road and there is no alternative to school roads. At ZoSS the required road safety facilities are zebra cross, warning signs, a guidance of crossing locations and many children's signs.

To limit the permitted speed at the ZoSS location of 20 km.jam, 25 km / h, and 30 km / h, the determination is done by the related survey:

a. The behavior of road users, including pedestrian behavior while crossing the street and pedestrian behavior down the street.
b. Traffic conditions include: Inventory of roads, traffic volume, pedestrians, and vehicle speed

The requirement of ZoSS is the unavailability of pedestrian bridges, schools that have direct access to roads with students over 50 students. According to the Ministry of Transportation of the Directorate General of Land Transport (2014), based on the school's location criteria, ZoSS can be classified by single and plural ZoSS. Single and plural ZoSS consists of 4 types of road segments ie 2/2 UD road types (two lanes, two undivided directions), 4/2 UD road type (four lanes, two undivided lanes), 2/2 D type road (two lanes, two divided directions), 4/2 D type road (four lanes, two divided paths).

a. Single Zoss

   It is a ZoSS assigned to 1 school in a location. Single ZoSS form and size consists of 4 types of road segments ie road type 2/2 UD, 4/2 UD, 2/2 D, and 4/2 D.

b. ZoSS plural

   It is a ZoSS defined for 2 schools or more that are located adjacent to the sailing. ZoSS plural fitted with the provisions *crosswalk* and outer distance ZoSS. Zebra cross is installed in every entrance/entrance to school and ZoSS outer distance is measured from the outermost schools. The distance between the access entrance to the school with other schools $\geq$ 50 meters, if <50 meters, then the zebra crossing merged into one.

   Facility of road equipment in ZoSS region consists of 3 parts:

a..Road markings

According to Warpani (2002: 95) " road markings are in the form of lines, images, darts, and the insignia on the road surface that serves to direct traffic flow and limit traffic areas of interest". The road marker contains command messages, warnings, directions and restrictions consisting of 3 colors:

1) The marka is white, stating that road users are obliged to follow orders or restrictions according to their form. The white marks consist of single intact white lines, single dashed white lines, double white lines, white lines and white dotted lines and white lines.
2) The marka is yellow, stating that road users are not allowed to stop in the area.
3) A red marka, a red road marker indicating a need or a special mark.

b. Traffic signs

Traffic signs are part of street equipment in the form of symbols, letters, numbers, sentences, and/or combinations that serve a warning, prohibition, command or instructions for road users (Kemenhub, 2014). Type of signs consists of 4 parts, namely:

1) Warning signs indicate the possibility of a road hazard to be traversed. Warning shaped rhombus, colored yellow base with black insignia or writing.

2) Prohibition signs that form the arrangement contain the prohibitions that can not be done by road users. A circular band with white base color and black or red writing or writing. Signs the special ban in the form of an equilateral octagonal.
3) Guidance signs that provide guidance on majors, road conditions, situation, the next city, the existence of facilities, and others. Signed instructions are rectangular.
4) Command signs are shaped clear and clear settings without any other interpretation that must be implemented by road users. Command signs are round, blue base color with white or red insignia or writing for oblique lines at the end of the command.

c. Road user safety device

Road user safety device that is in the form of ribbon band. Ribbon tape is an additional fixture on the road that serves to make the driver more alert. Rotting bands may be a road marker or other material mounted across a traffic lane with a maximum thickness of 4 cm. while the homecoming tape for the School Safe Zone is installed 50 meters from the outer ZoSS line. At very low traffic volume or kissing with a very good road, then the rider can run his vehicle in accordance with his wishes within the perceived safety, without feeling influenced by the existence of other vehicles. In this condition, it can be said that the traffic on the road is said to free flow, and the speed at this condition is called the velocity of free flow. However, if the road conditions are solid, there are many pedestrian factors crossing the road, affecting the road, causing conflict.Things that affect the road is called the side barriers that is a side activity that often leads to conflict and greatly affects the traffic flow. Side barriers such as; pedestrians crossing the road, pedestrians walking along the side of the road, parking vehicles or stopping by the side of the road, lambda vehicles, such as becak, bendi, bikes, and merchants selling on the sidewalk

3.Method

The type of this research is descriptive research to describe in fact the speed at the location using Zoss and location not using ZoSS. The data needed are primary and secondary data. Primary data is data obtained directly in the field is the time (seconds) and distance (km / h) at the location of SDN 3 Jl. Balam Ujung, SDN 68 Jl. Kesehatan, SDN 143 Jl.Tasykurun, SDN 18 Jl. Kulim, SDN 55 Jl.Kayangan and SDN 163 Jl. HR. Subrantas Pekanbaru. Time survey is sixty (60) minutes at the time of entry to and from school is at 6:30 to 7:30 pm and at 12:00 to 13:30 pm and observed on Monday, Tuesday and Wednesday. While the secondary SDN 68data obtained from related parties include Pekanbaru City Department of Education and the Department of Transportation in Pekanbaru, such as the type of road.

Speed can be measured as point velocity, travel speed, space velocity and velocity. (Oktaviani; 2014):
Speed formula in general:

$$V = \frac{d}{t} \tag{1}$$

V = speed (km /h)
d = mileage (km)
t = travel time (hours)

Speed is known for 5 kinds namely:

a. *Time mean speed* (average speed time)

It is the average speed of all passing vehicles at a certain point in certain periods of time.

$$Vt = \frac{1}{n}\sum_{i=1}^{n} vi \tag{2}$$

Vt = average speed of time (km /h)
n = number of speed data observed
vi = speed of each vehicle observed (km/h)

b. *Space mean speed* (average speed space)

It is the average speed of all vehicles that are in a road segment for a certain period of time. *Space mean speed* is calculated using the following formula. (Oktaviani; 2014):

$$Vs = \frac{1}{\frac{1}{n}\sum_{i=1}^{n}\frac{1}{Vi}} \tag{3}$$

Vs = average space velocity (km / h)

n = number of speed data observed

vi = speed of each vehicle observed (km /h)

c. *Spot Speed* (instantaneous velocity)

Speed is shown in the *speedometer.*

d. *Running speed* (moving speed)

It is the average speed of a vehicle during a move:

$$\text{Running Speed} = \frac{Mileage}{Running\ Time} \qquad (4)$$

e. *Journey speed* (traveling speed)

It is the average speed of the journey calculated from the distance divided by the travel time.

$$Journey\ Speed = \frac{Jarak\ tempuh}{Journey\ Time} \qquad (5)$$

4.Results and Discussion

The results of the study in the city of Pekanbaru, the average speed of motorists on schools that have facilities ZoSS of the survey results and the calculation speed of vehicles to SDN 3 Ujung Jalan Balam, SDN 68 and SDN 143 Health Jl.l Tasykurun can be seen in Table 1.

Table 1. Speed of vehicle in location using ZoSS in Pekanbaru City

Average vehicle speed at the site using Zoss					
Day	Time	School	n	ΣVi	Vt
Monday May 2017	06.30-07.30	SDN 3 Jl.Balam Ujung Pekanbaru	118	4233	36
	12:00 to 13:00		150	5534	37
Tuesday May 2017	06.30-07.30		167	6867	41
	12:00 to 13:00		187	7344	39
Wednesday May 2017	06.30-07.30		198	8977	45
	12:00 to 13:00		204	9345	38
Monday M ei 2017	06.30-07.30	SDN 68 Jl.Kesehatan Pekanbaru	377	12300	33
	12:00 to 13:00		392	13507	34
Tuesday May 2017	06.30-07.30		406	14355	35
	12:00 to 13:00		418	15663	37
Wednesday May 2017	06.30-07.30		423	15884	38
	12:00 to 13:00		438	15446	35
Monday May 2017	06.30-07.30	SDN 143 Jl.Taskurun Pekanbaru	110	3345	30
	12:00 to 13:00		121	3598	30
Tuesday May 2017	06.30-07.30		147	4325	29
	12:00 to 13:00		168	4453	27
Wednesday May 2017	06.30-07.30		179	5433	30
	12:00 to 13:00		197	5576	28
Average					35

With:

n = Lots of speed data

ΣVi = Total speed of vehicle (km / h)

Vt = Average speed (km / h)

From the data analysis only 143 Jl.Taskurun approaching SDN standard average speed of vehicles is permitted, namely 29 km / h but still exceeds the limit diz Inkan, is caused by the location of the category of the population and has a bustling location beside the huge obstacles. Causes the rider to adhere to ZoSS area rules. While in two different locations, motorists do not comply with the permit speed rules when passing ZoSS area, for SDN 3 Jl Balam Ujung Pekanbaru average speed of riders that is 41 km/h and SDN 68 Jl Health Pekanbaru average speed of rider that is 35 km /h. This is due to the fairly large road width, the less dense currents, the lack of side barriers and the very lack of awareness of motorists in passing through ZoSS territory. However, if averaged from the three locations, the speed of vehicles passing through the ZoSS area still exceeds the allowed speed of 35 km/h.

The results of the study in the city of Pekanbaru, the average speed of motorists on schools that do not have the facilities ZoSS of the survey results and the calculation speed of vehicles to SDN 18 Jalan Kulim, SD N 163 Jl. HR Subrantas and SDN 55 Jl Kayangan can be seen in Table 2.

Table 2. On-site vehicle speed does not use ZoSS in Pekanbaru City

Average vehicle speed at the site using Zoss					
Day	Time	School	n	ΣVi	Vt
Monday May 2017	06.30-07.30	SDN 18 Jl. Kulim Pekanbaru	121	4188	35
	12:00 to 13:00		135	4221	31
Tuesday May 2017	06.30-07.30		153	4323	28
	12:00 to 13:00		187	7134	38
Wednesday May 2017	06.30-07.30		203	7554	37
	12:00 to 13:00		245	8654	35
Monday M ei 2017	06.30-07.30	SDN 163 Jl.HR.Subrantas Pekanbaru	412	15663	38
	12:00 to 13:00		433	16773	39
Tuesday May 2017	06.30-07.30		389	16773	43
	12:00 to 13:00		418	17445	42
Wednesday May 2017	06.30-07.30		423	16889	40
	12:00 to 13:00		438	16773	38
Monday May 2017	06.30-07.30	SDN 55 Jl.Kayangan Pekanbaru	122	3877	32
	12:00 to 13:00		188	5885	31
Tuesday May 2017	06.30-07.30		204	6234	31
	12:00 to 13:00		226	8775	39
Wednesday May 2017	06.30-07.30		244	9877	40
	12:00 to 13:00		201	8775	44
Average					33

With:

n = Lots of speed data

ΣVi = Total speed of vehicle (km / h)

Vt = Average speed (km / h)

From the data analysis in Pekanbaru City average speed of motorists on schools that do not have the facilities ZoSS everything past the allowed limit is 28 km / h, 34 km / h, and 36 km / h. For SDN 18 Jl Kulim Pekanbaru the average speed of the rider is 34 km / h, SDN 163 Jl. HR.Subrantas Pekanbaru average speed of riders is 28 km / h and SDN 55 Jl.Kayangan Pekanbaru average speed of the rider is 36 km/hour. For SDN 18 Jl Kulim Pekanbaru and SDN 55 Jl. Kayangan Pekanbaru is caused by a large enough road width, less dense flow, and very lack of awareness of motorists in passing the area without ZoSS. For SDN 163 Jl. HR.Subrantas Pekanbaru However, if averaged from the three locations, the speed of vehicles passing through areas that do not have ZoSS facilities exceed the permitted speed of 33 km h.

5.Conclusions

From the survey results that have been done, it can be seen that:

1. The average speed of riders passing ZoSS exceeds the permitted standard speed limit (25 km / h for inner city roads) of 35 km / h.
2. The average speed of riders passing through school areas that do not have ZoSS exceeds the standard allowable speed limit (25 km / h for inner city roads) of 33 km /h

6.References

[1] Data Riau.com, 2014, *Jalan HR Soebrantas, Lintasan Maut di Pekanbaru*, Aviable from : http://datariau.com/read 13-829-2014-10-24-jalan-hr-soebrantas-lintasan-maut-di-pekanbaru.html, Accesed : 2016, April 13

[2] Kurniati, T ., dkk, 2010, *Evaluasi Penerapan Zona Selamat Sekolah di Kota Padang*, Jurnal Rekayasa Sipil Volume. 6, No.2, pp. 7 – 10Oktaviani. (2014). Teknik Lalu Lintas, Padang: Teknik Sipil Fakultas Teknik Universitas Negeri Padang.

[3] Peraturan Menteri Perhubungan Republik Indonesia Nomor PM 13 Tahun 2014 tentang Rambu Lalu Lintas

[4] Peraturan Menteri Perhubungan Republik Indonesia Nomor PM 34 Tahun 2014 tentang Marka Jalan

[5] Sari, N.M, 2015, "Tinjauan Kecepatan Kendaraan pada Wilayah Zona Selamat Sekolah (ZoSS) di Kota Padang", *ProsidingAnnual Civil Engineering Seminar*, ISBN :978-979-792-636-6, Universitas Riau, Pekanbaru

[6] Sari, N.M, 2015, "*Tinjauan Kecepatan Kendaraan pada Wilayah Zona Selamat Sekolah (ZoSS) di Kota Padang*", ejournal.unri.ac.id/ACES/article/download/2994/2926, Accesed : 2016, february 07

[7] Tamin, O.Z, 2008, "*Perencanaan, Pemodelan dan Rekayasa Transportasi",* Institut Teknologi Bandung
Wahyuni, Sri, 2012, "*Analisis Efektifitas Zona Selamat Sekolah (ZoSS) di Sekolah Dasar Kota Pekanbaru*", http://repository.unri.ac.id/xmlui/bitstream/handle/123456789/2050/SRI%20WAHYUNI%20%20(NIM%20070713 1855).pdf?sequence=1, Accesed : 2016, April 03

Comparison Pore Aggregate Levels After Extraction With Solvents Pertamax Plus And Gasoline

Muthia Anggraini
Universitas Lancang Kuning, Pekanbaru, 28265, Indonesia
E-mail:muthia@unilak.ac.id

Abstract: Loss of asphalt content extraction results become problems in Field Work For implementing parties. The use of solvents with high octane (pertamax plus) for the extraction, dissolving the asphalt more than gasoline. By comparing the levels of aggregate pores after using solvent extraction pertamax plus compared to gasoline could answer that pertamax plus more solvent dissolves the bitumen compared to gasoline. This study aims to obtain comparative levels of porous aggregate mix AC-WC after using solvent extraction pertamax plus compared to gasoline. This study uses the aggregate that has been extracted from the production of asphalt mixtures, when finisher and after compaction field. The method used is the assay of coarse and fine aggregate pores, extraction of bitumen content to separate the aggregate with bitumen. Results of testing the total absorption after extraction using a solvent preta max plus in the production of asphalt mixtures 0.80%, while gasoline solvent 0.67% deviation occurs 0.13%. In the finisher after the solvent extraction preta max plus 0.77%, while 0.67% gasoline solvent occurs deviation of 0.1%. At the core after extraction and solvent pertamax plus 0.71%, while gasoline solvent 0.60% 0.11% deviation occurs. The total water absorption after extraction using a solvent pertamax plus greater than gasoline. This proves that the solvent dissolves pertamax plus more asphalt than gasoline.

Keywords: Extraction, Asphalt, AC-WC, Pertamax Plus, Gasoline

1. Introduction

Aggregate is the result of natural stone processing, is the main component of pavement layer, which provides structural properties and contributes 9-95% of the weight or 70-85% of the volume of the pavement structure therefore aggregate properties greatly affect pavement performance. All aggregates are porous, aggregate porosity is the amount of pore content present on the aggregate. As a result, water will sink into the aggregate. Porous aggregates are useful for absorbing bitumen so that the bonds between asphalt and aggregate are good, but too much pore can lead to too much-absorbed asphalt resulting in a thin layer of asphalt.

Asphalt level extraction is often a problem in field testing, where there is a loss of asphalt content, the bitumen content in the field is not suitable or less than the bitumen content of the job mix formula. The use of a pertamax plus solvent which has a high octane content for extraction dissolves more of the asphalt than gasoline. Comparing aggregate pore levels after extraction using solvents pertamax plus and gasoline, is a way to prove it. Pertamax Plus is Pertamina's fuel oil which has an octane value of 95. The purpose of this study was to obtain a comparison of pore aggregate mixture of Asphalt Concrete - Wearing Course (AC-WC) after extract using solvent pertamax plus than gasoline. This study takes aggregates that have been extracted from the asphalt mixture production site, finisher and after compaction of the field.

Several existing studies, Toruan, et.al (2013), analyzing pore levels in several quartz locations. The larger the aggregate porosity the aggregate density becomes smaller so that the maximum density of the mixture becomes smaller and the smaller the aggregate porosity value the specific gravity of the aggregate increases, the greater the maximum density of the mixture becomes larger.[2] Achmad analyzed (2010), analyzing pore levels at several quartz locations in Gorontalo, mostly mixed asphalt using aggregate from quarry Pilolalenga, quarry Tangkobu, and quarry Molintogupo.[3] Negara and Putra (2010), analyzing the limestone pore content. State, et.al, analyzed the limestone pore content of Nusa Penida as a pavement aggregate.[5] Pertamax plus as a solvent in the extraction of bitumen content dissolves bitumen than gasoline solvent. Where is the deviation in AMP -0,03%, from finisher -0,01%, and core -0,02%.[6] Soehardi did an extraction analysis of asphalt content of AC-WC mixture of coarse gradation with gasoline solvent.[7]

2. Literature Review

2.1 Aggregate Porosity

Water absorption by aggregates can be used to estimate the amount of bitumen that can be absorbed by aggregates in the mixture. The porous aggregate will absorb the asphalt, so the mixture tends to be dry or cohesive. in the asphalt-aggregate mixture (hot mix) there is a slight addition of asphalt content to satisfy the asphalt absorption by aggregates. Very porous aggregates, when used in the mixture, should be added asphalt quite a lot. Aggregates with very high porosity are not used in asphalt-aggregate mixtures unless the aggregate has excellent properties. Aggregate porosity is generally characterized by the amount of water that can be absorbed by the aggregate when immersed in water.

2.2 Method of Weight Types and Absorption of Crude Aggregate Water

Grain aggregate specific gravity testing was carried out in accordance with the Indonesian National Standard, Weighted Aggregate Water Type and Absorption Water Method using SNI 03-1969-2008. How to calculate it is:[8]

1) Bulk Specific Gravity (SNI 03-1969-2008)

Specific gravity calculates the weight of the dry aggregate and all aggregate volumes, using equations:

$$Sd = \frac{A}{(B - C)} \quad (1)$$

with:

Sd : Bulk density
A : Weight of dry matter oven test (gram)
B : Weight of specimen saturated surface dryness (gram)
C : Weight of specimen in water (gram)

2). Saturated Surface Dry (SNI 03-1969-2008)

Specific gravity calculates the weight of the aggregate in the dry state of the surface and the aggregate volume, using the equation:

$$Ss = \frac{B}{(B-C)} \quad (2)$$

with:

Ss : Specific gravity of dry saturated surface
B : Weight of dry surface saturated test object (gram)
C : Weight of specimen in water (gram)

3). Apparent Specific Gravity (SNI 03-1969-2008)

Specific gravity taking into account the weight of the aggregate in the dry state and all aggregate volumes not permeated by water, using equations:

$$Sa = \frac{A}{(A - C)} \quad (3)$$

with:

Sa : Apparent Specific Gravity
A : Weight of dry matter oven test (gram)
C : Weight of specimen in Water (gram)

4) Water Absorption (SNI 03-1969-2008)

Absorption is the percentage of water weight that can be absorbed by pores on the weight of the dry aggregate, using the equation :

$$Sw = \left[\frac{B - A}{A}\right] x\ 100\% \qquad (4)$$

with:

Sw : Water absorption
A : Weight of dry matter oven test (gram)
B : Weight of test surface saturation surface (gram)

2.3 Pertamax Plus

Pertamax Plus is Pertamina's fuel oil, has an octane value of 95. Pertamax Plus, like Pertamax and Premium, is a petroleum product from petroleum processing, generated by adding additives to the process in the oil refinery. An octane number is a number indicating how much pressure it can give before the gasoline burns spontaneously. The name octane comes from octane (C8), because of all the molecules of the gasoline compound, octane which has the best compression properties. Octane can be compressed to a small volume without spontaneous combustion, unlike for heptane, for example, which can be spontaneously burned. The heptane function is as the solvent in the extraction. By having a high octane the fuel will burn completely.[10]

3. Methodology

The study was conducted by laboratory testing. Refers to the tests of Specific Weight and Absorption of Crude Aggregate Water using SNI 03-1969-2008 and Tests of Smooth Aggregate Type and Water Absorption using SNI 03-1970-2008. Aggregates used from quarry Solok, West Sumatra. The aggregate after extraction is taken from the aggregate that has been separated from the asphalt using the solvent pertamax plus on each specimen from Asphalt Mixing Plant (AMP), back finisher, and core. Sampling on Marpoyan-Lipat Kain road segment. The results were compared with testing of specific gravity and absorption after being extracted using a gasoline solvent.

4. Results and Discussion

4.1 Comparison of aggregate pore content after extraction of AMP solvent pertamax plus with gasoline

Comparison of aggregate pore content after extraction of AMP using the solvent pertamax plus and gasoline is shown in the Table 1 :

Table 1. Comparison of Pore-Level After Extraction Of AMP Solvent Pertamax Plus With Gasoline

No	Type of Aggregate	Notation	Water Absorption of Aggregates After Extraction (%)		Max Specific Terms (%)
			Solvent Pertamax Plus	Solvent Gasoline *	
1	*Coarse agregat*	BIN III	0,99	0,88	3
2	*Medium agregat*	BIN II	0,74	0,57	3
3	Sand	BIN I	0,84	0,76	3

* Fitri analysis (2014)

From the table above shows that the aggregate water absorption after extraction from AMP with solvent pertamax plus and gasoline on BIN III (coarse agregate) deviation 0.11%., In BIN II (medium agregate) 0.17% deviation, in BIN I (Sand) deviation occurs 0,08%.

4.2 Comparison of pore content of extraction aggregate after extraction of solvent core pertamax plus with gasoline

The results obtained are in the Table 3:

Table 2. Comparison of Pore Level After Extraction Of Core Solvent Pertamax Plus With Gasoline

No	Type of Aggregate	Notation	Water Absorption of Aggregates After Extraction (%)		Max Specific Terms (%)
			Solvent Pertamax Plus	Solvent Gasoline *	
1	*Coarse agregat*	BIN III	0,94	0,86	3
2	*Medium agregat*	BIN II	0,61	0,48	3
3	sand	BIN I	0,77	0,69	3

* Fitri analysis (2014)

4.3 Comparison of total aggregate water absorption in the mixture after extraction of solvent pertamax plus and gasoline from AMP, finisher, and core

The recapitulation of a moisture content of AMP, finisher, and core is in the Table 4 :

Table 3. Comparison of Total Aggregate Absorption In Mixed After Solvent Extraction of Pertamax Plus and Gasoline From AMP, Finisher¸ and Core

No	Sample	Total water absorption in the mixture (%)		Syarat Spek Maks. (%)
		Solvent Pertamax Plus	Solvent Gasoline *	
1	*AMP*	0,80	0,67	3
2	*Finisher*	0,77	0,67	3
3	Core	0,71	0,60	3

* Fitri analysis (2014)

From the table above shows that total water absorption in the mixture after extraction from AMP with solvent pertamax and gasoline deviation 0.13%. At the finisher, there is a deviation of 0.1%. And at the core deviation 0.11%. The results obtained are still included in the specification requirement that is a maximum of 3%. The pore content of the extraction aggregate test with the solvent plus solvent is higher than that of the gasoline solvent. This proves that the solvent pertamax plus more dissolves the asphalt compared to the gasoline solvent. Because the solvent pertamax plus has a high octane content of 95 compared to gasoline which is only 88. The higher the octane value makes the fuel burned perfectly.

5 Conclusions

Pertamax plus as a solvent in the extraction of bitumen content produces more bitumen content on the same quarry aggregate because it dissolves the asphalt more than the gasoline solvent. Where the pore rate deviation from extraction aggregate testing with solvent pertamax plus and gasoline are 0,13% in AMP, 0,1% in finisher, and 0,11% in core.

6 References

[1] Sukirman, Silvia. 2003. *Beton Aspal Campuran Panas.* Bandung. Granit.

[2] A.L.Toruan,O.H. Kaseke,L. F. Kereh, dan T. K. Sendow. 2013.*"Pengaruh Porositas Agregat Terhadap Berat Jenis Maksimum Campuran", J.Sipil Statik*, Vol. 1, no. 3, pp.1-6.

[3] F.Achmad.2010.*"Tinjauan Sifat-Sifat Agregat Untuk Campuran Aspal Panas (Studi Kasus Beberapa Quarry Di Gorontalo)".J.Saintek,*Vol.5, no.1, pp.1-10.

[4] I.N.Widana Negara dan T.G. Suwarsa Putra.2010." *Potensi Batu Kapur Nusa Penida Sebagai Agregat Perkerasan Jalan", Jurnal Ilmiah Teknik Sipil*, Vol.14,no.1, pp 1-5.

[5] I. N. Widana Negara dan T. G. Suwarsa Putra .2010. *Potensi Batu Kapur Nusa Penida Sebagai Agregat Perkerasan Jalan", Jurnal Ilmiah Teknik Sipil.* Vol.14, no.1, pp.1-5.

[6] M.Anggraini, S.Wiyono, dan A.Wanim.2015.*"Kajian Kadar Aspal Hasil Ekstraksi Penghamparan Campuran AC-WC Gradasi Kasar Dengan Job Mix Formula", Prosiding Seminar Teknik Sipil Annual Civil Engineering Seminar ,* ISBN 978-979-792-636-6, pp 96-103.

[7] F.Soehardi, S.Wiyono, dan A.Wanim. 2015. *"Kajian Perbandingan Kadar Aspal Hasil Ekstraksi Campuran AC-WC Gradasi Kasar Dengan Cairan Ekstraksi Menggunakan Bensin", Prosiding Seminar Teknik Sipil Annual Civil Engineering Seminar ,* ISBN 978-979-792-636-6, pp 137-144.

[8] Departemen Pekerjaan Umum (1969-20080). *Standar Nasional Indonesia Cara Uji Berat Jenis dan Penyerapan Air Agregat Kasar.*SNI 1969-2008.

[9] Departemen Pekerjaan Umum (1970-2008). *Standar Nasional Indonesia Cara Uji Berat Jenis dan Penyerapan Air Agregat Halus.* SNI 1970-2008.

[10] M.Anggraini, 2014, *"Kajian Kadar Aspal Hasil Ekstraksi Penghamparan Campuran AC-WC Gradasi Kasar Dengan Job Mix Formula", Tesis*, Program Magister Teknik Sipil Universitas Islam Riau, Pekanbaru.

Inventory of Heritage Building in Kampung Bandar Senapelan Pekanbaru City, RIAU

Rika Cheris[1] and Repi[2]
1. Universitas Lancang Kuning, Pekanbaru, 28265, Indonesia
Telp:+6275152324, Fax: :+6275152324
email: rika.cherish@unilak.ac.id
2.Universitas Lancang Kuning, Pekanbaru, 28265, Indonesia
Telp:+6275152324, Fax: :+6275152324
email: repi@unilak.ac.id

Abstract: Heritage building is a building that has a character value of local history and culture in a village and city. This heritage building is usually located in city center and into the early development of a city. In Pekanbaru, there is a group of heritage buildings with some characters. The area named Bandar Senapelan on the edge of the Siak river which has been developed before the colonization of the Netherlands.This research was conducted with field observations and interviews with local residents and then conducted measurements on the building. The data collected are compiled into a table to make it easier to conduct an inventory of the identification of the structure and construction of the building as well as the types of responses. Some of the heritage buildings in the area of Bandar Senapelan were influenced by some ethnic groups such as Minangkabau, Malay and Chinese. Those buildings have distinctive character although they do not use too many elements of architectural art, but their existence is quite able to prove past glory in the city of Pekanbaru. In this research, a core zone area was created, then the buildings' remains were calculated. After that, the calculations and sincronization were done with a map of virtue as many as 52 heritage buildings made of wood and stone. The buildings have different tipology, as a single home (individual), the home of the fisherman, merchant's house, warehouse, shop and home stores.The output from this research will generate a mapping of the heritage buildings and it is expected that it can provide useful information for the development of the city stakeholders to determine the direction of future development of Bandar Senapelan
Keywords: *Heritage building, inventory and typology*

1.Introduction

Bandar Senapelan is an area in Central of Pekanbaru City that is attached directly to the Siak River. At the region there` was the Centre of trade and transport links between central Sumatra and the lands on the opposite side of the Mallaca Straits . The existence of cultural heritage buildings are still found in the area. The Government of the Republic of Indonesia firmly has mentioned in the law on cultural heritage No. 11 of 2010, that more than 50 years old buildings are objects of cultural heritage and must be preserved. However, to reach the stage of preservation, it needs to go through the process of inventory and documentation as well as in the assessment of these buildings. Inventory of historic old buildings aims to give public awareness especially local authorities about how important it is maintaining and keeping the visual quality of the city.The new development is one of the factors of economic development and improvement of a city. However, new development should not sacrifice the cultural heritage values in one place.

The research method is to do a field survey to each heritage building. Data taken in field such as existing photos, history, ownership, construction structures, architectural details and typology. After collecting all the data through interviews and the depiction of the measurement, then the process of designation or classification action on the heritage building preservation was done. For that data were summarized in a table in order to facilitate the understanding of the condition rather than the object of research.

2. The theory of cultural heritage buildings

Old buildings establish the identity of the architecture of the later origins of the past i.e. traditional architecture, (Antariksa, 2015). This statement is an affirmation to the existence of traditional architecture. Traditional architecture provides visual richness for the community of the city and at the same time reflects the local wisdom of traditional society in the past such as the cleverness and precision of traditional community breakfast buffet on the buildings they build. Every structure and construction of the building support mutually and are able to survive decades of even hundreds of years. While in the carvings, color, detail architecture there are meanings of such a high position, rank, as well as advice for the next generation. Traditional societies have a high sensitivity to the environment around them. When they intend to build a house or other building, then they always pay attention to power support a stable land as land for housing, land for farming, as well as a source of clean water.

Modern societies concern for the environment, so Interested to do the maintenance of the architectural cultural heritage buildings as a comparison to the businessmen in the field of property in order to join in observing the environment when it will open a new land. Specifically, the objects are essential for building preservation and there are several criteria. Some general criteria are used to determine the objects preservation, among others:

1. Aesthetic, is assessment of beauty and artistic value of a building in part of the city as it represents a special achievement in a particular historical style. The consideration of the decisions taken vary so it is quite difficult to prove a building is more important than the other, this aesthetic benchmarks associated with aesthetics in terms of form, structure, and space.
2. The plurality, is a group of buildings that represent a special type of class or period of a particular time which forms the specific character. So the benchmark of the plurality emphasizes on how far the work of the architecture represents a range or a special kind of specifity.
3. The scarcity of a building that consists of one type of latter examples, or that still exists is rare or even that is the only one in the world, it does not belong to other areas.
4. The role of history, is a building or group of buildings with an important historic event to preserve as a symbolic bond between previous and present events.
5. Strengthen the area nearby, namely buildings and parts of the city that will affect urban character with a very meaningful presence to improve the quality and image of the surrounding environment.
6. Privileges, i.e. buildings which are protected because they have privileges, for example the first, longest, tallest, oldest, largest, and so on.

After doing classificationthe above, then the data carried out the conservation action types. There are three categories of types of actions, namely:

a. the first category is based on the age of the building. Through the study of the history of the development of the city it will be retrieved timeline (time series) which is important in the formation of the city. Time series with this approach will make it easier to determine the population of buildings based on time. How prevalent is done through the choice of the population which has certain specifications in determining the group. This can be done when it has been known years of development.
b. The second category; based on the flavor of local residents regarding the uniqueness of the form or the motives of the ornament that is owned by heritage building, the influence of different cultures.
c. the third Category; the buildings after the second world war. Usually not much uniqueness of cultural values. In this population in the buildings that have no importance against this can be removed from the list check (check list)

After doing the above, then the data collection carried out conservation action type of separation process. There are three categories of types of actions, namely:

a. the first category; is based on the age of the building. Through the study of the history of the development of the city will be retrieved timeline (time series) are important in the formation of the city. Time series with this approach will make it easier to determine the population of buildings based on time. How prevalent is done through the choice of the population which has certain specifications in determining his group. This can be done when it has been known years of development.
b. The second categoryis based on the flavor of local residents regarding the uniqueness of the form or the motives of the ornament found in the heritage building which may show the influence of different cultures.
c. the third Category is the buildings after the second world war. It does not show much uniqueness of cultural values. In this case,the buildings have no importance, thus can be removed from check list.

After the selection was done the buildings on the list should be assessed through specific interests based on historical values, interesting architectural characteristics, circumstances and structures as well as the influence of the surrounding area against the building. Assessment of heritage architecture in this area can be done by analyzing the existing building typology. Building typology is describing a group of objects on the basis of the similarity of character and basic forms. Grouping multiple objects is performed based on the nature of the similarities in the form type, model and idioms.

Classification of buildinhg typology is done based on understanding the object, function and symbolic perception. Typology of traditional Malay houses is the home stage with high poles. (Mahyudin Al Mudra, 2003). The Overview of Malay architecture Building typology in Bandar Senapelan shows that there have been changes or substitution of some elements of the building. This is caused by the traditional typology of the building because of the influence of Malay culture and external cultural influences that developed at that time. Malay architecture building in the Bandar Senapelan is built based on the function of building, the House, warehouse, trade accounts, and places of worship (Titin, Rika, Repi, 2016).

Minangkabau ethnic residential typology is determined by the number of Poles, the floor and the end of the building, while the other characteristics remain the same (Ibenzani, 1985). Minangkabau architecture also emphasizes the functional aspects rather than on the symbolic ones. A home is not just a place to live, but also as a place of discussion and traditional ceremony. Room is spacious enough built to accommodate guests on certain occasions. The bedroom is arranged in a linear-shaped in order to bring more impression on the house. Minangkabau house has several types and characters that have been mutually agreed by the experts. The second type is the basis of development of the minangkabau house so it undergoes transformation from time to time. The house of Chinese there has also undergone many changes and is almost disappeared in this area. Anyhow, it can be seen in the area surrounding the location of research. The cultural heritage of ethnic Chinese that is still found is the intangible heritage which is always in use by the Chinese community.

2) The process of determination of cultural heritage buildings
In the case of the research area of Kampung Bandar Senapelan, determination of the building is done by several stages through the principles below:

1) Making a list of the records against the heritage buildings by doing an inventory buildings that have plurality, aesthetics, rarity, history, reinforcing the role of the regions, and privileges. Action was taken based on the quality of the building, maintenance and the contributions on the face of the city. All buildings contain cultural heritage in kampung Bandar Senapelan marked in the map. Then to the environment, the selection can be made by:
   a. Environmental conservation which can be selected based on the grouping of heritage buildings. This means buildings which are not worth the environmental legacy but included in the heritage building will get special attention.
   b. The environment is considered an important contribution to the quality of the face of the city as a whole (including oldand new buildings), will be considered as environment conservation. There are times when there are regions that make up a good angle and give the quality of the image of the city as a whole. the vegetation is one of the things that makes the space on a particular view that gives good quality. This atmosphere should also be maintained.

Figure 1. The guidelines limit the area at kelurahan Bandar Senapelan.

2) Classifying the environmental and cultural heritage building by making the outline zoning district. The area which has already been in the zoning would be a major concern to do an inventory process. By grouping the environment and buildings in zoning, it will facilitate all actions on the location. Said Zoning is better known in archaeology while in architecture it is known as deleniazation. Deleniazation and zoning have already been agreed by all preservation experts to use in grouping zones that must be protected. Zoning is therefore divided into 3 namely core zone, buffer zone and developed zone. The zoning research stresses only on the core zone, due to limited time. As for the buffer zone and the developed zone will be continued on subsequent research. The core zone is a group of buildings that have been classified. In this core zone of protection of heritage building have been preserved in accordance with theories that have been there. Then to protect the core zone guidelines on building maintenance and environment setup and development as a reference in the future, should be created. In Figure 01 below. It is clear that using the core zone limits the street name as well as the edge of the River, that is North of the River with unbounded Siak, South with the Kampung Bukit, to the East with Mohamadd Yatim street and the West with the Waqf. Decision making of this limitation is intended to make it easier to control and maintain.

IOP Conf. Series: Earth and Environmental Science **97** (2017) 012021 doi:10.1088/1755-1315/97/1/012021

Table 1.
Numbered buildings of cultural heritage in Kampung Bandar Senapelan

No	Nama bangunan	alamat
01	*Surau Al-Irhaash*	Jl.Senapelan no.57 Kampung bukit
02	*Rumah Pribadi*	Jl.Senapelan no.55 Kampung bukit
03	*Rumah Pribadi*	Jl.Senapelan no.53 Kampung bukit
04	*Rumah Pribadi*	Jl.Senapelan no.51 Kampung bukit
05	*Rumah Pribadi*	Jl.Senapelan no.50, Kampung bukit
06	*Rumah Pribadi*	Jl.Senapelan Gg. Perdagangan
07	*RumahTuan Kadi*	Jl.Senapelan Gg. Perdagangan
08	*Kios*	Jl. Kota Baru
09	*Gudang*	Jl. Kota Baru
10	*Rumah Tinggal*	Jl. Kota Baru
11	*Gudang Garam*	Jl. Kota Baru
12	*Kios*	Jl. Kota Baru
13	*Gudang Garam*	Jl. Kota Baru
14	*Gudang*	Jl. Kota Baru
15	*Gudang*	Jl. Kota Baru
16	*Gudang*	Jl. Kota Baru
17	*Gudang*	Jl. Kota Baru
18	*Toko Veleg*	Jl. Kota Baru
19	*Kios*	Jl. Kota Baru
20	*Kios*	Jl. Saleh Abbas
21	*Kios*	Jl. Saleh Abbas
22	*Kios*	Jl. Saleh Abbas
23	*Kios*	Jl. Saleh Abbas
24	*Kios*	Jl. Saleh Abbas
25	*Kios*	Jl. Saleh Abbas
26	*Rumah Tinggal*	Jl. Saleh Abbas
27	*Rumah Tinggal*	Jl. Saleh Abbas
28	*Kedai Kopi Kim Teng*	Jl. Saleh Abbas
29	*Gudang*	Jl. Kampar
30	*Toko Ban Mobil*	Jl. Kampar
31	*Gudang*	Jl. Kampar
32	*Toko Ban Mobil*	Jl. Kampar
33	*Rumah Tinggal*	Jl. Kampar
34	*Rumah Singgah Sultan*	Jl. Pergadangan
35	*Rumah Tenun*	Jl. Pergadangan
36	*Rumah Tinggal*	Jl. Pergadangan
37	*Rumah Tinggal*	Jl. Pergadangan
38	*Rumah Tinggal*	Jl. Pergadangan
39	*Rumah Tinggal*	Jl. Pergadangan
40	*Rumah Tinggal*	Jl. Pergadangan
41	*Gudang/Tempat Tinggal*	Jl. Pergadangan
42	*Rumah Tinggal*	Jl. Pergadangan
43	*Rumah Tinggal*	Jl. Pergadangan
44	*Rumah Tinggal*	Jl. Pergadangan
45	*Rumah Tinggal*	Jl. Pergadangan
46	*Rumah Tinggal*	Jl. Pergadangan
47	*Rumah Tinggal*	Jl. Pergadangan
48	*Rumah Tinggal*	Jl. Pergadangan
49	*Rumah Tinggal*	Jl. Pergadangan
50	*Rumah Tinggal*	Jl. Pergadangan
51	*Rumah Tinggal*	Jl. Pergadangan
52	*Mesjid Dagang*	Jl. Pergadangan
53	*Rumah Tinggal*	Jl. Pergadangan
54	*Rumah Tinggal*	Jl. Pergadangan

3) Buildings which will bring the history of both the history of the building as well as the history of its surroundings, 2) flavor of the local population with respect to uniqueness of the shape or the motives of the ornament that is found in heritage building, the influence of various cultures, 3) buildings after the second world war do not have uniqueness of cultural values. From the survey results and interviews with residents who still reside in the location of the median – median research that the building stands before the time of the second world war. So the category assignment on a building that looks in the table is the category one and category two.

4) Determination of the building, this assignment consists of 4 types such as, tipe A is the building with the maintenance of preservation through action on the exterior and the interior, The exterior maintenance, tipe B is yet on the inside may be modified in accordance with the functions of the present, tipe C is the maintenance may change the exterior and interior, while tipe D is the building that can be destroyed as stalls that stick to the walls of buildings of cultural heritage or glue that is deliberately attached so it may damage the original building and so
5) Determination of the typology of the building, grouping shapes, characters and styles on a group of buildings. In the area of research in German calculation, in accordance with current function. As mentioned in the introduction that the typology of the building of the cultural heritage found in kampung bandar senapelan are several buildings namely; a) residential Building - housing, b) Shop House/kiosk/stall, 3)mosque, 4) Warehouse,
    a. Housing = 48 %
    b. Shop house/stall = 22,2 %
    c. Warehouse = 18,5 %
    d. Mosque = 3,7 %

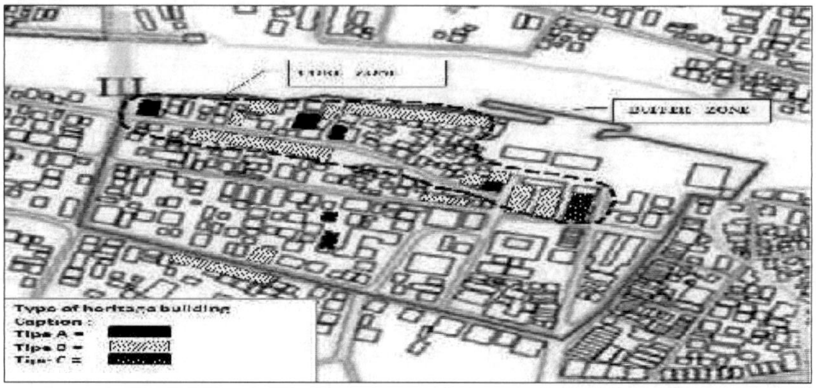

Figure 2. The determination of the core zone, buffer zone, and the preservation of cultural heritage building types

Comparison of the function of buildings found in the kampung bandar senapelan caused by the activities of the trade. Different function will affect the appearance of the building as a whole. Likewise with the surrounding environment. Until now, the port activity is quite high especially for the heavy stuff. Several buildings that served as the House turned into a goods shed weight. This will bring the impact on the existence of the building and if it is not treated well, it will then fall by itself. The specialty of the building in this area is the average use of materials from wood to start from the floor to the roof, for the foundation stones and limes were used. This indicates that most of the Malay population live in this region including to (have a high economic level). The average of the society work as traders, only a small fraction that become fishermen. The number of elites left the house being left to others, they are fixed.Most of the buildings remain to this day are more than a hundred years and the damage does not occur only because of bad maintenance but also because of the location of the fishermen's homes which are located along the lip of the river which cause damage and moldiness of building materials. Quite worrying is when these damaged buildings fall away with age. For this reason, special handling is required so that the diversity of the old buildings in the area is still preserved. .

4.Conclusion

From the results of the analysis and the discussion above, it is inferred several things. The first, the need to complete the process of documenting this further to a building that has not yet been identified. Then for keeping this area, the Government is expected to control and evaluate the rules enforcement of construction or building permit. The necessity to do the rejuvenation of existing buildings is to provide a clean, comfortable visual experience for all citizens. The research will be continued to manufacture managed plan for handling the preservation areas which will deepen the process of determination of the zoning area. It is hoped the input of this article for self development.

5. Reference

[1] Antariksa. 2015. *Preservation of architecture and an integrated city.*Cahaya Atma Pustaka Press.
[2] Anne and John Summerfield, 1999. *Walk In Splendor Ceremonial Dress and The Minang Kabau.*
[3] BWSB. 2003. *Inventory of Heritage Building in Sawahlunto West Sumatra.* Goverment of Sawahlunto Publishers.
[4] Hamidy, UU. 2009, *The universe is "Malay Cultural Trajectory in Riau*, Bilik Kreatif Press.
[5] Ismael, Sudirman. 2007. *Minangkabau traditional architecture.* Bung hatta University Press.
[6] Mudra, Mahyudin Al, 2003, *Melayu housing*, Adicita
[7] Titin, Rika, Repi, 2016. *The potensial of Bandar Senapelan To be the culture and historic Tourism.*

Approach of Urban Design Elements in Preservation Area Kampung Bandar Senapelan Towards Pekanbaru Metropolitan City

Yose Rizal [1], Rika Cheris[2], Repi[3]
[1,2,3] Universitas Lancang Kuning, Pekanbaru, 28265, Indonesia
Telp:+6275152324, Fax: :+6275152324
email: yose@unilak.ac.id

Abstract: The development of Pekanbaru City is very rapid, consequently is constantly experiencing changes in buildings, areas or cultural objects that need to be preserved to be disrupted, replaced by economic-oriented development - commercial. The contradiction between the construction of the metropolis will be the beginning of the problem for urban areas. Kampong Bandar Senapelan is an early town of Pekanbaru town located on the banks of the Siak River. The settlement has a typology of Malay and vernacular Malay architecture. The existence of these villages experienced concern as a contradiction of the city's development toward the metropolis which resulted in degradation of the historical value of urban development in this region. This study was conducted to make an important assessment of preserving Kampung Bandar Senapelan as the oldest area and its great influence on the development of metropolis. Preservation of historical and cultural heritage with conservation and preservation measures is one of the urban design elements to be considered for all city stakeholders to safeguard the civilization of a generation. Considerations that will become a benchmark is the history, conservation and urban development towards the metropolis. The importance of awareness of the conservation of the city through conservation and preservation in this area can lead to new characters and values to the building and its environment and will create an atmosphere different from the rapid development (modern style). In addition, this preservation will be evident in a harmonious life with a high tolerance between multi-ethnicity that co-existed in the past.

Keywords: *Preservation, Conservation, Urban Design Element and Metropolis Development*

1. Introduction

Pekanbaru City is currently experiencing a rapid growth of physical development. This very significant development spread to all sectors such as economy, politics, religion and culture. All this can not be circumvented because it has become a consequence of the vision of Pekanbaru City as a Metropolitan City Madani. The pace of development The city's development towards the Metropolitan makes all stake holders of the city strive to increase energy in achieving it. Especially from the economic side, that the economic benefits become the symbols of the glory of a city so that the construction of multi-storey buildings began to be present on every side of Pekanbaru City.

The construction of spectacular buildings became a strategic part of City Marketing, a city effort to stand out in the context of global and tight competition (Laksono, 2013, 41). This principle has swept across cities in some countries, not only in Indonesia and even other countries have started in the first place, and as a result most of the city realized the mistake in organizing their metropolitan city. History has proved that the great cities of the world at this time just want to re-arrange their city by creating a humane civilization and more identity. Learning from our neighboring country, Singapore, they have succeeded in creating a smart city by not leaving the local culture and becoming a high-tech city. Singapore is building a fairly balanced city between modernization, and nature and culture conservation. All get the attention full of involving the citizens of the city so it becomes a fun place for the citizens of the city.

2. Existing Condition of Kampung Bandar

Learning from several metropolitan cities in the world, the Government of Pekanbaru really need to pay attention to the urban aspects if you want to achieve the vision of the city mentioned above. According Shirvani (1985) there are eight aspects in the planning of a city, one of which is the conservation aspect. Conservation of the city is one of the things that play a role in determining the identity of the city, as well as the identity of the community also as a distinction between one city with another city in the world. This is due to differences in climate, geography and local culture that will create a local culture of its own. Identity is a disorder that belongs to a place, so as to distinguish it from other places. The things that shape the image of a place need to be known for sure to be driven maintenance efforts against the place. Component in question consists of physical elements of the city such as buildings that often attract public attention because of the uniqueness and properties of excellence, the atmosphere of the place that shows the lifestyle of the local population, the characteristics of landscape and plants that exist, the daily activities of human Show local lifestyle.

Pekanbaru City seems to have to find an identity for their city to be easy to know and has its own characteristics. This identity search is easier through the local culture aspect because as mentioned above, local culture is something different from other regions. Seeing from the formation of Pekanbaru City that this city comes from a village that is on the edge of the river built by the King of Young Muhammad Ali in 1784 and was named Pekan Baharoe.

Kampong Senapelan is growing rapidly due to the presence of ports with large ships to sail down to the Melaka Strait. So that the port of Senapelan become a means of society to conduct trade and sailing to neighboring country. Thus indirectly the development of this area becomes a magnet for the entire community, especially the Malay, Kampar and Minangkabau.

These three tribes inhabit Kampung Bandar Senapelan by bringing their own culture and building their housing. But with the limitations of artisans, materials then the construction of residential homes in doing improvise without using the intricate details like their original home. So the building is more of a function than the beauty and splendor as in their area of origin. House of stilts became their main alternative to wake up because this area is very close to the edge of the river. In the Dutch colonial period then one by one Ethnic Chinese also come to this area and abide by taking the profession as a trader. And they also build their housing with all the limitations.

2.1 Decline in Environmental Quality

Along with the passage of time and the development of Pekanbaru City forces Kampung Bandar Senapelan have to adjust to the condition of society that has been very diverse. The economic and regulatory pressure of the region's function has made Kampung Bandar Senapelan has changed considerably. Buildings with Traditional Architecture have been replaced with modern buildings deliberately. According to observations in the field turnover of this building is caused by several things such as, 1). Stilt house with wood material is not able to survive for the present function, 2). Difficulty to find wood that can be used to repair the building, 3) Ignorance of the community about the importance of maintaining their buildings which can actually be the identity of the city and culture, 4). The absence of government awareness of the importance of maintaining buildings that have cultural values. The reasons are so basic to the community that they unwittingly replace the structure and construction of their buildings. But in reality this village has decreased urban visual quality especially towards the community buildings with traditional architectural typology. One by one this building disappeared and replaced with a store house building with regulations that have not been established. If the building is no longer in the village then it is feared Pekanbaru City will lose the urban identity. Though the identity of a city reflects the civilization of its people.

3. Research Method

In this study, the authors use a qualitative research methodology comparative. Comparative qualitative research is an objective research because it interacts with the facts studied and provides a description of a precedent study as a comparison of research objects. In this study, the authors attach some precedent studies that are considered successful in the application of the concept of conservation. In addition, this study uses an inductive approach that is an approach that collects data in advance to create a hypothesis. In other words, the inductive approach is as a process of concluding (or hypothesis formation) based on one or more facts (evidence) obtained. And in this study the authors use qualitative historical research methods and descriptive because it takes some data historically and data information analyzed and the results of its analysis is a picture of a phenomenon that occurs.

4. Conservation Theory

In Urban Design, several elements of urban design are land use, building arrangement, circulation and parking, open space, pedestrian path, supporting activities, information governance, preservation and conservation (Hamid Shirvani, 1985). In this theory conservation becomes an important thing to note.Indications of a region of value as a historical area of urban heritage include:

1) Urban Heritage is an area once a city centers that have the complexity of functions and economic, social and cultural activities that accumulate historical significance. The richness of urban heritage typology and morphology such as the historic site, traditional district and colonial district is generally a Locus solus that once served as economic, social and cultural "in various environmental scales (distric, sub distric, neighborhood, area and sub area).
2) Urban Heritage accumulates the values of cultural significance (cultural significance). The cultural meaning of a place manifests itself in its physical fabric, its place and its contents, the contents accumulated in urban heritage have significant values such as: aesthetics / Architectonic, / typical, scarcity, history, influence on the environment and Privileges (Attoe in Catanese and Snyder, ed., 1984).

Typology of Cultural Heritage City (Urban Heritage) Typology of cultural heritage areas (urban Heritage) in Indonesia are generally grouped into three models (BWSB, 2002, 12), namely:

1) Traditional area; a) It is a locus solus area that accumulates the cultural meaning of the region with its traditional character. b) Typology and cultural significance of traditional areas are defined again in several regional scales, ranging from neighborhoods (traditional villages, kauman villages, ethnicity villages, Chinatowns, etc.) to the kadipaten. c) Generally some typology, cultural meaning and living culture of traditional areas in Indonesia are still conserved fairly well.
2) Colonial area, a) It is a locus solus area that accumulates the cultural meaning of the region with the colonial character. Generally located in large cities or areas that are considered important to be the center of office activities, industry, settlements and trade VOC in the colonial period. b) Typology and cultural significance of the colonial region is defined again in several regional scales of the region and the district. c) Generally, existing living culture has gone along with the occurrence of decolonization.
3) Historical Site, While the historical footprint is meant for areas with very high historical value in the form of palaces and religious monuments

Some of the conservation principles that need attention are:

1) Conservation is based on appreciation of the original state of a place and as little as possible to physical intervention of the building, so as not to change the historical evidence it has.
2) The purpose of conservation is to recapture the cultural significance of a place and should be able to secure its security and preservation in the future.
3) Conservation of a place should be considered in all aspects related to its cultural meaning, without emphasizing one aspect only and sacrificing other aspects.
4) A building or a historic work must remain in its historical location. The removal of all or part of a building or work is not permitted, unless it is the only way to ensure its sustainability.
5) Conservation maintains a suitable visual background such as shape, scale, color, texture and development. Any new changes will have a negative impact on the visual background should be prevented.
6) Appropriate conservation policies for a place should be based on an understanding of the cultural significance and physical condition of the building.

Successful Adaptive Re-Use Strategies in Conservation Areas

This success cannot be separated from the strategy applied in it. These strategies include: The availability of strong political will from the government in promoting the acceleration of the revitalization process. The existence of political will of the government that became the main key to the success of the concept of revitalization. The city government as the holder of political authority should see revitalization as an opportunity. The Urban Redevelopment Authority (URA) in Singapore and Hong Kong is the most active government agency in developing revitalization concepts to revive old economically dead areas (Kamil, 2008). Having a mixed-use concept of integrated and integrated development The success of revitalized areas in Asian and European countries is, among others, reinforced by an integrated Master Plan concept. This Master Plan document contains a comprehensive range of regional planning strategies. In the Clarke Quay and Boat Quay areas, the concept of land use uses a high & best use and dynamic tenant mix approach, complemented by regional spatial design guidance and attractive streetscape device design (Kamil, 2008).

Having a unique and competitive district economic identity strategy to compete with other urban areas One of the reasons for the demise of economic activity in urban areas is the inability of the region to adapt to new economic challenges. Therefore, one of the most important revitalization strategy concepts is to reposition economic identity oreconomic re-positioning. Examples include the Far East Square area in Chinatown and the Mohamed Sultan region, both in Singapore, successfully revitalized from the area of general trade and abandoned shelter into an active and successful dining and entertainment resort. (Kamil, 2008)

5. Results and Discussion

After reading some of these theories and in pairs with the current state of the building and environmental groups in the village of senapelan, there can be several strong approaches to support the efforts to preserve this kampong, namely;

- From typology and morphology Kampong Senapelan classified to traditional area or traditional district because its position is still in an area,
- Has historical value with the original typology of the river basin is the stage house as the most appropriate alternative to living around the river edge,
- Some typology presents Living Culture because of ethnic heterogeneity such as Minangkabau ethnic, Malay enik, ethnic Chinese, ethnic Ocu Kampar and others which their number still sangt slightly like Batak ethnic,
- With still not much change of heritage building so it can still to restore cultural meaning of this area,
- Maintenance of the visual background of the city that will bring the city to cultural identity, especially Malay culture.

Figure. 1. Port is still active. Private doc

These five reasons will be able to place Kampung Bandar Senapelan in the first rank position as an area that must be preserved. So much historical significance and cultural significance inherent in this village that will be able to grow the identity of local communities in the midst of metropolitan development.
Comparative Studies

Figure. 2. Building and environmental maintenance in Clare Quay Singapore as one of the comforts of the current Singapore community. (Source: Kompasiana)

The Clarke Quay and Boat Quay areas of Singapore Clarke Quay and Boat Quay are areas along the Singapore River that adopt the concept of conservation that is well known and considered successful. This area was originally a slum area and has no eyesight or reuse. This situation forced the Singapore government to tear down old buildings along the river and replace them with new, more modern and useful buildings. However, on the various inputs and insistence of the planners, it was taken a policy to preserve the old buildings into a new function.

After the Singapore government committed to clean up the river for 10 years and completed in 1987, this river basin area can eventually be fully utilized. Utilization of the maximum one of them juggle old buildings with a little reshuffle on the façade and the inside. The old functions of this building were transformed into a new function in order to better benefit the Singapore government in the field of tourism. In addition, by closing the path for motor vehicles so that open space in front of the building becomes like a pedestrian expanse for pedestrians and can be used for open dining for some cafes.

Figure. 3. The atmosphere of conservation areas in Clarke Quay Singapore (source: Kompasiana)

Urban Revitalization Strategy :

From the experience of several areas in major cities in Asia that succeeded in revitalization, at least there are 5 development strategies that can be used as a case study in the success of revitalizing an urban area. These strategies include:

5.1. Political initiative of the city government.

Many cities in developing countries in Asia are not aware that the urban areas have an economic age or economic life cycle that in a certain period must be recycled. It is the government's political will that is the key to the success of the concept of revitalization. The city government as the holder of political authority should see revitalization as an opportunity. The Urban Redevelopment Authority (URA) in Singapore and Hong Kong is the most active government agency in developing revitalization concepts to revive the economically dead old areas. URA in Hong Kong even conceptualized its vision and mission with four pillars of city planning strategy: Redevelopment for newly developed areas of dead or vacant land. Revitalization to restore the economic pulse in the old urban areas and die economically. Rehabilitation for buildings that have deteriorated their physical quality. Preservation for areas and buildings that have historical significance or architectural quality that must be preserved.

5.2. Have a regional management agency

In some cases, the urbanized areas are revitalized, the complexity of the problems and the scale of the area often requires a special area management strategy. For the first model we can see in China. In China where the land is wholly state-owned, the management of Xin Tian Di area in Shanghai and the historical area of Shamian Island in Guangzhou is given entirely to developers who are given business concessions to revitalize and develop these areas. Xin Tian Di is managed by developer Shui On Properties. Shamian Island by Swire Properties. The second model is like in Singapore and Hongkong, where the governing body of the revitalized area remains from the government. This happens because this government body has enough solid experience and a vision of a competitive global economy. Clarke Quay revitalization project in the revitalized area of Singapore River 2 times is the result of urban tourism-based revitalization concept that is directly conducted and driven by URA. For cases in Indonesia where generally government agencies are not as active as URA Singapore or Hong Kong, the establishment of a regional management body as a government partner consisting of local stakeholders becomes crucial. If the role is not as a technical decision-making body like the example in China, at least this body has a strong role in determining the strategies and concepts that are appropriate for the area to be revitalized.

5.3. Has a new competitive economic identity
One reason for the demise of economic activity in urban areas is the inability of the region to adapt to new economic challenges. Therefore, one of the most important revitalization strategy concepts is to reposition economic identity or economic re-positioning. Examples include the Far East Square area in Chinatown and the Mohamed Sultan region, both in Singapore, successfully revitalized from the area of general trade and abandoned shelter into an active and successful dining and entertainment resort. Bukit Bintang Street in Kuala Lumpur, formerly known only as a moderate trading area, is now the icon of urban tourism in Malaysia known globally.

5.4. Have a concept of integrated regional development
The success of the revitalized areas in Singapore, Kuala Lumpur and Shanghai is, among others, reinforced by the integrated Master Plan concept. This Master Plan document contains comprehensive regional planning strategies. Examples include the Master Plan for revitalizing the Singapore River in an effort to reposition this region in capturing the global economic opportunities of potential urban tourism. In the area of Boat Quay and Clarke Quay along the Singapore River, the concept of land use is based on the concept of high & best use and dynamic tenant mix which is complemented by spatial design guidance of the area and the design of an attractive streetscape device. Usually for areas where land ownership is quite complex and owned by many parties, the Urban Design Guidelines (UDGL) document as a control device for the design of the area or building becomes important. When investment begins to enter and physical construction begins to be done then UDGL serves as a guide in aligning the physical concept with the overall economic theme of the region.

5.5. Have a pragmatic phasing strategy
In general, urban areas are revitalized is a large area, so it is impossible to develop the entire region at the same time. Therefore, the phasing strategy becomes crucial. The initial phase or pilot project is generally selected in the area that invites the most investment and areas that are able to represent strongly the new image of the revitalized area.

6. Conclusions
History is the best bridge between current events and events in the past, therefore every region or historic building is fitting for us to preserve. Preservation application of a region or building that is currently often used is conservation, because this way is considered quite successful in maintaining a historic heritage in Indonesia. Learning on urban pilot areas in the above discussion shows that the concept and strategy of revitalization must be done innovatively and comprehensively. Innovative means the revitalization area planning approach that the main issues vary should be solved with creative concept innovations and not solely based on general theories of urban planning / design. Comprehensive means all aspects of the problems and involvement of all parties must be completed and involved from the concept process until the implementation process. The success of revitalizing these troubled urban areas proves indirectly to be one of the solutions in responding to urban issues in a larger context as a result of urbanization and rapidly changing urban culture in urban areas in major Asian cities. Conservation efforts An area or a building can be successful if done with a mature and directed strategy. Therefore, the right strategy in conservation efforts is to transform a region or building into a place with a new function that is more profitable than the old function, without changing the character and historical value contained in it. Changes in the function of a region or building should use the concept of mix-used development, so that each element is mutually filling and integrated so as to produce many benefits in all fields. In the economic field, the results obtained can be used for maintenance and maintenance costs so that the building can be maintained and its historical value. While in the field of tourism, social and culture is able to invite young people to better appreciate the historical heritage and able to attract local and foreign tourists while maintaining the identity of a city.

7. References
[1] Shirvani, Hamid. 1985. *The Urban Design Process Van Nostrand Reinhold*. Michigan University.

[2] Attoe, W. 1989. *Perlindungan Benda Bersejarah*. Dalam Catanese, Anthony J. Dan Snyder, James C. (Editor). *Perencanaan Kota*: 413-438. Jakarta: Erlangga.
[3] Badan Pelestarian Pusaka Indonesia. 2003. *Piagam Pelestarian Pusaka Indonesia*.
[4] Budiharjo, Eko. 1997. *Arsitektur Pembangunan Dan Konservasi*. Jakarta: Djambatan.
[5] Budiharjo, Eko. 1997. *Arsitektur sebagai Warisan Budaya*. Jakarta: Djambatan Dinas Kebudayaan & Permuseuman Provinsi Daerah Khusus Ibukota Jakarta.
[6] --------. 2007. *Sejarah Kota Tua*. Jakarta: Jakarta Culture & Heritage.
[7] Dinas Kebudayaan & Permuseuman Provinsi Daerah Khusus Ibukota Jakarta. 2007. *Guidelines Kota Tua*. Jakarta: Jakarta Culture & Heritage.
[8] Dobby, A. 1978. *Conservation and Planning*. London: Hutchinson.
[9] Fitch, J. M. 1992. *Historic Preservation: Curatorial Management of The Build World*. New York: Mc Graw Hill Book Company.
[10] Irpansa. 2014. *Re-Use Sebagai Konsep Eko Urban Pada Kawasan Kota Tua*. Jakarta. http://id.scribd.com/doc/200121635/Arling-Kota-Tua
[11] Kamil, Ridwan. 2008. *Strategi Revitalisasi Kota-kota Asia dalam Konteks Persaingan Global*. Dalam wordpress blog.
[12] Karmila. 2013. *Pengkajian, Penelitian Arsitektur dan Upaya-Upaya Kegiatan Pelestarian Bangunan/ Kota/ Kawasan Bersejarah*.
[13] Martokusumo, Widjaja. 2008. *Revitalisasi Sebuah Pendekatan Dalam Peremajaan Kawasan*. Bandung: Jurnal Perencanaan Wilayah dan Kota ITB vol. 19 no. 3 Desember 2008, 5773.
[14] Purwantiasning, Ari Widyati. 2009. *Konversi Bangunan Tua Sebagai Aplikasi Konsep Konservasi*. Jakarta: Jurnal Ilmiah Arsitektur NALARS Volume 8 Nomor 2 Juli 2009.
[15] Purwantiasning, Ari Widyati. 2013. *Kajian Konsep Adaptive Reuse Sebagai Alternatif Aplikasi Konsep Konservasi*. Jakarta: FT UMJ.
[16] Purwantiasning, Ari Widyati. 2005. *Konservasi Bangunan di Singapura*. Published-04-2005-konservasi.

The Influence of Addition of Plastiment-VZ to Concrete Characteristics in Riau Province

Shanti Wahyuni Megasari[1] and Winayati[2]
1. Universitas Lancang Kuning, Pekanbaru, 28265, Indonesia
Telp: +6281275254771
E-mail: shanti@ unilak.ac.id
2.Universitas Lancang Kuning, Pekanbaru, 28265, Indonesia
Tel/Fax: +681275138910
E-mail: winayatimt@gmail.com

Abstract: Riau Province has an area of 8,702,000 ha consisting of 7,121.344,00 ha of forest and 3,867,000 ha in the form of peatlands. Peat structures are soft and have pores that make it easy to hold water. Peat water has a high color intensity, low pH, high organic content and has an acidic properties So it does not qualify as a mixture of concrete. To meet the needs of water in the concrete mix then water should be obtained from another place but it will require a greater cost and time. To resolve the issue, the advancement of concrete technology has resulted in admixture that can help in maintaining the quality of concrete. Plastiment-VZ is a plasticizer material that can increase workability of concrete without adding water. However, for the use in the field, the selection of admixture must be adjusted to the planned concrete situation and condition. Excessive use of admixture will also result in uneconomical concrete. The design of the job mix using the Department of Environment (DOE) method with compressive strength concrete plan fc '= 25 MPa. The percentage of Plastiment-VZ addition is 0%, 0,05%; 0,10%; 0,15% and 0,20% to the weight of cement. The reduction of the amount of water in this study is 10% of the total amount of water. Specimens in each variation were made using cylinder mold with 15 cm in diameter and 30 cm high. After specimens are created and maintained, testing of compressive strength concrete held in 28 days. The test results show that the trend of average compressive strength has increased along with the addition of Plastiment-VZ percentage. The equation resulting from the average compressive strength is $y = -362,7x^2 + 133,3x + 28,10$ with value $R^2 = 0,969$. The highest average compressive strength value was obtained in the addition of 0,20% Plastiment-VZ at 40,76 MPa. Statistical testing with Analysis of Variance - ANOVA states that there is a very real interaction or treatment between the compressive strength of the concrete with the addition of Plastiment-VZ. So it can be concluded that the reduction of the amount of water with the addition of Plastiment-VZ has an effect on the increasing of concrete compressive strength characteristics.
Keywords: Anova, Concrete, Compressive Strength, Influence, Plastiment-VZ.

1.Introduction

Peatland is a wetland ecosystem formed from the accumulation of organic matter in forest land from vegetation debris over a long period of time. Peat structures are soft and have pores causing it easy to hold water. Peat water has a high color intensity, low pH, high organic content and acid properties.

T. Mulyono (2004), the water required in the manufacture of concrete to trigger a chemical process cement, aggregate wetting and provide convenience in concrete work. Water containing harmful compounds, contaminated with salt, oil, sugar or other chemicals will degrade the quality of concrete and even change the properties of the concrete produced.

Based on these conditions, the condition of peat water does not meet the requirements for use as a mixture of concrete. To meet the water requirement in concrete mix then water must be obtained from other place but it will require cost and bigger time.

To overcome these problems, the advancement of concrete technology has resulted additives which can help in maintaining the quality of concrete. One product output from PT. Sika Indonesia is Plastiment-VZ. Plastiment-VZ is a plasticizer material that can improve the workability of concrete without adding water.

However, for use in the field, the selection of admixture or must be adjusted to the planned concrete situation and condition. Excessive use of additives will also result in uneconomical concrete, given the expensive price of additives.

These conditions, complicate the community both service providers and users of construction services, especially in Riau Province, because it is not yet known clearly the dosage that is suitable with the condition of both cement and aggregate materials available in Riau Province. So to get maximum results that can be applied widely by the community both service providers and users of construction services in Riau Province, research will be needed to analyze the effect of the addition of Plastiment-VZ on the characteristics of concrete in Riau Province.

2. Literary Review

2.1 Water Peat

Peat is an accumulation of organic materials composed of the remains of plants and decaying plant tissue. The peat deposition process generally occurs in the basin area and ultimately forms the deposition of peat water.

Kusnaedi, (2006), Peat water is a surface water that is commonly found in tidal, swampy and lowland areas, with the following characteristics :

a. High color intensity with brownish red
b. Low pH
c. The organic content is very high
d. Turbidity and low suspended particle content
e. Low cation content

2.2 Additive

P. Nugraha dan Antoni, (2007), additive is a material other than water, aggregate and hydraulic cement mixed in concrete or mortar added before or during stirring.

The addition of added ingredients in a concrete or mortar mix does not change the large composition of the other ingredients, Because the use of this added material tends to be a substitute or substitution of the concrete mixture. Because the goal is to improve or change certain properties and characteristics of the resulting concrete or mortar, then the tendency to change the composition in volume weight is not felt directly compared with the initial composition of concrete without added ingredients.

In general, the added materials used in the concrete can be divided into two, namely chemical admixture and mineral additive.
S. Aprilianti and Nadia (2012), o produce concrete with low fas but still easy to work then needed added materials. One of them is using the superlasticizer admixture. With the ability to reduce high water use, concrete mixes with added materials are capable of producing higher compressive strength of concrete.

2.3 Plastiment-VZ

According to Technical Data PT. Sika Indonesia (2011), Plastiment-VZ is a concrete plasticizer material and water reducer in large quantities, in accordance with ASTM with accelerated hardening effect, in accordance with ASTM C 494-92 type D.

Plastiment-VZ is used as an additive for various concrete uses where the condition of casting requires high quality concrete. The advantages of Plastiment-VZ will produce the following properties:

a. Accelerate hardening time in hot weather
b. Accelerate hardening after setting
c. Improve workability without adding water
d. Reduces water without loss of workability
e. Improve the quality of concrete
f. Reduce shrinkage
g. Slump loss control lasts longer
h. Better surface finish
i. Chloride free, so as not to damage reinforcement
j.

The dosage of Plastiment-VZ is 0,15 – 0,40 to the weight of the cement. The effective dose of Plastiment-VZ depends on cement type and aggregate quality with a general dose of 0,15% - 0,25% for low absorbent concrete to fine aggregate. This dose can be increased up to 0,60% to overcome difficulties in terms of cement or aggregate quality, high temperature and difficult casting conditions. Plastiment-VZ must be mixed first directly into the water mixture after it is aggregated. Characteristics of Plastiment-VZ can be seen in table 1.

Table 1. Plastiment-VZ characteristics

Technical Data	
Form	Polyhydroxy Carbon Salts
Colour	yellow
Specific gravity	1,17 – 1,19 kg/ltr
Age and Storage	Minimum of 1 year if kept in original unopened packaging in a

	dry, cool and shady place
Packaging	Drum 240 kg, Bulk 1000 kg

(Source : PT. Sika Indonesia, 2011)

2.4 Analysis of Variance

S. Yitnosumarto, (1991), Analysis of Variance (ANOVA) is one hypothesis test on parametric statistics, to test the interaction between two factors in an experiment by comparing the mean of more than two samples. ANOVA was developed by Ronald Fisher (1918) to analyze the diversity of a response and divide it into sections dealing with a known source of diversity and is associated with random error. The known sources of diversity will be associated with the independent variables of the tried (treatment) factors.

Procedure of variance analysis Using single numerical variables measured from a number of samples to test the null hypothesis of a population that (estimated) has the same mean. The variable should be a quantitative variable. This variable is sometimes referred to as a dependent variable.

In the ANOVA test, sample evidence is taken from each population under study. The data obtained from the sample is used to calculate sample statistics. The sampling distribution used to make statistical decisions, ie rejecting or accepting the null hypothesis (H0), is the F Distribution.

In this test it is assumed that all populations under study have the same variant, regardless of whether the populations have the same or different mean count.

3. Data Analysis

3.1 Materials Research

The material used is :

a. Type I Cement (Ordinary Portland Cement - OPC), production of PT. Semen Padang
b. Coarse aggregate or crushed stone from Pangkalan area
c. Fine aggregates from the Teratak Buluh area
d. Water at Batching Plant Laboratory PT. Mekar Abadi Mandiri Jalan Lintas Pekanbaru - Duri Km.62
e. The additives Plastiment-VZ production PT. Sika Indonesia

3.2 The Design of The Test Object

Concrete design using Department of Environment (DOE), with cylinder molds of 150 mm x 300 mm. The compressive strength of the plan is fc '= 25 MPa.

The admixture used are Plastiment-VZ. Percentage of Plastiment-VZ addition that is 0%, 0,05%, 0,10%, 0,15% and 0,20% to the weight of cement. The reduction of the amount of water in this study as much as 10% of the composition of the mix design.

Initial research begins by conducting preliminary testing on the material, manufacture of specimens, slump test, stages of concrete treatment for 28 days and compression concrete testing.

3.3 Analysis of Concrete Compressive Strength

Calculation of compressive strength value can be used formula :

$$f'_c = \frac{P_{max}}{A_C} \quad (1)$$

where :

f'_c = Concrete Compressive Strength (MPa)
P_{max} = Maximum load (N)
A_c = Surface area (mm^2)

3.4 Data Analysis Method

Analysis of variance would be a valid statistical technique to be applied using the following assumptions:

a. The studied population has a normal distribution.
b. Sampling is done randomly and each sample is independent or not bound by another sample.
c. The populations in which the sample values are obtained have the same population variance values.

Table 2. Distribution treatment

No.	% Plastiment-VZ	Speciment			
		1	2	3	Total
1.	0	$Y1_1$	$Y1_2$	$Y1_3$	$\sum_{j-1}^{p} Y_{ij}$
2.	0,05	$Y2_1$	$Y2_2$	$Y2_3$	$\sum_{j-1}^{p} Y_{ij}$

3.	0,10	$Y3_1$	$Y3_2$	$Y3_3$	$\sum_{j-1}^{p} Y_{ij}$
4.	0,15	$Y4_1$	$Y4_2$	$Y4_3$	$\sum_{j-1}^{p} Y_{ij}$
5.	0,20	$Y5_1$	$Y5_2$	$Y5_3$	$\sum_{j-1}^{p} Y_{ij}$

$$F_{count} = \frac{KT_{treatment}}{KT_{experiment}} \quad (2)$$

Where :

$KT_{treatment}$ = Middle square $_{treatment}$

$KT_{experiment}$ = Middle square $_{experiment}$

Distribute F at a significant level, $F_{critical}$ $\alpha = 0{,}05$ or 0,01. Jika $F_{count} > F_{table}$, then there is a very real treatment. With note if $\alpha = 0{,}05$, called different or significant effect, and if $\alpha = 0{,}01$ called different or very significant effect. Data analysis of F test result can be seen in table 3.

Table 3. Data analysis of F test result

SK	Db	JK	KT	F_{hitung}
Treatment	$(p\text{-}1)$	$n\sum_{i}^{P}(Y\text{-}Y..)^2 = JK_P$	KT_P	KT_P/KT_G
Experiment error	$(pn\text{-}1)$	$\sum_{i}^{P}\sum_{i}^{n}(Y_{il}\text{-}Y_{i}..)^2 = JK_G$	KT_G	
Total	$(pn\text{-}1)$	$\sum_{i}^{P}\sum_{i}^{n}(Y_{il}\text{-}Y_{i}..)^2 = JK_T$	-	-

(Source : S. Yitnosumarto, 1991)

4. Results and Discussion

4.1 The compressive strength of the concrete

After testing the specimens with the addition of Plastiment-VZ, then the test result of the compressive strength can be seen in table 4. In Figure 1 shows the value of compressive strength of concrete on each specimen with variation of Plastiment-VZ, while In Figure 2 pada gambar 2 menunjukkan nilai kuat shows the average compressive strength value with the addition variation of Plastiment-VZ.

Table 4. The compressive strength of the concrete

% Plastiment-VZ	Speciment	Weight (g)	Area (cm^2)	Load (kN)	Compressive Strength (MPa)	Average Compressive Strength (MPa)
0	I	12680	176,79	520	29,99	27,49
	II	12810	176,79	460	26,53	
	III	12860	176,79	450	25,96	
0,05	I	12910	176,79	590	34,03	35,18
	II	12980	176,79	620	35,76	
	III	12860	176,79	620	35,76	
0,1	I	12840	176,79	660	38,07	37,49
	II	12890	176,79	670	38,65	
	III	12290	176,79	620	35,76	
0,15	I	12950	176,79	700	40,38	39,03
	II	12890	176,79	660	38,07	
	III	12980	176,79	670	38,65	
0,2	I	12850	176,79	720	41,53	40,76
	II	12665	176,79	680	39,22	
	III	12750	176,79	720	41,53	

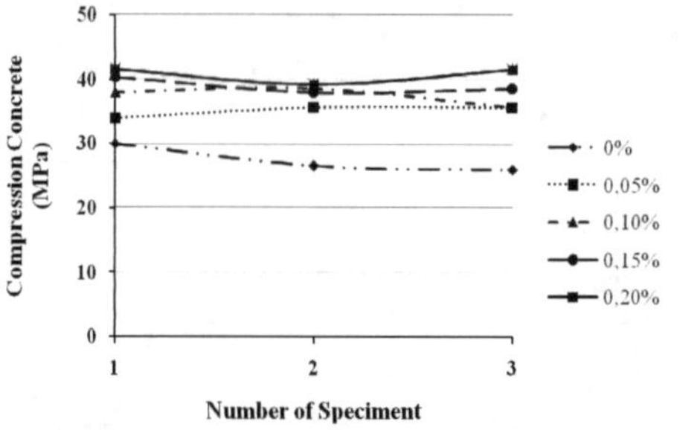

Figure 1. Compressive strength of concrete on each specimen with the addition of Plastiment-VZ

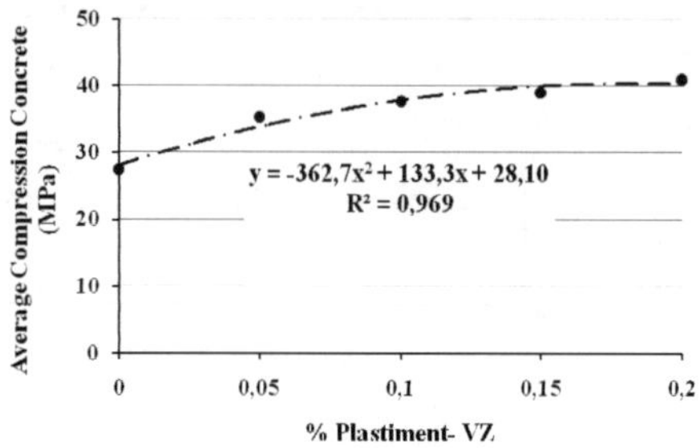

Figure 2. The relation between average compressive strength of concrete with the addition of Plastiment-VZ

4.2 Analysis of Variance – ANOVA
Treatment design with mathematical notation can be seen in table 5, while the analysis of variance can be seen in table 6.

From Table F it can be seen that $F^{0.05}_{Table}$ = 3,48 and $F^{0.01}_{Table}$ = 5,99, while F_{count} = 35,55. Because $F_{count} > F_{Table}$, there is a very real interaction or treatment on compressive strength of concrete with the addition of Plastiment-VZ.

4.3 Discussion
From the test result on all specimens, the compressive strength of the concrete is higher than the compressive strength of the plan is 25 MPa. The test results show that the trend of average compressive strength of concrete has increased along with the addition of Plastiment-VZ. The equation resulting from the average compressive strength of the concrete is $y = -362{,}7x^2 + 133{,}3x + 28{,}10$ with $R^2 = 0{,}969$. The highest average compressive strength of the concrete was obtained in the addition of 0,20% Plastiment-VZ at 40,76 MPa.
Statistical testing with Analysis of Var

iance - ANOVA states that there is a very real interaction or treatment between the compressive strength of the concrete with the addition of Plastiment-VZ. So it can be concluded that the addition of Plastiment-VZ influences the increase of concrete compressive strength characteristics.

Table 5. Mathematical Notation

No	% Plastiment-VZ	Sampel 1	Sampel 2	Sampel 3	Sum
1.	0	29,99	26,53	25,96	$\sum_{j-1}^{3} Y1_j = 82{,}48$
2.	0,05	34,03	35,76	35,76	$\sum_{j-1}^{3} Y2_j = 105{,}55$

3.	0,10	38,07	38,65	35,76	$\sum_{j-1}^{3} Y3_j = 112{,}48$
4.	0,15	40,38	38,07	38,65	$\sum_{j-1}^{3} Y4_j = 117{,}10$
5.	0,20	41,53	39,22	41,53`	$\sum_{j-1}^{3} Y5_j = 122{,}28$
				Total	**539,89**

Table 6. F Test Results Using Compressive Strength of Concrete

SK	DB	JK	KT	F_{count}
Treatment	4	321,36	80,34	35,50
Experiment error	10	22,63	2,26	
Total	14	343,98	82,60	

5. Conclusion

The conclusions are :

a. Statistical testing with Analysis of Variance - ANOVA states that there is a very real interaction or treatment between the compressive strength of the concrete with the addition of Plastiment-VZ.
b. The addition of Plastiment-VZ influences the increase of concrete compressive strength characteristics.
c. The test results show that the trend of average compressive strength of concrete has increased along with the addition of Plastiment-VZ.
d. The equation resulting from the average compressive strength of the concrete is $y = -362{,}7x^2 + 133{,}3x + 28{,}10$ with $R^2 = 0{,}969$.
e. The highest average compressive strength of the concrete was obtained in the addition of 0,20% Plastiment-VZ at 40,76 MPa.

6.References

[1] T. Mulyono. 2004. *Teknologi Beton, Edisi II.* Yogyakarta : Andi Offset.

[2] Kusnaedi. 2006. *Mengolah Air Gambut dan Kotor Untuk Air Minum.* Jakarta : Penebar Swadaya.

[3] P. Nugraha and Antoni. 2007. *Teknologi Beton.* Yogyakarta : Andi Offset.

[4] S. Aprilianti and Nadia. 2012. "*Analisis Pengaruh Beton dengan Bahan Admixture Naphtalene dan Polycarboxilate Terhadap Kuat Tekan Beton Normal*", *Jurnal Konstruksia.* Vol. 3, No. 2, April 2012. pp. 33 – 40.

[5] PT. Sika Indonesia. 2011. *Data Teknis Know-How From Site To Shelf.*

[6] S. Yitnosumarto. 1991. *Percobaan Perancangan, Analisis Dan Interprestasinya.* Jakarta: PT. Gramedia Pustaka Utama.

IOP Conf. Series: Earth and Environmental Science **97** (2017) 012024 doi:10.1088/1755-1315/97/1/012024

Blended Design Approach of Long Span Structure and Malay Traditional Architecture

Titin Sundari
Universitas Lancang Kuning, Pekanbaru, 28265, Indonesia
Tel/Fax: +62761 53108
E-mail:titin@unilak.ac.id

Abstract: The growing population in the world is so fast, which is followed by the increasing need of some new and large activities. Architects face the problem on how to facilitate buildings with various activities such as for large meeting, conference, indoors gymnasium and sports, and many others. The long span structure of building is one of the solutions to solve that problem. Generally, large buildings which implemented this structure will look as a technological, modern and futuristic ones or even neo futuristic performance. But on the other hand, many people still want to enjoy the specific and unique senses of local traditional architecture. So is the Malay people who want an easy pleasant large facilities which can be fulfilled by implementing modern long span building structure technology. In the same time, their unique sense of Malay traditional architecture can still be maintained. To overcome this double problems of design, it needs a blended design approach of long span structure and Malay Traditional Architecture.
Keywords: blended design approach, long span structure, Malay traditional architecture.

1.Introduction

The growing population of Riau Province during recent decade is very fast. Based on the Statistics of Riau Province in 2016, Riau has 2.56% growth rate in 2014 - 2015. Besides this growth of people, Riau also becomes one of tourist destinations. Many seminars and conferences were held as part of MICE (Meeting, Incentive, Convention and Exhibition) activities in Riau. For example in 2012 Riau became the host of a national sport event (Pekan Olah Raga Nasional XIII) and National Sport for Difable Person XIV. With the high rate of population growth Riau tend to be a high density region . One of the consequency of that growth is the need of many new larger public facilities, such as terminal buildings, harbors, indoors sports and gymnasium buildings, stadions, tourism facilities and others. Such buildings refer to the structures of large scale building to cover an open clear space without any obstractions, which are called long span structures.
Malay people in Riau Province have a great pride of their traditional architecture as they often perform it in their buildings, either private or public. The regional building regulation also ask the builder to implement traditional architecture as the component of the special building.

\Architects have to do their design with an approach to both the long span structure and the Malay traditional architecture. This article will explore a four step design approaches. Firstly there will be a glance review to know the variations of structure of the large buildings in Riau Province which already exist. Secondly, a brief review of long span structure, and especially on shell structure and flat truss structure based on the reference from Heinrich Engel. Thirdly, a general review to the Riau Malay traditional architecture will be written based on the reference of Tenas Effendy. Fourthly, a brief concept of a blended design approach by choosing the structure system that is suitable with the need of performing Riau Malay traditional architecture. At the end of the article, it will show a 3D picture of an alternative pre-designs.

2. Blended Design Approach

Blended design approach is a process of designing by synthesizing two or more design aspects simultaneously to achieve the design goal. The building structure design approach done by architect is not exactly the same with what a civil engineer does. They do the same object, but with different goal and method. A civil engineer's goal is to produce the whole system as well as the detail of building structure designed by calculating , drawing and even doing the tests. But the building structure they calculate was early designed by architects who estimated the structure system mostly by rule of thumb not by calculating it. Building structure design is always merely approached on the structural aspects, such as loads and moments mechanism, deflections, strength, durability etc. Whilst in this blended approach, the building structure will be analyzed by two aspects simultaneously, i.e. structural aspect and traditional architecture aspect. In a building with conventional structural approach, the traditional architecture is generally built as accessories, whilst in a blended approach the traditional architecture will be the structural element itself, so that it will occur as long as the building exist.

3. Large Buildings in Riau Province

As mentioned earlier, Riau Province has already had several large buildings as venues, which especially were constructed for facilitating the event of PON XIII in 2012. These buildings are mostly placed in Pekanbaru, but some of them are outside. This review will look at some of them which to be considered as representatives, they are main stadium, basketball hall, gymnasium hall and Youth Center (figure 1). In general all of these buildings are long span structure buildings. They are modern buildings which have interesting form and look as sporty huge performance. But none of them show a strong sense of Riau Malay traditional architecture, except a very small touch of accessory like a *lipat kajang*. When this accessory is uninstalled, the building will lose its sense of Riau Malay traditional architecture and looks like other buildings outside Riau Province. In order to become permanent identity, a traditional architecture should be a part of the building structure itself.

Figure 1. Main Stadium, Gymnasium Hall, Youth Center and Basket Ball Hall.

The main stadium reinforces concrete in below structure and the roof structures are steel trusses which are placed in two directions so that it gives a vector active structure mechanism. While the other three buildings applied reinforce concrete as middle structures and space frame as roof structures. Gymnasium and basketball halls reinforce concrete structure as middle structure and space frame as roof structure.Youth Center reinforces concrete structure as middle structure and space frame as roof structure. The building has two masts tapered up to joint in its peak. The masts are accessories, not a structural mast. Another venue is the Unilak Student Center (*Pusat Kegiatan Mahasiswa Unilak*) in Rumbai. This building has a shell-like structure. Its roof is a spherical form which is very impressing combined with the line of steel main structure. In fact it is not a really shell structure, but space frame back bones as main structure combine with planar trusses to support the roof. Steel pipe columns stand along the periphery of the roof. But there is no identity of Malay traditional architecture at all, something that often be forgotten to place.

Figure 2. Unilak Student Center in Rumbai Pekanbaru

The last monumental building in Pekanbaru is Regional Library of Riau Province (fig.2) this is not a building with specific traditional architecture of Malay Riau. No element which has a sense of Malay traditional architecture is found there. It is supported by reinforced concrete combine with steel as the middle structure and steel frame with metal sheet cladding as roof structure.

Figure 3. Regional Library of Riau Province I Pekanbaru.

From the six buildings above, it can be concluded that it is rare to find of large buildings in Pekanbaru, Riau, which applies Riau Malay traditional architecture. When the traditional element is used such as *lipat kajang*, it is usually placed as an accessory in front of the main entrance of the building. But it only can be seen by people who come into the gate. It is not useful to give an identity of the whole building.

4. Long Span Structures

There are several terms of long span building which are made by individuals and institutions. In this article the term will be understood from Encyclopedia Britannica and American Institute of Architect (AIA). According to the Encyclopedia Britannica a long span building create an unobstructed, column-free spaces greater than 30M (100feet) for a variety of functions. These include activities where visibility is impotant for large audiences (auditorium and covered stadium), where flexibility is important (exhibition halls and certain types of manufacturing facility), and where large movable objects are housed (aircraft hangars). AIA didn't define any specific definition of long span building. A 60 feet office building is considered as a long span, but not to a gymnasium. Whereas a 60 feet is a long span to timber structure, but not to steel structure.

To create a long span building, it can apply many types of structure. As the material of structure developed fast, the long span building is much easier to build. In this article, the long span structure pre-design will apply shell structure as main structure combined with flat truss structure as secondary structure. Sometimes a building applies a shell-like structure but actually it is not a real shell structure, which means that the building has a curve shape like a shell but the structure is not surface active structure system. The design approach of real shell structure in this case will be based on structure knowledge in Heinrich Engel's book, *Structure System.* The shell structure concept here will cover 3 aspects, they are understanding shell structure as a surface active structure system, generating a geometrically defined basic shell shape of shell structure and developing the basic shell shape.

4.1 Shell Structure as Surface Active Structure System

Before talking about the concept of shell structure, it should review the relation of shell structure and surface active structure system. Surface active systems can be understood as structural surfaces which can be composed to form mechanisms that redirect forces. Structural continuity of the elements in two axes, i.e. surface resistance against compressive, tensile, and shear stresses are the first prerequisite and first distinction of surface active structure

The shell structure is one of the types of surface active systems, which will be developed in this article as a pre-design long span building, i.e. a harbor. The concept of this structure is to unite the logically and geometrically defined of long span structure with the artistic sense of Malay traditional architecture. It verily needs good consideration on choosing the type of structure. In designing a long span habour, its important to create an unobstructed space to facilitate many people and vehicles with goods to be transported therein. The space also needs large openings, ventilation and natural lighting as well. Shell structure offers all of that is needed. This structure also has a clear interior performance and can be adjusted artistically. Later the concept of logically and geometrically defined will perform in pint 4.2 below.

4.2 Generating a Geometrically Defined Shape of Shell Structure

It is very important to generate a shell shape that geometrically defined. This form is to ensure that the builder can do the construction easier and without any difficulties that sometimes occur on a geometrically undefined shape. A geometrically defined shape is very important in a shell structure. A shape that is geometrically undefined in a shell structure will come to difficulties on its construction. Based on the book of Heinrich Engel, here will develope the generation of the basic shape. The generation is based on a rotate method with two main elements, i.e. a directrix and a generatrix (see fig. 4).

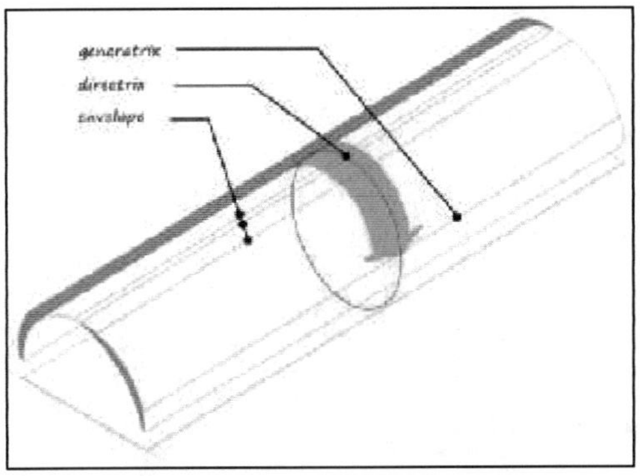

Figure 4. Generate the basic shell shape

A straight line here as a generatrix will form the shape by rotating it along a directrix that is a half circle in a plane that lies at right angles to the generatrix. The action will form an envelope, half cylindrical shape, that's a basic unit of shall shape. The shape is geometrically defined, because it has a defined radius of directrix, i.e. 4.0 meters. That will make the contractor's job easier.

4.3 Developing Basic Shell to a Larger Shape

The shell structure in this paper is designed to cover a large space, which needs several units of basic shell shape to be connected. As mention in Heinrich Engel's book, there are several methods of juxtaposition for cylindrical surface connection. One of the methods which is appied here is a continous connection (see fig 5). Another connection type that will be applied is transverse folding.

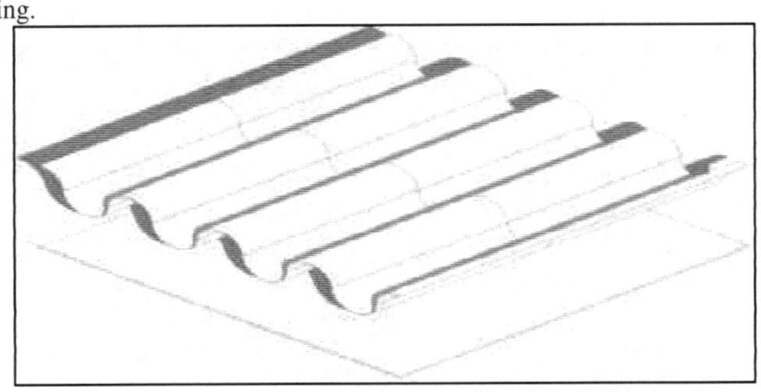

Figure 5. Continous connection of juxtaposition for cylindrical surface

5. Riau Malay Tradisional Architecture

Traditional architecture is always an interesting object to be related with MICE and tourism. A harbor as a tourist gate of the region, must be designed with a special local tradition sense attract the guests' attention who come through that gate. In this paper the harbor is designed with a Riau Malay tradional architecture component, that is *singap* or bidai layer (see figure 6). *Singap* is a layer to cover a front end of the roof, it is usually ornamented with carving timber. This ornament originally has a function as ventilation.

Figure 6. *Singap* or bidai layer

(source: Mayuddin, Yogyakanta, in Tenas Effendy)

In the blended design here, bidai layer will be transformed to be a structural element with steel pipe material. It is impossible to use timber as structural material that will combine with fiber concrete shell material. The transformation form will change the ornament to a structural form of a vector active structure system. This system applies small trusses to be arranged as both structural and aesthetic. It is structurally elements which are dominated by triangles as stable form. It also still performs an aesthetic component of Riau Malay traditional architecture, even though it has been transformed (see figure 7).

Figure 7. The transformed bidai layer

6. The Design

The pre-design of harbor building structure will be based on the concept of blended long span structure and Riau Malay traditional architecture. The structural concept can be explained as follow.

a) The whole structure system is the combination of surface active structure system, i.e. shell structure and vector active structure system, i.e. planar trusses structure.

b) Shell structure as the first structural component consists of several basic units of cylindrical shape which are connected in juxtapose connection and folded connection methods.

c) Bidai layer is transformed to be the second structural element of triangle trusses that act as a vector active structure system.

The pre-design consists of plan, elevation and isometric (see figure 8, 9, 10 and 11). Because it is structural pre-design, so all of the figures are merely to show the building structure. There is no architectural space analysis and design.

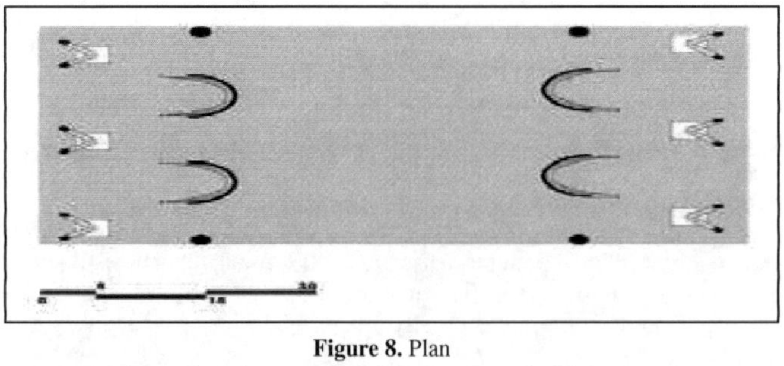

Figure 8. Plan

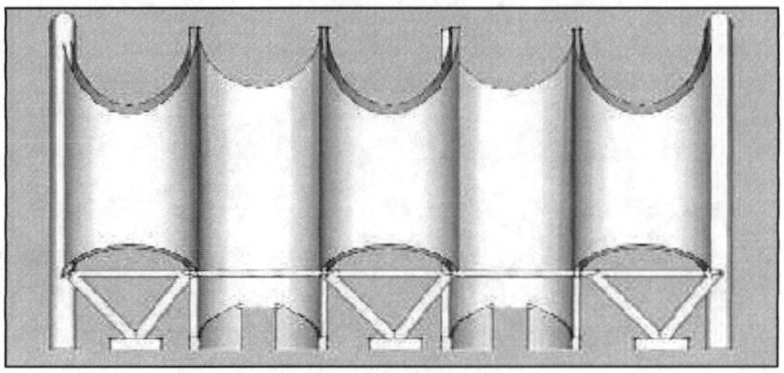

Figure 9. Side Elevation

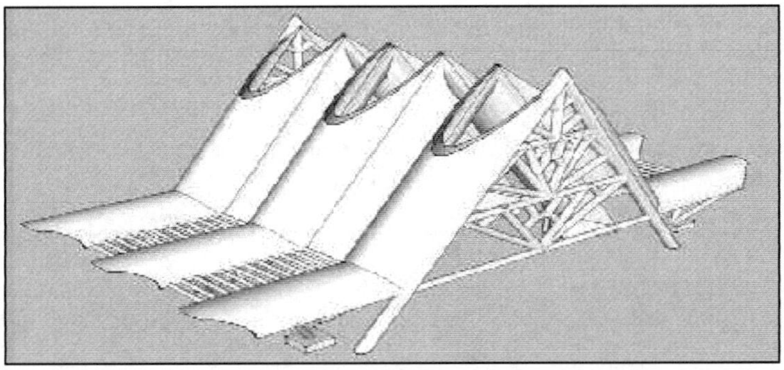

Figure 10. Isometric

The building is a symmetrical form with structure that consists of shell structure and flat truss, it is stable to support vertical, horizontal and vibration loads. In the part of shell structure, there are five basic units that connected in juxtapose method at each side, let say right and left side. The two sides connected in folded method at the peak, while the shells are cut off in interval manner, as sky lights to allow a free orientation view to the sky. The shells are also folded at each side in interval manner, in order to form a Malay roof shape. Then in the part of flat truss, that consists of two huge flat trusses located at each end of the shell assemblage. The flat truss functions to stabilize the whole shell structure.

The building is a symmetrical form with structure that consists of shell structure and flat truss, it is stable to support vertical, horizontal and vibration loads. In the part of shell structure, there are five basic units that connected in juxtapose method at each side, let say right and left side. The two sides connected in folded method at the peak, while the shells are cut off in interval manner, as sky lights to allow a free orientation view to the sky. The shells are also folded at each side in interval manner, in order to form a Malay roof shape. Then in the part of flat truss, that consists of two huge flat trusses located at each end of the shell assemblage. The flat truss functions to stabilize the whole shell structure.

Figure 11. The Harbor

Finally, figure 11 shows the harbor building as a large building with a modern, futuristic and technological performance but still has the aesthetic of local traditional architecture of Malay Riau. The structure itself is simple with every part of the structureis honest and easy to understand.

7. References

[1] Effendy, Tenas.
[2] Engel, Heinrich.1981. *Structure System.* Van Nostrand Reinhold Company
[3] Siegel, Curt. 1962. *Structure and Form in Modern Architecture.* USA: Van Nostrand Reinhold Company.
[4] Statistics of Riau Province, https://riau.bps.go.id
[5] Encyclopedia Britannica https://www.britannica.com/technology/building-construction/Long-span-buildings

Experimental Study of Axially Tension Cold Formed Steel Channel Members

Widya Apriani[1], Fadrizal Lubis[2], and Muthia Angraini[3]
[1,2,3]Universitas Lancang Kuning Pekanbaru, 28265, Indonesia
E-mail:widyaapriani@unilak.ac.id

Abstract: Experimental testing is commonly used as one of the steps to determine the cause of the collapse of a building structure. The collapse of structures can be due to low quality materials. Although material samples have passed laboratory tests and the existing technical specifications have been met but there may be undetected defects and known material after failure. In this paper will be presented Experimental Testing of Axially Tension Cold Formed Steel Channel Members to determine the cause of the collapse of a building roof truss x in Pekanbaru. Test of tensile strength material cold formed channel sections was performed to obtain the main characteristics of Cold Formed steel material, namely ultimate tensile strength loads that can be held by members and the yield stress possessed by channel sections used in construction. Analysis of axially tension cold formed steel channel section presents in this paper was conducted through experimental study based on specificationsAnnualBook of ASTM Standards: Metal Test methods and Analitical Procedures, Section 3 (1991). The result of capacity loads experimental test was compared with design based on SNI 03-7971-2013standard of Indonesia for the design of cold formed steel structural members. The results of the yield stress of the material will be seen against the minimum allowable allowable stress range. After the test, the percentace of ultimate axial tension capacity theory has a result that is 16.46% larger than the ultimate axial tension capacity experimental. When compared with the load that must be borne 5.673 kN/m it can be concluded that 2 specimens do not meet. Yield stress of member has fulfilled requirement that wass bigger than 550 MPa. Based on the curve obtained ultimate axial tension capacity theory, results greater than experimental. The greatest voltage value (fu) is achieved under the same conditions as its yield stress. For this specimen with a melting voltage value fy = 571.5068 MPa has fulfilled the minimum melting point value of 550 MPa required for standard mild steel materials in accordance with the code SNI 03-7971-2013 about Cold formed steel.

Keywords:cold formed channel sections, axially tension members, SNI 03-7971-2013.

1. Introduction

Failure of roof truss structure of the building can lead to large losses for building owners as well as threaten the security and safety of building users. Based on surveyed**,** In 2016 there is a failure in the roof truss structure X building in Pekanbaru which resulted in nearly the entire framework of the roof of the building collapsed. Based on visual observation the failured at the connection and bend at a couple of bars.The collapse of structures can caused by low quality materials. Although the material samples have passed laboratory tests and the existing technical specifications have been met but there may be undetected defects and known material after failure. Changes in cross-sectional shape will affect the strength of the structural elements and their bending behavior. In addition, in the field found the case of the roof structure more quickly collapsed than the results of the analysis performed, because the calculation analysis is done using the specifications of the actual lightweight steel while in practice used zincalume which has much lower strength than mild steel (Amalia, 2012) .

From the above description it was necessary to analyze the factors causing the failure of the structure include how the strength of lightweight steel plate material used in the roof framework of Building x Pekanbaru?

IOP Conf. Series: Earth and Environmental Science **97** (2017) 012025 doi:10.1088/1755-1315/97/1/012025

2. Cold Formed Steel Design Method

The Cold formed steel design method used is SNI 03-7971-2013 which is a modified adoption of AS / NZS 4600: 2005 because the majority of cold rolled steel industry uses this standard. Several modifications are made to this SNI to suit the conditions in Indonesia.design for axial tensionto A member subject t a design axial tensle force (N)according to SNI 03-7971-2013shall satisfy based on eq 1:

$N \leq N_t$ (1)

where

= capacity reduction factor for members in tension (0,9)

N_t = nominal section capasity of the member in tension.

based on SNI 7971 2013, the nominal section capaity of a member in tension shall be taken as the lesser of :

$N_t = A_g.f_y$ (2)

$N_t = 0{,}85\ k_t.A_u.f_u$

where:

A_g = gross area of the corss section

f_y = yield strength used in design

k_t = correction factor for distribution of forces .

A_n = net area of the cross-section, obtainer by deducting from the gross area of the cross section, the sectional area of all penetrations and holes, including fasteners holes

f_u = tensile strength used in design

3. Material Properties and Test Specimens

The material tensile test is performed to obtain Ultimate Axial Tension Capacity Experimental-kN as well as the main characteristics of lightweight steel material. Specimens and tests refer to the Annual Book of ASTM Standards: Metal Test Methods and Analytical Procedures, Section 3 (1991). Tensile test specimens were taken from the Channel profile portions of C 150.75.80 and 150.75.75. Which has a nominal thickness of 0.75 mm. The shape and dimensions of the material tensile test specimens are given in Figure 1.

Tensile testing is performed by providing static loads that gradually increase at a rate of 1 mm per minute until a breaking condition occurs in the specimen. The number of specimens tested were 4 pieces from 2 different light steel rods: Stem-1, Stem-2 is light steel 150.75.80 and 3 and 4 test pieces are mild steel 150.75.75.

Experimental procedure to carry out Tension test and determine modulus of elasticity is as given below: a) A piece of standard size was cut from the C section using Shear off machine b) Gauge length 50.2 mm was marked on the specimen and is fixed on the testing machine. c) Initially the thickness of member is 0.75 and 0.8 mm. d) Tensile test is carried out by using loads e) Tensile load was applied gradually on the specimen and observations were recorded f) Photograph shows the specimen (before and after test). g) A graph is plotted on the basis of recorded value.

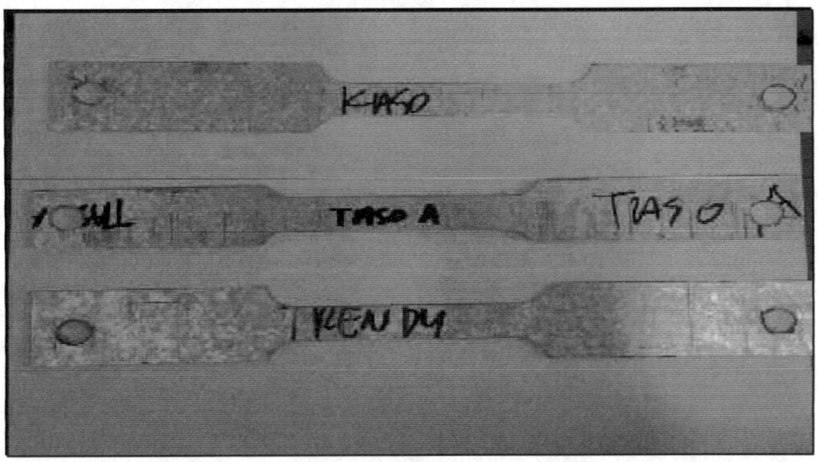

Figure 1: Test Specimens before Test

Figure 2: Test Specimens After Test

Figure 3: Test Specimens before Test

4. Analitycal Investigation

After the analysis was done, the inner force of the axial tensile force is obtained in the maximum structure of the lightweight steel frame structure in the form of a tensile force. From the above roof frame load analysis, the maximum tensile load to be retained is 5,673 kN/m. The experimental tension capacity loads (figure 3) was obtained based on specificationsAnnualBook of ASTM Standards: Metal Test methods and Analitical Procedures, Section 3 (1991).

Tabel 1.
Ultimate Axial Tension Capacity Theoritical and Experimental

Specimens	Fy (Mpa)	Ultimate Axial Tension Capacity Experimental-kN	Ultimate Axial Tension Capacity -Theoritical -kN)
1	571.51	5,578	6.001
2	559.22	5,458	5.872
3	632.88	4,633	6.645
4	509.62	4,850	2.201

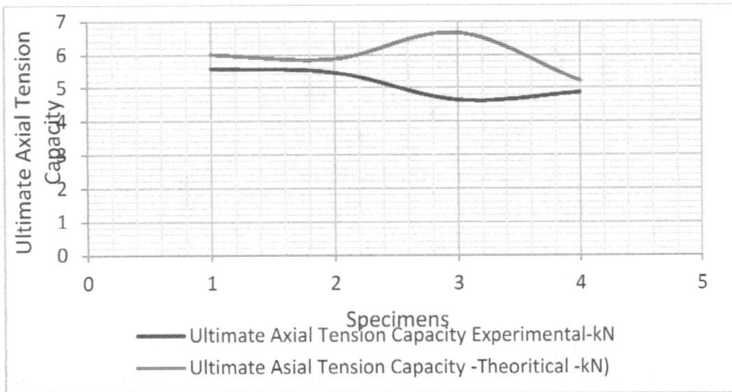

Figure 4. Ultimate Axial Tension Capacity Theoritical and Experimental

After the test, from the table1 we had observed thefollowing result 1. The percentace of ultimate axial tension capacity theory Has a result that is 16.46% larger than the ultimate axial tension capacity experimental. When compared with the load that must be borne 5.673 kN/m it can be concluded that 2 specimens do not meet. Yield stress of member has fulfilled requirement that is bigger than 550 Mpa. Based on the curve obtained ultimate axial tension capacity theory results greater than ultimate axial tension capacity experimental of figure 4.

4.1 Curve Stress and strain cold formed steel

The various material failures that occur in all test specimens show the same result that the breaking condition occurs suddenly without preceded by a significant increase in the length of the specimen visible. The approximate average length increase measured from the post-tensile specimen length is Δ = 12.8 mm. At the disconnected nor a reduction in cross-sectional area (necking) such as those encountered in the tensile test ductile material. The material stress-strain curve obtained from the tensile test of all specimens shows the same curve pattern. As shown in Fig. 5, at the beginning the stress and strain loading increases almost linearly to a value of 568.3 MPa; This stress can be taken as the value of the yield stress of the material (fy). Furthermore, the strain value increases at a relatively equal value of stress until the specimen breaks out. The greatest tension value (fu) is achieved under the same conditions as its yield stress. For this specimen with a yield stress value fy = 571.5068 MPa has fulfilled the minimum yield stress value of 550 MPa required for the standard cold formed steel material in accordance with the SNI 03-7971-2013. In this rule it is also established that the ultimate voltage value (fu) is equal to the voltage value yield stress (fy) if the minimum steel quality of 550 MPa is met. Material elasticity (E); The results obtained close to the standard elasticity modulus value set in the regulation. Recapitulation of tensile test results can be seen in Table 2.

IOP Conf. Series: Earth and Environmental Science **97** (2017) 012025 doi:10.1088/1755-1315/97/1/012025

Table 2.
Characteristic of axially tension test material

specimen	fy(Mpa)	fu (Mpa)	El(%)
1	571,5	571,5	5,6
2	559,2	559,2	8,0
3	474,7	474,7	7,6
4	496,9	496,9	12,8

(source: Testing laboratory of Materials Technology University of Riau)

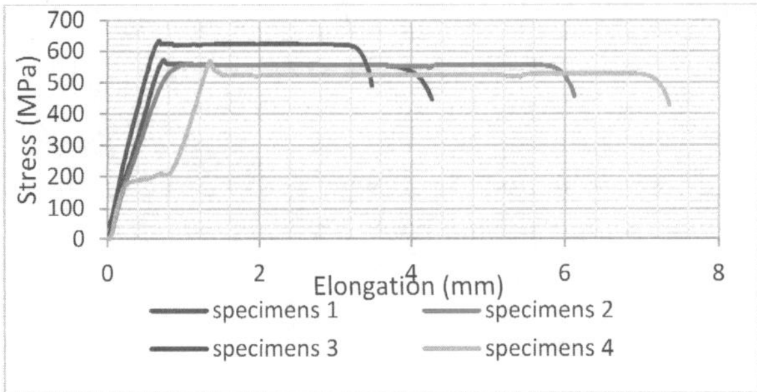

Figure 5. Characteristic of axially tension test material

5. Conclusions

Axiallytension tests on cold-formed channel sections were conducted. A total of 4 lipped channel specimens were tested. This paper has outlined current approaches to the design of lipped channel sections using SNI 03-7971-2013as well as the test results. The experimental test results arethe percentace of ultimate axial tension capacity theory has a result that is 16.46% larger than the ultimate axial tension capacity experimental. Yield stress of member has fulfilled requirement that is bigger than 550 Mpa. Based on the curve obtained Ultimate Axial Tension Capacity Theory results greater than Ultimate Axial Tension Capacity experimental.

6. References

[1] Amalia R. 2012. *Studi Eksperimental dan Analisis Rangka Atap Tipe Howe Menggunakan Profil Baja Ringan*. Tugas Akhir Program S1 Teknik Sipil Universitas Sriwijaya. Sumatera Selatan.

[2] Akmal I. 2009. *Rangka Atap Baja Ringan, Majalah Rumah Ide edisi 10/IV*. Jakarta: Gramedia.

[3] Ariyavinanta, E. 2015. *Pengembangan Spreadsheet untuk Perhitungan Kapasitas Baja Canai Dingin dengan Pengaku Berdasarkan SNI 7971:2013*. Tugas Akhir Program S1 Program Studi Teknik Sipil Universitas Kristen Petra. Surabaya.

[4] ASTM. 1991. *Annual Books of ASTM Standards, Section 3, Metals Test Methods and Analytical Procedures. American Society for Testing and Materials (ASTM)*. Philadelphia, US.

[5] Badan Standarisasi Nasional. 2013. *Tata Cara Perencanaan Struktur Baja Canai Dingin, SNI 03-7971-2013*. Bandung.

[6] Badan Standarisasi Nasional. 2012. *Tata Cara Perencanaan Ketahanan Gempa Untuk Bangunan Gedung, SNI 03-1726-2012*. Bandung.

[7] J. Rhodes. 2013. *Cold Formed Steel Structure*. Cambridge: Cambridge University Press.

[8] Joint Technical Committee BD-082, Cold-formed Steel Structures. 2005 AS/NZS4600-2005. Australian/New Zealand Standard Cold Formed Steel Structures.

[9] Kurniawan. 2008. *Analisis Baja Ringan dengan Baja Konvensional*. Tugas Akhir Program S1 Teknik Sipil Universita Diponegoro. Bandung.

[10] Nur, K. 2012. *Analisis Stabilitas Elemen Baja Ringan Sebagai Bahan Alternatif Pengganti Baja Konvensional Pada Rangka Batang (Studi Kasus Rangka Atap Gedung Fakultas Teknik UNG)*. Laporan Penelitian BNBP Teknik Sipil Universitas Negeri Gorontalo.

[11] Rahmat, S., B. 2010. *Stabilitas Kuda-Kuda Baja Ringan Star Truss Type C (studi kasus: Pengujian kuda-kuda Baja Ringan Bentang 6 m)*. Tugas Akhir Program S1 Teknik Sipil Fakultas Teknik Universitas Gadjah Mada. Yogyakarta.

IOP Conf. Series: Earth and Environmental Science **97** (2017) 012026 doi:10.1088/1755-1315/97/1/012026
***** Formatting Issue - Best Available Paper/Graphic *****

Breakdown Voltage of Carbon Ionised by Impurity of Butane and Air at Slightly Atmospheric Pressure

Ikhsan Rahman Husein[1*] and Saktioto[2]
[1]Department of Physics, Faculty of Mathematics and Sciences University of Riau
Kampus Bina Widya Jl, HR Subrantas KM 12,5 Pekanbaru, 28293, Indonesia
Email: Ikhsanrahman.h@gmail.com

Abstract : This paper investigates the breakdown voltage of Carbon ionised affected by butane and air gas impurities at slightly atmospheric pressure. The ionization process is conducted by electrical discharge using graphite electrode with the gap in millimeter scale to generate plasma by the direct current voltage less than 3.5kV. The experimental results show the breakdown voltage and current of plasma. Carbon plasma density and temperature are calculated from the current and voltage. These results show the ionization of air at slightly low pressure having a maximum voltage of 570V. The breakdown voltage and current are respectively 530-570V and 0.27 to 0.45mA. At high pressure it has a maximum voltage of 1160 V, the voltage and the current breakdown are respectively 900-1100V and 0.46 to 0.6 mA. When gas impurities are injected at slightly low pressure, it has maximum voltage of 772V, the breakdown voltage is 536-775V and the current is 0.03 to 0.45mA, whereas at high pressure it has a maximum voltage of 1044V, the breakdown voltage is 675-105 V and the current is 0.03 to 0.69mA. The results are interesting phenomena and can also be applied to develop a level of purity of Carbon Nano Tube formation at the cathode end of the effects of the density and temperature of the plasma generated.
Keywords: Arc Discharge, DC Voltage Ionization, Plasma Carbon,

1. Introduction

The development of atmospheric plasma and technology research is nowadays increasingly important in industries such as steel cutting, welding, as a source of energy and the development of materials, especially to grow Carbon Nano Tubes (CNTs) in the cathode surface as well [1]. Particularly in the growth of CNTs, the effort to ionize the gas, produce and grow CNT remains a research choice today. Many research results are found, such as a mixture of Helium-Argon gases has been carried out to generate plasma by arc-discharge technique at atmospheric pressure using a mixture of Nickel and Yttrium catalysts. This ionization process produces Single-walled-nanotubes (SWNTs) purified by oxidation and centrifugation techniques [2]. CNTs have attracted great attention due to the demands of nanotechnology, conservation and storage energies, and large nanosensors. CNTs have been produced effectively by arc-discharge methods at atmospheric pressure using a gas mixture and are essential for producing high density Carbon species [3]. The degree of purity of ionization is very important to be observed in forming CNTs, especially ionization products with high purity Carbon species in generating CNT density in their structural and geometric diversity. The gases used are Methane, Argon, Helium and Argon + Helium mixtures [4]. Another example is the method of arc discharge method, laser ablation, HIPCo, Molten Salt Technique and chemical vapor deposition (CVD) [5]. The arc discharge method is

IOP Conf. Series: Earth and Environmental Science **97** (2017) 012026 doi:10.1088/1755-1315/97/1/012026
***** Formatting Issue - Best Available Paper/Graphic *****

the first method used in finding CNTs and this method is the easiest and most efficient method of generating carbon ion ionization [6].
Given the impurity of ionized gases causing source and media factors greatly affect CNT formation. The source of the voltage breakdown on ionization will experience some contribution and resistance to the formation of Carbon ions. At a small electrode distance and diffusion length will also deliver ionizing effects that require varying energy. The DC voltage source is expected to deliver a positive Carbon ion to the negative cathode so that its mobilization can be formed in large quantities and regular structures and geometries. Under the same conditions high voltage power and the impurity of air in the Alkane group gas will increase the higher the breakdown voltage. In the case of a high-density gas, the collisions of gases mostly produce elastic collisions or electron excitations in atoms amongst particles, and polarization of dielectric gases by an electric field. This direct current voltage source (DC) will inhibit ionization while the movement of the gas particles will also be hampered not only because of the gas crossover, but also the atomic and molecular polarization effect reduces the kinetic velocity of the particles so that the kinetic energy and potential energy for ionization will decrease.

Therefore, although experimentally the phenomenon of breakdown voltage has been successfully understood because it works at the state of before and after the ionization occurs, but the understanding of the processes and effects of the particle gases diversity, high pressure and DC voltage remain the research concern for an increase in the number of CNTs formed. In this experiment, a small plasma reactor tube contains a pressurized atmospheric carbon gas with two electrodes is connected to a voltage source. The carbon gas used will be ionized and produce plasma discharge so that the density and temperature can be calculated. This density and temperature analysis can be utilized to predict the growth of CNTs, their purity, geometry, structure and chain formation produced.

2. Ionization of Carbon Plasma on Butane and Air Mixture

Ionization of gas by electron impact collision is very significant to produce plasma at high pressure [7]. Kinetically, plasma can occur when the temperature or energy of the gas is increased by the electrons out of the orbit. The magnitude of the ionization energy of an atom or molecule can be written in Equation 1:
(1) Where the particle mass, m is the velocity of the particles, v is effort because the electric field, W_E is the electrical potential energy, m_e is the mass of the electron, v_e is the velocity of the electron, e is the charge of the electron, and v_i is the ionization potential.

The process of plasma is influenced by several factors including the kinetic energy of atomic, molecular and photon collisions; Thermal and acoustic mechanical energy from external sources that accelerate atomic/molecular motion to excite electrons to ionization. Another factor is the potential energy source of the electric field and the external magnetic field thus accelerating the ionization process. The ionization process is also accelerated by involving both kinetic energy and potential energy. The ionization process, arc discharge can occur with two electrodes and high electric potential difference through the medium gas ($(N_2+O_2+CO_2+\text{trace})_{air}$ + C_4H_{10} ionisasi (Air + Butane)). The high voltage causes the polarization current and the kinetic collision leads the excited electrons to ionize to form ions [8]. Although the ambipolar diffusion process of polarization inhibits kinetic motion, high electrode voltage difference, electrode distance and short diffusion length will help the ionization process to occur fast along the degree of ionization of Butane which is of low ionization energy. Butane has a large density, large latitude appearance, relatively small velocity and ionization energy of 10.6eV which facilitates ionization. The ionization energy required for each molecule in the butane gas for ionization is a vector quantity that can be defined as follows, (2) The energy distribution function of each atom or molecule will collide for each particle and contribute diverse ionization energy. Therefore, by knowing the distribution of this energy, the ionization potential will easily form the plasma until the breakdown voltage occurs.

IOP Conf. Series: Earth and Environmental Science **97** (2017) 012026 doi:10.1088/1755-1315/97/1/012026

***** Formatting Issue - Best Available Paper/Graphic *****

3. Experiment

The ionization process is carried out by increasing the voltage gradually so as to form plasma with voltage and electric current shown in Figure 1.

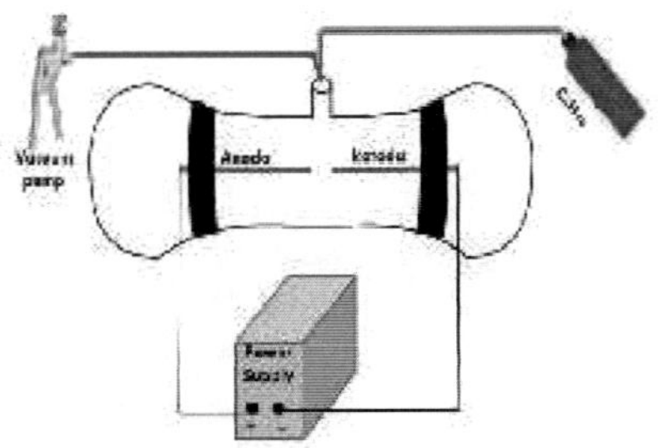

Figure 1. Design and operation of directional carbon plasma

Figure 1 is an electric discharge schematic of the gas ionization tube. A direct voltage source is provided by connecting a carbon electrode (commercial pencil) about 1mm away. Voltage is given until the electric spark is reached and maintained until it is in arc discharge condition. The air in the gas cylinder is reduced in the order of 0.1Atm which then occurs the breakdown voltage of any reduction in air pressure. The electric current is maintained only at the order of 10mA in order to maintain arc discharge conditions. The butane gas are also alternately streamed until 1Atm is subtracted to obtain a voltage drop phenomenon until the air density becomes smaller as a form of impurity of ionized gas, dominated by Alkane gas. Butane are used interchangeably and not mixed in the same process. The butane gas which is flowed with the air gas is in a random velocity. This speed can be known statistically by using the Maxwell Boltzmann Equation:

(3)The velocity distribution of gas particles in the tube can be shown in Figure 2.
Figure 2. Gas particle velocity distribution at atmospheric pressure

(4)Physically, the kinetic and thermodynamic contributions strongly dominate the event of the breakdown voltage with the electric field effect that atomic and molecular polarization speeds up the ionizing process. Plasma density and plasma temperature of carbon species with general air impurities can be expressed as follows:

(5)In the event of this thermal plasma, not all particles are fully ionized so that the density and temperature of the plasma depend on the source of the electric voltage, the effect of the electric field between the electrodes, the high particle velocity and the mean free path of the short particles distance facilitates recombination and particle association, although this is highly desirable in the formation of CNTs on the cathode surface. Therefore the voltage drop varies greatly with the gas pressure of the Alkane by the impurity of the air mixture.

4. Results and Discussion

Observation of ionization on air is done by providing pressure variations. At 12 cmHg (570 V) pressure it produces the breakdown voltage of 0.53-0.57 kV and a current of 0.27-0.45 mA. The pressure of 16 cmHg (580 V) produces the breakdown voltage of 0.49-0.56 kV and a current of 0.31-0.48 mA. The pressure of 26 cmHg (920 V) produces the breakdown voltage of 0.86-0.88 kV and a current of 0.54-0.6 mA. Pressure 36 cmHg (1160 V) produces the breakdown voltage of 0.9-1.1 kV and 0.46-0.6 mA. This data shows an almost linear relationship between the pressure with the current and the resulting drop voltage. The DC voltage given is high because the gas ionization is at a relatively high pressure and the

IOP Conf. Series: Earth and Environmental Science **97** (2017) 012026 doi:10.1088/1755-1315/97/1/012026

density of the particles inside the tube is at > 10^{24}particles/m^3. Electric current from electron particles and positive ions moves successively to the anode and cathode.

Movement of particles to the anode is assumed to be faster than that to the cathode because the electrons are lighter with much density, while the Carbon positive ions move slowly toward the cathode as the CNT builder. This movement is due to the influence of the potential difference of each electrode. When the pressure inside the tube is reduced then there is also a reduction in the density. The particle density also affects the resulting fall stress as shown in Figure 3.

High densities result in collisions occurring predominantly at the excitation level of atoms and electrons so that ionization is difficult to reach at high pressure because some of the energy is absorbed for even superelastic collisions. Some of thermodynamic data of plasma is shown in Figure 4.

Figure 3. Air ionization with variation of air pressure

Figure 4. Air ionization with some pressure

Figure 4 shows a nearly linear kinetic relation, although not all of all particles are ionized, this kinetic quantity is in thermodynamic equilibrium. Increased plasma density is followed by increased pressure so that plasma temperature will also increase. This temperature is the movement of electrons and ions in which both particles remain at the same temperature. At 13cmHg pressure, the density of electrons and ions is relatively low, the pressure is small so that the collisions are small and the temperature decreases.

The ionization of the butane gas is subjected to the variation of pressure to illustrate the resulting ionization difference. At a pressure of 20 cmHg (1199V) it has a particle density of 5.78x10^{24} particles/m^3, the breakdown voltage of 536-775V and the current of 0.03-0.45mA. At pressure of 20 cmHg (866V) having a particle density of 2.77x10^{25} particles/m^3 produces the breakdown voltage of 587-871V and the current of 0.05-1.10mA. Pressure of 26 cmHg 1007V has a particle density of 2.79x10^{25} particles/m^3, resulting in breakdown voltage of 703-1007V and a current of 0.04-0.55mA. The pressure of 27 cmHg 1044V has a particle density of 2.80 × 10^{25} particles/m^3, resulting in breakdown voltage of 675-1055V) and a current of 0.03-0.69mA.

The applied voltage for ionization of butane gas with air mixture tends to be greater than the total air gas due to increased butane density, relatively slow particle velocity, and the area of the gas particle cross section which tends to be large. These results indicate to ionize butane gas having a high voltage because butane has high density and large cross-sectional area.

Figure. 5, 6, and 7 describe the breakdown voltage to the change in the density of the gas by the ionization occurring between the mixture of and air + butane.
Figure 5. Ionization of butane + air and propane + air ionization

Figure 5. shows the ratio of breakdown voltage generated by the butane propane gas. Butane drop voltage is not only contributed to the high electric field, but also the greater carbonized ionization opportunities. Butane has an ionization energy of 10.6 eV. The ionization of butane gas requires a high voltage even though the ionisation energy is large. From Equation 1, ionization Alkanes lies in the kinetic energy regime, the speed and the heat of butane.

Figure 6. Discharge of propane and butane gas with air mixture. Figure 6 shows the electrical discharge process between the two electrodes. Electrons move at speeds that are affected by electric potential, electric fields, drifting collisions, and collision frequencies that occur during the ionization process. The energy that is given to ionization is made up of mechanical energy and electromagnetic energy. The electric current is maintained from the effects of the system in order to maintain the balance of the arc discharge so that the plasma conditions remain in a few seconds.

Figure 7. Density of alkane particles in thermal plasma
The greater the plasma density occurs in the tube, the higher the plasma temperature is reached as shown in Figure 7. This is also followed by the addition of the gas pressure. This result is smaller than the ionization in air due to the relative butane mass greater than the air particles. The carbon chains of butane gas are also large so that ionization gets easier. Ionization of butane has a low temperature, due to the effect of un-ionized hot gas making it easy to recombine the load, greater butane cross-sectional effect.

5.Conclusions
The ionization of the carbon produced by the butane gas has a linear tendency. The design and operation of carbon plasma has been successfully carried out. The ionization of butane gas. At a pressure of 20 cmHg (1199V) has a particle density of $5.78x10^{24}$ particles/m^3, the breakdown voltage of 536-775V, a current of 0.03-0.45mA, density plasma $1.01x10^{23}$ particle/m^3 and plasma temperature is $2.66x10^4$ K. At a pressure of 20 cmHg (866V) having a particle density of $2.77x10^{25}$ particles/m^3 produces the breakdown voltage of 587-871V, a current of 0.05-1.10mA, density plasma $1.13x10^{23}$ particle/m^3 and plasma temperature is $2.39x10^4$K. Pressure of 26 cmHg, 1007V has a particle density of $2.79x10^{25}$ particles/m^3, resulting in the breakdown voltage of 703-1007V, the current of 0.04-0.55mA, density plasma $1.32x10^{23}$ particle/m^3 and plasma temperature is $2.65x10^4$ K. The pressure of 27 cmHg, 1044V has a particle density of 2.80×10^{25} particles/m^3, resulting in the breakdwon voltage of 675-1055V), the current of 0.03-0.69mA, density plasma is $1.34x10^{23}$ particle/m^3 and plasma temperature is $2.72x10^4$ K. The plasma density and temperature of butane have been calculated. The ionization has an additional voltage when the gas density increases. The butane + air mixture requires high pressure for ionization because of the cross-sectional, ionisation energy, and relative mass of the gas. The plasma ionisation in this research is the reactive Alkane gas with impurities of air mixture so that the resulting CNT product is not pure to the carbon composition.

6. References
[1] Yamaguchi, T. 2014. *Synthesis and Characterization of Carbon Nanotubes by Arc Discharge with Methane Gas*. Universiti Teknologi Malaysia

[2] Jum'h, I.Y., Sekhaneh, W.AL., Al-Thaani.H., Albiss.B.A. 2016. Preparation and Characterization of High Qualitysingle-Walled Carbon Nanotubes Using Arc Discharge Technique. *Digest Journal of Nanomaterial and Biostructure*. Vol 11. No 2. 517-523.

[3] Su, Yanjie., and Zhang, Yafei. 2015. Carbon Nanomaterials Synthesized by Arc Discharge Hot Plasma. *Carbon*. 83. 90-99

IOP Conf. Series: Earth and Environmental Science **97** (2017) 012026 doi:10.1088/1755-1315/97/1/012026
***** Formatting Issue - Best Available Paper/Graphic *****

[4] Raniszewski, Grzegorz. Szymanski, Lukasz, dan Kolacinski, Zbigniew. 2012. *Carbon Nanotubes Synthesis By Electric Arc Plasma With External Magnetic Field.* Technical University Of Lodz, Institute of Mechatronics And Information Systems.

[5] Rashaad, A.A., Noaman, R., R. Mohammed., S.A, Yousif. 2016. Synthesis of Carbon Nanotube. *Journalof Nanoscience and Technology*. 2(3). 155-162

[6] Kaushik B.K and Majumder M.K. 2015. *Carbon Nanotube Based VLSI Interconnects*, Springer Briefs inApplied Sciences and Technology.

[7] Chen, F. F. 1984. *Plasma Physics and Controlled Fusion.* Los Angeles: Plenum Press.

[8] Tufail, Kashif. 2014. *Synthesis And Characterization Of Carbon Nanotubes By Arc Discharge With Methane Gas*. Universiti Teknologi Malaysia.

IOP Conf. Series: Earth and Environmental Science **97** (2017) 012027 doi:10.1088/1755-1315/97/1/012027

Indigenous Knowledge of Petalangan in Conserving Tesso Nilo National Park, Riau, Indonesia

Muhammad Syafi'i [1], Auzar [2], Sukendi [3], and LN. Firdaus [2]
[1]Enviromental Sciences Postgraduate Program Universitas Riau, Pekanbaru, 28131, Indonesia
E-mail : forsyafii@gmail.com

Abstract: Environmental conservation based on the Indigenous knowledge (IK) was increasing become the attention of Indonesia in line with the declining quality the environment, especially in the national park area. Tesso Nilo is one the national park conservation areas in Riau Province with an area of 38,576 ha has experienced almost 66.2% depreciation in the period 2004-2013. This study aims to explore the values of Indigenous knowledge from peoples in the region Tesso Nilo Pelalawan, Riau Province. The survey method has been used for all indigenous communities domiciled in 7 villages within the Tesso Nilo National Park. 47 respondents consisting of village head, *perbatinan*, and *tokoh adat* were determined by purposive sampling. Data were collected through *depth interview*, observation, and documentation, then analyzed descriptive-qualitative. There are nine *IK* Petalangan society in TNTN, that is *adat menyusuk kampung, adat menyusuk dusun, adat Berladang, adat rimbo larangan, adat rimbo kepungan sialang, adat rimbo simpanan, adat beramu kayu, adat berburu, and adat berikan*. It can be concluded that *IK* Petalangan community a strategic value in environmental conservation TNTN. Integration values *IK* Petalangan community into the TNTN environmental conservation micro policy is urgent in the perspective of sustainable development.

Keywords: Conservation, *Indigenous Knowledge,* Petalangan, Tesso Nilo National Park

1. Introduction

Environmental conservation based on the values of local wisdom (indigenous knowledge,IK) progressively become attention of indonesia in line with progressive quality of environment crop area of national, report of Wold Wide Fund[1], wide 10 the last year 2004-2013 have losing of its forest equal to 48.139 ha or it Tesso Nilo owning wide 83.068 ha since extension in year 2009 owning 72.720 ha forest and have experienced of losing of equal to 4.813,9 ha or 5,79% TNTN in this time which still forested remain 24.581 ha.

Community have an important role in management of TNTN. Three tribe group which compose Petalangan, Logas tanah darat, Gunung Sahilan have custom claim to natural resources area. Although their settlement beyond area proposal of Tesso Nilo , but they have interaction direct with this area . Tribe group of existing in environment of TNTN the which claiming each other custom to this area natural resources make TNTN have uncared in meaning less is getting attention in the effort continuation of and custody of environmental habitats .Its effect issue expanding TNTN have at the verge of destruction . though a lot of natural potency able to taking its benefit make importance of human life , especially society exist in this area . have ought to this TNTN taking care of by society area of TNTN because sense of belonging for the sake of have offspring continuation.

All society member have traditional ecological knowledge : old fellow, woman, men, just children only is amount of *IK* had by the individual vary . Age , education , gender , social status , and economic, everyday experience , influence from outside , responsibility and role at home and

community , profession , availibility of time , and attitude of kapabilitas intellectual , degree and skill of observation , ability to wander and independence and also operation of natural energy seumber represent a number of factor which very influence IK , Firdaus[2].
Indigenous knowledge (IK) is used at the local level by communities as the basis for decisions pertaining to food security, human and animal health, education, natural resources management, and other vital activities. IK is a key element of the social capital of the poor and constitutes their main asset in their efforts to gain control of their own lives. For these reasons, the potential contribution of IK to locally managed, sustainable and cost-effective survival strategies should be promoted in the development process, Gorjestani [3].

Indigenous peoples are a group of people who have historical support and its existence is not in doubt. According to tombo they come from Johor using a boat, and open the forest in their present settlement. They became *kawula kerajaan Kampar* now better known as Pelalawan. Under the rule of the *Pelalawan Sultan* they have recognized the right to their forest territory *Hutan Tanah Perbatinan Kurang Satu Tiga Puluh* , led by an adat head known as the *batin*. The legal protection rights over customary law and customary land in the Petalangan community have received legitimacy, Tenas E [4].

One of the most important *IK* systems to be explored and developed in the Petalangan community is the traditional unique knowledge of indigenous peoples in the management of the natural environment which in this case is the TNTN area. *IK* is thought to be a very effective model in TNTN conservation, because in *IK* there are local values and wisdom. These values contain the traditional / local knowledge of the Petalangan community since ancient times, especially in preserving the forests, both the flora and fauna and the ecosystems within TNTN. This research aim to for the mengeksplorasi of value - assess indigenous knowledge of roof-gutter society residing in area of TNTN in the effort continuation of TNTN of Pelalawan regency Riau provinsi.

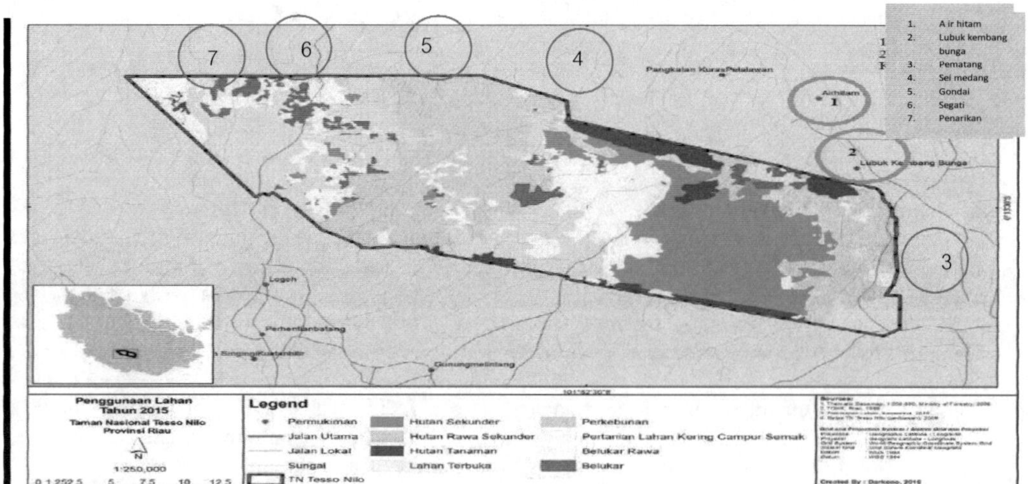

Figure 1. Map of Research location

2. Method

The determination of the respondents from the community is done by *purposive sampling* with the criteria that apatrepresents,that is the indigenous peoples and petalang community related to the utilization of TNTN pelalawan as many as 47 people in 6 pebatinan in 7 village located in TNTN. The method used in this research was survey. Data collecting to interview depth , observation , and documentation. Technique analyse data in by deskriptif-kualitatif.

3. Result.

To roof-gutter Malay society , setlement or countrified have to awaking up the full calculation , because over there they remain to and lengthen clan . Settlement awaking up with basis for custom / cultural and also trust which embracing of , later then completing with prohibition prohibit which goning into effect tightly .Old people roof-gutter Malay remind " in menyusuk or membangun kampung , custom holding institute hold highing .(interview : Kijing, Batin Hitam, on 26 September 2015)

Reference above giving points what a tighten and is careful him rule of custom about developing countrified or settlement . Old fellow affirm in *adat menyusuk dusun* wearing institute hold highing , or telling if kampung will founded , regulation and custom become guidance , prohibition and prohibition order becoming hold , general consensus become base .Even in course of development of house , roof-gutter Malay society have to consider very important several things for the shake of taking care of dweller security from a having the character of mystique and also physical . start from ground construction, environmental or neighbour , land ground status and entire/all him have to get statement of custom deliberation and permit (interview elite figure Nantan , Batin hitam , on 26 september 2015)

From place facet there three land category to develop,build house among him its land construction and is rust colored , plain but heel rear , bush land , close to source of water . While land type which must avoiding among him of place bleed and sand of peat land , bird droppings land, oblique land easterly go out to sea and communal ownership land (interview : Bondo , Batin hitam elite figure on 10 december 2015) .

Andya [5] , saying that each tradition have typical kampung characteristic , identify completely cousin kampong, kampung of Sumaresmi have complete karateristik culture which enough strength earn also influencing by social environment from outside .

Doing tradition have land have value see the condition of the land do either in making countrified land, for example hilly not marshland . Besides seeing also from facet apart to deal with other countrified , apart with economic resources (place paint maintenace) , and distance with source of clean water . if condition of the land have up to standard good to making is countrified , hence study also come land.

Further *mengaji asal tanah* determines and ensures the origin of the land whether included in *tanah ulayat pebatinan / pesukuan* that will open the village or included in the land ulayat pebatinan / tribal others. When entering the communal land of other people's *pebatinan* or in a state of dispute with other parties then the land is not feasible and not be used as a village land. If included in the inner self-mortification of the soil, it should also be ascertained whether the forest is indeed intended as a forest land settlement. If you already meet the new requirements performed ritual activities called to *menengok tuah tanah.*

Activity *menengok tuah tanah* which to see situation of land forest in its communication with various being of is including occult natural being . This as according to trust and confidence in place society tradition that there is land owning tuah tanah (luck).

Ritual *sema tanah* to move supernatural beings to other places so that people who own the land avoid the disaster and are expected to bring tuah.Ini make the village Petalangan have the characteristics of care about the surrounding environment with tradition.

Aims to create a dusun dedicated to the plantation garden, as an extension of the village land, by custom used by the institution is upheld or custom and the rules become guidelines. Deliberation by way of discuss about making a village. In make the hamlet has two elements, namely 1.dusun old, 2. young hamlets, hamlets where overgrown with hard trees, hard trees as a fence between the hamlets one with other hamlets, perennials are also defined as the fence of wildlife, fence so that people do not disturb the habitat of wild animals.

In determining the village to see some provisions, based on interviews with (Mr Kijing, Batin Hitam, on September 26, 2015) it is said that the village is an old hamlet and a young hamlet. The place grows hard trees so there is a fence between one hamlet and Other hamlets, there are old hamlets and young hamlets adar hierarchy of old hamlets and newly formed hamlet, where the village fence is defined with hard plants so that wild animals do not disturb the people of Petalangan and society also does not interfere with zoning wild animals.

Ritual *menilik tanah* to see the type and state of the soil, soil fertility, and plant species, is not justified to make a hamlet in the soil there are plants belonging to the category of abstinence forbidden to be felled, the soil that made the hamlet does not interfere with the survival of wild animals.Selanjutnya ritual Set the ground by measuring the land according to the existing community with the basis of the principle of justice, land pegging considering all the plots of land have the same type and type. The last held the division of land that has been pegged with the customary institutions through ritual or custom tradition of dug the hamlet in the lead by the mind and assisted by the man.

Shifting is done to move on a certain stretch and should not move outside the expanse, shifting cultivation is done with a period of 5 years long enough - 15 years circulation. Land that has been used will be abandoned and overgrown with shrubs to become secondary forest with the condition of hereditary fields In one family group only.

Tradition in *adat berladang*: to get rid of permits to the magical spirits of the land keepers. Make the path, make the boundary or border by cutting down small trees. *Menobas* is an activity of clearing by cutting small woods, cutting roots, Bamboo and lush clumps using a machete or an ax. Buckets are done by knocking down large woods and cutting medium-sized timbers. Midfielders set fire borders around fields to be burned or relocate so that fire does not burn elsewhere. Further Membake is The technique of burning the field, *Memoun* cleans the wood or roots that are not consumed by the fire, Mencacak bonih done the day before menugal done ritual down *mencocak bonih* by using incantation, Besiang do maintenance by making smoke burn woods for animals and animals- Creatures that are believed to exist know that there are people as its owner. Pele effort Stariandat fields experience narrowing of the land and due to the massive logging of TNTN resulting in the pattern of cultivation of cultivated land undergoes a change. It is also revealed by Asysyifa [6], that with the increase of population in the three communities elbow Dayak elephants caused the narrowness of land and declining fertility So that the pattern of cultivation of community farming is changing.

According to adat is a forest / forest that should not be destroyed .Rimba / forest that is passed down from generation to age should not have any activity including the field. Rimba also should not be used for settlements. Rimba maintained indigenous with protected from any threat from any party. Rimba larangan always has a large
enough area so that thousands of hectares that are customary rights, have biodiversity. Customary leaders are responsible for the prohibition of the forest.

The extinct customary rights have made the defame of the spirit concern about the prohibition of the ban, this is different from the situation presented by Elfriandi [7], that through the cultural values of indigenous peoples Rumbio enthusiasts provide encouragement and awareness of environmental awareness *imbgho larangan* to every individual who is inside Their communities.

The *adat rimbo kepungan sialang* which is located between some farming land, as forest / *kepungan* where bee nesting in sialang tree, kind of sialang trunk is vine *rimbo kepungan sialang* which is located between some farming land, as forest / *kepungan* where bee nesting in *sialang* tree, kind of *sialang* trunk is vine stem, house of fighting and jack water consisting of child *sialang,* wood *sialang* and *sialang*. *Sialang* trees are kept by cleaning around the trees and doused with diamond water. *Sialang* bosses serve as a water source, accelerate forest conservation, food sources for the community

IOP Conf. Series: Earth and Environmental Science **97** (2017) 012027 doi:10.1088/1755-1315/97/1/012027

and potential honey economic places. A sialang can produce 200 honeybee honeycomb with 26 kg honey / beehive. Honey bee taking is done in a traditional way known as *manumbai* custom which has organization of honey bee harvesting procedure done by custom.

Utilization of forest products in TNTN by Petalangan community is very important in the sustainability of farmers' life. This dependence needs to be given a good legal reinforcement. According to Rachmawaty [8] that the functions and benefits of forests are dependent on individuals or groups of people who have an interest in forests.

Adat rimbo simpanan oriented to the management of natural resources. In certain times, the community of Petalangan provides or supplies food stocks in case of rainy season and flood. The public is prohibited from excessive storage of the forest. The management of the savannah forest designated by a single bridegroom may only manage the savings for the community in the pebantin only with the management not being undertaken in a personal way.

According to Fadhilah Amir [9], the forms of local wisdom of the community of molamahu Pulubala Gorontalo can basically be categorized in the form of local agricultural system, local food production techniques, local food production motives, processing patterns and storage of local food sources, and values / ideas (1) the community has food independence, does not depend on a single food commodity, (2) fosters the active participation of the community in the diversification of food sources that are not dependent on rice, (3) the public is open to Change, but not eliminate the tradition itself, (4) maintaining social relations through customary events.

The activities of collecting forest materials for the fulfillment of daily needs, clothing, food and medicines by way of gathering done by individual men and entourage into the forest for a day. Before *kemantan / dukun* is asked to provide clues about good days and good fortune because in concocting they face many dangerous things like tigers and other venomous animals. The customary ritual of mixing wood is done with a diamond ceremony. Concern for wood by likening wood to something sacred.
Conducting activities chasing, stalking and catching and killing the animals that exist in the jungle to make a meat or fish, done by individual or group. Extension is done by individuals or more, done at night, bringing traditional gobok / lance / rifle. Known by one person's direction.Strategy of finding traces or seeing traces of animal food to be searched. Before departing prepare to arrange the gear and recite the mantra (prayer) ask the forest dwellers to help to tame the animals. Animals hunted are usually large ones such as deer and deer. Do not hunt animals like wild pigs, other than haram wild boar also as a wild animal food in the forest like tiger for tiger ecosystem awake. Large number of extension agents are usually animals such as lindak, kancil, napuh should not be arrested because it is not sufficiently divided.

Menapak, done during the day with a large number of entourage, usually by looking for a trail that is still new by sneaking.Headed by someone who has experience in tread. Tracing is also done to search for hunted animals that have been shot dead while tearing at night by following the trail and blood splatter. Animals that do not die must also continue to be sought and should not be allowed to meet the carcass because kalua left will be believed to torture and also if not sought it will be hit by reinforcements. *Menjoat* and capture is one of the customs of hunting by trapping the animal buran. The process takes a long time with the installation of the first snares of the jarring installed and abandoned in a relatively long time. *Menjoat* is not justified in many places and in the area of pebantin only.

By *adat berburu* it is believed that it can suppress the market demand for animal meat production because only hunting animals are hunted for the needs of petalangan communities, according to PattiselanoFredy [10], that one of the causes of excessive use of wildlife hunting is increasing market demand Production of meat from animal origin.

Adat berikan is nailed to a stream. The river becomes sacral because the pulse of the life of the community of Petalangan for daily life one of them catch fish as food. Must not take the whole type of fish and is prohibited from taking small fish. The process of fishing is done by way of catching by catching fish / shrimp using a cud. Woven cane made of woven rattan, small hole made small size measurement so that small is not taken. *Sempiyai*, small fish trap, *miawai*, by holding back taju or *menaju*. *Melukah* is a fishing gear made of reed or rattan made and woven in such a way that is round and long, the front of the fish entry is made somewhat larger than the back. The backside is made somewhat closed so that the incoming fish can not get out. The length of the painting is approximately 1.25 m- 1.50 m.

In the adat berikan there is a ritual "tapa naik tapa turun" where the value in the implementation of this tradition gives a message for people to pitch fish *tapa naik* can be done fishing until tapa season down. And will wait about a year later to take a fish big enough to harvest. According to PrahamaSusyanthi [11] said in a fish harvest conducted once a year for the welfare of the local community to be fair divided by the family's head for the realization of equity as a tradition of society pangakan indarung cultural fishing in Lubuk larangan.

4. Conclusion

IK of Petalangan community has a strategic value in the perspective of the conservation of the environment TNTN. The value of *IK* is always guarded by the community of petalangan aims for the environment able to provide assurance of the implementation of values *IK* good value to all communities of petalangan and with nature. Nine value and the culture of the peasant *adat menyusuk kampung, adat menyusuk dusun, adat berladang, adat risa mba larangan, adat rimba simpanan, adat rimba kepungan sialang, adat beramu kayu, adat berburu, adat berikan* ensuring the social sustainability of the peasant society Which are inherited from the puppets left to their children and grandchildren.

Integrating the values of *IK* of Petalangan communities into the TNTN conservation micro-life policy is urgent in a sustainable development perspective. As a follow-up to the development of national identity through the revitalization of *IK* culture of Petalangan community in preserving TNTN, the understanding of the local nation's philosophy must be done through a zoning system that integrates the value of *IK* Petalangan community.

5. References

[1] Wold Wide Fund Riau (2014). *Laju Deforestasi Taman Nasional Tesso Nilo. Provinsi Riau . Indonesia.*

[2] Firdaus L.N, 2009 .*Kewaspadaan Nasional terhadap Perlindungan Indigenous Knowledge dalam Rangka Good Governance.* Lembaga Ketahanan Nasional Republik Indonesia.

[3] Gorjestani, Nicolas. (2000). Indigenous knowledge for development: Opportunities and challenges. [This paper is based on a presentation made by the author at the UNCTAD Conference on Traditional Knowledge in Geneva, November 1, 2000.]. Retrieved 16 February 2012 from: http : // www. Worldbank .org /afr /ik / ik paper_0102.pdf

[4] Tenas. E, 1998. *Orang Talang dan Kebudayaannya.* DewanKesenian Riau

[5] Andya,2004. *Karakteristik Lanskap Kampung Tradisional di Halimun Selatan dan Faktor-Faktor yang Mempengaruhinya.* Institut Pertanian Bogor.

[6] Asyisyifa, 2009. *Karakteristik Sistem Perladangan Suku Dayak Meratus Kecamatan Loksado Kalimantan Selatan.* Universitas Lambung Mangkurat.

[7] Elfiandri, 2014. *Model Komunikasi Lingkungan dalam Pelestarian Imbho Laghangan (Rimba Larangan) di Kabupaten Kampar Provinsi Riau*, Universitas Riau.

[8] Rachmawaty, 2004.*Hutan Fungsi dan Peranannya bagi Masyarakat.*Fakultas Pertanian Program Ilmu Kehutanan Universitas Sumatera Utara, Digital Library USU.

[9] Fadhilah Amir, 2013. *Kearifan Lokal dalam membentuk Daya Pangan Lokal Komunitas Molamahu Pulubala Gorontalo.* Al Turas Vol. XIX, No. 1, januari 2013

[10] Pattiselanno Fredy (2010). *Kearifan Tradisonal Suku Maybrat dalam Perburuan Satwa sebagai Penunjang Pelestarian Satwa.* Makara Sosial Humaniora. Vol. 14. No 2. Desember 2010

[11] Prahama Susyanthi (2016), *Budaya Menangkap Ikan di Lubuk Larangan(Studi Kearifan Lokal Masyarakat Desa Pangkalan Indarung Kecamatan Sengingi Kabupaten Kuantan Sengingi.* Universitas Riau.

The Effect of Coloring and Compacting Pressure Paving Block by Adding 5 Wt.% Fly Ash in The Compressive Strength

Nurzal and Aries Nursyuhada
Mechanical Engineering Department, Padang Institute of Technology, Padang, 25143,, Indonesia
E-mail: nurzall@gmail.com

Abstract: This research aims based on SNI 03-0691-1996 to investigate the effect of coloring and compacting pressure with the addition of 5 wt.% fly ash (Fa) on compressive strength. Fa derived from waste material coal-fired *Sijantang Sawahlunto* thermal power plant. The growing production of Fa caused negative environmental impact. So, one of the solutions to overcome that effects is to use the Fa as a raw material for paving block mixture that can reduce the cost of raw material and increase its strength. Paving blocks are gray and red with 0 wt.%, 5 wt.% Fa + Pb composition. Compaction pressure variations 55, 65, 75, 85 and 95 Kg/cm^2. The drying time for 35 days. Specimens were produced in the form of rectangular bar (length, L = 20 cm, width, B = 10 cm, thickness, W = 6 cm). The test results showed that the addition of 5 wt% FA has a compressive strength value higher than 0 wt%. The red color has a compressive strength lower than the gray color paving block caused the red color (Iron Oxide) is less binding at the time of mixing the material. Gray and red Paving blocks both increase in each additional compaction pressure, because the higher the compaction pressure will increase the bond between the particles so porosity is reduced increased compressive strength. The overall data, the gray paving block with the composition of 5 wt% FA at compaction pressure 95 kg/cm^2 with the optimal compressive strength value of 36.1 MPa and the lowest value is found in the red color paving block at 0 wt% FA at a pressure of 55 kg/cm^2 with a value of 6.5 MPa. Gray and red Color paving blocks has a compressive strength quality based on SNI 03-0691-1996.

Keywords: fly ash, paving block, SNI, colouring and compressive stength.

1. Introduction

Fly ash is residual coal combustion is widely used in the combustion process in a steam power plant. The increasing demand of coal in power plants for the combustion process leading to an increase of production new utilization of fly ash were little, which is approximately 20 to 30%, thus causing environmental pollution such as air pollution and groundwater. Therefore, it is necessary to find a solution to overcome the problem by the use of fly ash as a raw material for mixtures paving blocks that can reduce the cost of raw material and increase its strength. In developing the technical field of fly ash has superior properties, such as: hardness, high strength and capable of good work, so that it can be applied in the fields of construction, mechanical and chemical industry [4]. The utilization of fly ash as an additive can also improve the quality of paving blocks. On the composition of the fly ash 10% - 40%, paving block is watertight aggressive medium, which is resistant to industrial waste water, brackish water and sea water [3].

This study is a continuation of previous research. On the research found 5 wt% the composition of the fly ash is the best composition for the absorption of water by the value of quality A & B Quality Compressive Strength [10]. Binder best is mineral water [11,12] while the drying time is best obtained during 35 days [11]. The research aimed to determine the effect of coloring and compacting pressure with the addition of 5 wt.% fly ash (Fa) on compressive strength and determine the classification of the quality of paving blocks with SNI 03-0691-1996.

2. Paving Block

Paving block is the building material products of cement are used as an alternative ground cover or surface hardening. Paving blocks also known as concrete bricks (concrete block or cone block). Paving block is a building material composition made from a mixture of Portland cement or hydraulic adhesive material like, water and aggregates with or without other material that does not reduce the quality of concrete bricks. Among the various alternatives of land surface covers, paving blocks over has a lot of variety in terms of shape, size, color, pattern and texture of the surface as well as strength.

2.1 Advantages of Paving Block

Absorption of water through Block Paving can preserve the balance of groundwater to support concrete/ houses on it. Weight *Block Paving* relatively lighter than concrete/ asphalt and support the foundation of a stable house. The house has good water absorption, save soil water availability to be drilled and used for daily purposes. Easy iimplementation and does not require heavy equipment. Easy Maintenance and can be replaced after be demolished.

2.2. Weaknesses of Paving Block

Easy wavy when the foundation is not strong and less comfortable for the vehicle at high speed, thus paving block pavement is suitable for controlling the speed of the vehicle environment in residential and the crowded urban. Application of paving blocks can be classified based SNI 03-0691-1996 [4]

Table 1.
Quality Paving Block (SNI 03-0691-1996)

Class	Compressive Strength (MPa)		Max average water absorption (%)	Application
	Average	Min	Average	
A	40	35	3	Road
B	20	17	6	Parking area
C	15	12,5	8	Pedestrian
D	10	8,5	10	Parks and other users

Paving blocks manually produced concrete quality is usually included in class D or C is for the purpose of non-structural use, such as for parks and other uses that are not required to hold the load weight on it. Quality paving block the process by using a press machine can be categorized into concrete quality class C to A with a compressive strength above 125 kg/cm^2 depends on the mixing ratio of the materials used. Sightings of paving blocks are manufactured by hand and pressed paving block machine in plain view are relatively similar, but the surface of paving blocks are manufactured with a pressing machine looks denser than manually created [6].

2.3 Shapes and Sizes

Based on the shape of paving block can be divided into two rectangles and polygons. The thickness of 6 cm, 8 cm, and 10 cm. The color is generally gray or according to customer orders. Tolerance size is ± 2 mm for the size of the width and ± 3 mm of thickness and weight loss when tested sodium sulfate maximum 1 %.

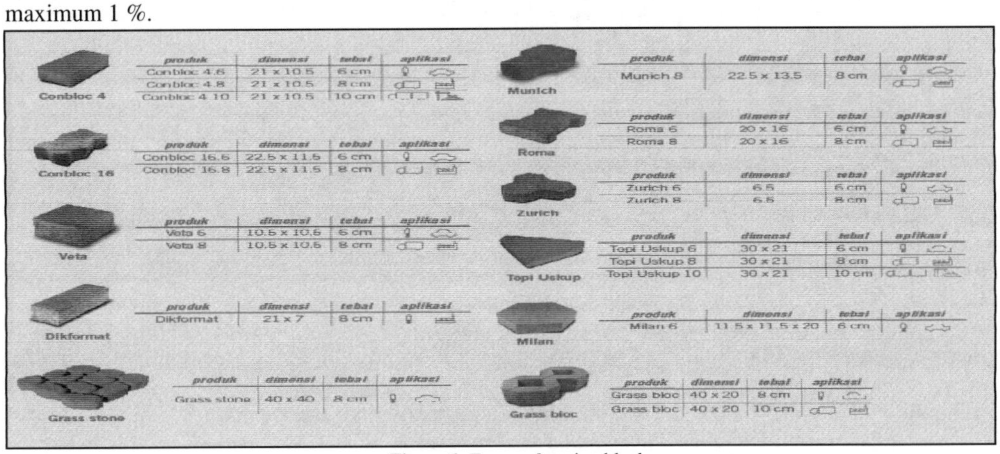

Figure 1. Forms of paving block

2.4. Fly Ash

Fly ash is a coal combustion residue in the form of amorphous fine particles; an inorganic material is formed from mineral material changes due to the combustion process. From the coal combustion process in a steam generating unit (boiler) will form two types of ash are: fly ash and bottom ash. The composition of the coal ash produced consists of 10-20% bottom ash, while the remaining approximately 80-90% in the form of fly ash captured by the electric precipitator before being discharged into the air through the chimney.
According to ACI Committee 226, explained that the fly ash has a fairly smooth grain, which passes a sieve No. 325 (45 mile micron) 5-27% by the specific gravity of between 2.15 to 2.6 and blackish-gray color. Coal ash contains silica and alumina about 80% with partial-shaped amorphous silica. The physical properties of coal ash include a density of 2.23/cm^3, about 4% water content and mineral composition is predominantly α-quartz and mullite. Besides coal ash containing SiO_2= 58,75 %, Al_2O_3= 25,82%, Fe_2O_3= 5,30% CaO= 4,66 %, alkaline= 1,36 %, MgO= 3,30% dan other material = 0,81%.. Actually, fly ash does not have the ability to bind as well as cement, but with the presence of water and the size is fine, oxides of silica contained in the coal ash will react chemically with the calcium hydroxide formed from the hydration process of cement and will produce a substance that has the ability of binding .

3. Method

A specimen was performed in Industrial paving block *Sumber Intan Betung Tebal Padang*. Paving block material consisting of cement and sand sized 0.0625 to 2 mm (matrix), while for reinforcing materials used fly ash with a size of 80 meshes. Binder is mineral water.

Table 2.
Paving block composition for 10 specimens

composition (wt.%)	cement (kg)	Sand (kg)	FA (Kg)	Water (liter)	Iron Oxide (gr)
100	4,7	23,3	0	0,8	133,33
95	4	22,6	1,4	0,8	133,33

Figure 2. Material used: (a) Fly ash, (b) Cement, (c) sand,(d) Red iron oxide coloring and (e) Mineral Water

Based on SNI 03-0691-1996, Specimens were produced in the form of rectangular bar (length, L = 20 cm, width, B = 10 cm, thickness, W = 6 cm). Paving block color is gray and red (thickness ± 1.3 cm of 6 cm). Based on the Composition of Materials paving block (Table 2). The materials are manually mixing to homogeneous and flush with mineral water until it forms a paste after that into a mold paving blocks. Compaction using a press machine with pressure variations: 55, 65.75, 85 and 95 Kg /cm^2. After finish compacting, Paving was placed above the board and let it harden for 1 night and not exposed to direct sunlight to avoid cracking, after that paving block is dried in direct sunlight for 35 days.

(a) (b) (c) (d) (e)

Figure 3. The compacting process: (a) manual mixing, (b) mold, (c) the press machine, (d, e) Paving

The Compressive Strength Test:
The process employ Measuring compressive strength by equation:

$$\sigma = \frac{F}{A}$$

Where:
σ = compressive strength (kg / cm^2)
F = Force press (kg)
A = sectional area (cm^2)

4. Results And Discussion

Gray and red paving block are both experienced an increase in each additional compaction pressure, because the higher the pressure molding the density is also higher because of the higher compacting pressure will increase the bonding between the particles so that the gaps between the particles/porosity is reduced through the new particle contact and leads to better packing and increased compressive strength. The thickness of the paving blocks are also influential because it is a parameter that must be maintained is as thick as 6 cm [4] so that its increases weight.

The compressive strength higher than the overall data is the gray paving block with a composition of 5% by weight of fly ash which is at a pressure of 95 kg/cm^2 with a value of 36.1 MPa and the lowest value contained in the paving block of red in the composition of 0 wt % fly ash that is at a pressure of 55 kg/cm^2 with a value of 6.5 MPa.

Comparison of the results of the compressive strength test of the gray paving blocks higher than the color red, this is due to dye red iron oxide used in the manufacture of paving blocks to contain 95 % Fe_2O_3 which is a ceramic material that bonding process be better if aided by the heating process.

Table 3.
The test results Compressive Strength refers to the SNI 03-0691- 1996

Compaction Pressure (kg/cm^2)	Compressive Strength (MPa)			
	Grey	Quality	*Red*	Quality
5% wt. FA				
55	17.8	B	11.8	D
65	26.9	B	13.7	C
75	27.9	B	15.6	B
85	32.2	B	18.0	B
95	36.1	A	22.5	B
0% wt. FA				
55	14.8	C	6.5	<D
65	18.3	B	7.1	<D
75	20.1	B	10.8	D
85	24.8	B	13.1	C
95	30.3	B	15.1	C

All the test results included in the SNI 03- 0691-1996:Quality A: for roads, Quality B: the Court parking Quality, C: Pedestrians Application and Quality D: Applications Parks and other .

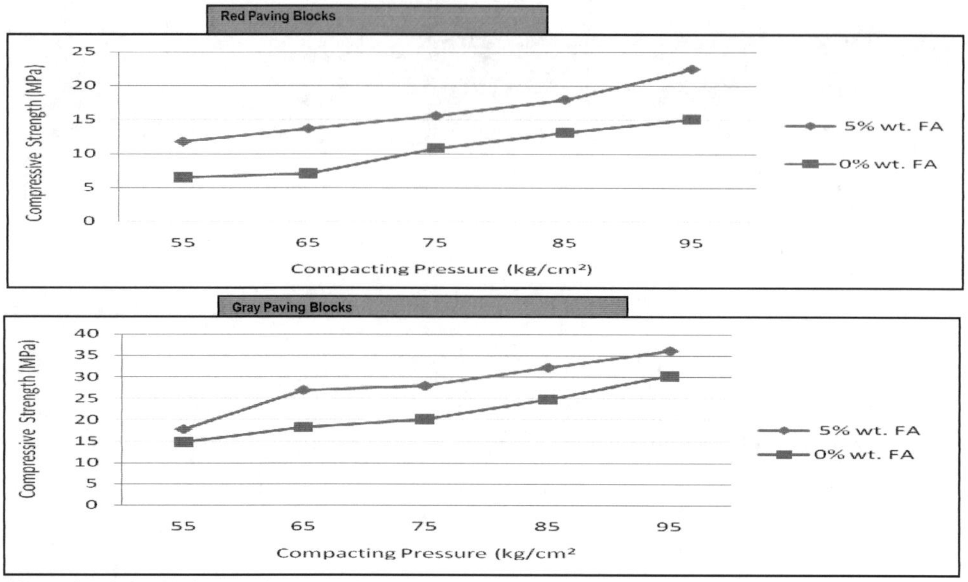

Figure 4. Relationship between the compacting pressure paving blocks with compressive strength of paving blocks

5. Conclusions

Based on the results, it can be deduced as follows:
The higher Compacting pressure paving block make the increased compressive strength. Compressive Strength is highest in pressure 95 Kg/cm^2 in both compositions. Compressive strength between the gray higher than the red color paving blocks. The compressive strength between gray and red paving blocks have quality according to the standard SNI 03-0691-1996.

6. References

[1] Aswin, B. 2007. *Strong Push and Pull Strong Concrete with Fly Ash as Cement Replacement*. Yogyakarta: Universitas Islam Indonesia.

[2] Barsoum, MW. 1997. *Fundamentals of Ceramics*. New York: Mc Graw-Hill Book Co.

[3] Eridani, Zeta. 2004. *Paiton Coal Utilization Waste as Materials Added to the concrete mix. Thesis Jurusan Teknik Sipil*: Universitas Gadjah Mada.

[4] Benavides, EG. 2003. *Densification of Ash from a Thermal Power Plant*. Ceramics International Journal, 29, 61-68.

[5] Boccacini, AK. 1995. *Glass-Ceramics from Filter Dust From Waste Incinerators*. Ceramics International Journal, 21, 231-235.

[6] Claudia Muller, EF. 2006. *Making Training Module Block and Paving tile or brick*. International Labor Organization.

[7] Djumari, ES. 2009. *Technical and Economic Assessment of Coal Waste Utilization (Fly Ash) on the Production Block Paving*. Media Journal of Civil Engineering, Vol IX, ISSN 1412-0976.

[8] Hima, AK. 2003. *Use of Fly Ash in Preparation Paving Block*. Surabaya: Final Department of Civil Engineering, Kristen Petra University.

[9] Karasawa, A. d. 2003. *Application of Fly Ash to Concentrate Paving Block. The7th Proceedings of the International Conference on Concrete Block Paving (Pave AFRICA)*, ISBN: 0-958- 46091-4.

[10] Nurzal, Zepriady Zakir. 2014. *Effect of Fly Ash Composition on Paving Blocks Compressive Strength*. ITP Engineering Technical Journal, Vol 4, No. 1, Pages: 15-21, ISSN 2089-4880.

[11] Nurzal and Ardiansyah. 2015. *Effect of Drying Time By Adding 5 wt% Absorption on Paving Blocks Compressive Strength (Binder PT.X)*. ITP Engineering Technical Journal, Vol 5, No. 2, Pages: 127-132, ISSN 2089-4880.

[12] Nurzal and Taufik. 2016. *Effect of Drying Time By Adding 5 wt % Fly Ash on Paving Blocks Compressive Strength (Binder mineral water)*. ITP Engineering Technical Journal, Vol 6, No. 1, Pages: 43-49, ISSN 2089-4880.

The Characteristics of Peats and Co_2 Emission Due to Fire in Industrial Plant Forests

Ambar Tri Ratnaningsih[1] and Sri Rayahu Prasytaningsih[1]
[1]Universitas Lancang Kuning, Pekanbaru, Indonesia
E-mail : ambar_trn@yahoo.com
srirahayu_pn@yahoo.co.id

Abstract: Riau Province has a high threat to forest fire in peat soils, especially in industrial forest areas. The impact of fires will produce carbon (CO_2) emissions in the atmosphere. The magnitude of carbon losses from the burning of peatlands can be estimated by knowing the characteristics of the fire peat and estimating CO_2 emissions produced. The objectives of the study are to find out the characteristics of fire-burning peat, and to estimate carbon storage and CO_2 emissions. The location of the research is in the area of industrial forest plantations located in Bengkalis Regency, Riau Province. The method used to measure peat carbon is the method of lost in ignation. The results showed that the research location has a peat depth of 600-800 cm which is considered very deep. The Peat fiber content ranges from 38 to 75, classified as hemic peat. The average bulk density was 0.253 gram cm^{-3} (0.087-0,896 gram cm^{-3}). The soil ash content is 2.24% and the stored peat carbon stock with 8 meter peat thickness is 10723,69 ton ha-^{1}. Forest fire was predicted to burn peat to a depth of 100 cm and produced CO_2 emissions of 6,355,809 tons ha^{-1}.

Keywords: peat, forest fire, carbon, bulk density

1. Introduction

The area of peatland in Indonesia is more than 20 millions ha, 6.29 million ha is in Sumatra, while 4.044 millions ha are in Riau Province. Riau Province contributes 45% peatland in Indonesia where most of it is for Industrial Timber Plantation (HTI), in which at present its condition is threatened. The hot spot data from the last 5 years shows the source of forest and land fires in Riau has shifted from mineral land to peatland. Most of the fires occurring in peat forests are classified serious due to the characteristics of peat which is composed of litters of organic materials with vegetation above and are potential for fuel. The forest fires that occur in peatlands will increase the amount of CO_2 in the atmosphere so that the availability of carbon in peat declines.

Peatlands store carbon in plant biomass, litters beneath peat forests, peat layers and mineral soil under the peat substratum. The range of peat carbon content is about 30-70 tons m^3. This amount is equivalent to 300-700 tons $ha^{-1}m^{-1}$ carbon stock. If peat has a thickness of 10 m, then the carbon stored in it is about 3000-7000 tons ha^{-1} [1] . Peat soils store C much higher than mineral soils. Each one gram of dry peat stores about 180-600 mg of carbon, whereas every one gram of mineral soils contains only 5-80 grams of carbon. In the tropics, the carbon stored by soil and plants on peatlands can be 10 times the carbon stored by soil and soil in mineral soils (Agus and Subiksa, 2008).

Peat with a depth of 1 m has a carbon content of 600 tons ha^{-1} [2] , while peat forest biomass contains only about 200 carbon tons ha^{-1}

This study aims to determine the characteristics and reserves of peat carbon and CO_2 emissions from fires in industrial plantations. The benefits of this research can provide information on the characteristics and reservation of peat carbon and CO2 emissions generated during the fire in industrial plantations so as to facilitate the management of peatland forests in climate

2. Method

The research was conducted on peatlands of industrial plantation of PT. Sekato Pratama Makmur, Bengkalis Regency, Riau Province. It is conducted in April to July 2016. Measurement of peat carbon reserves at the research sites is done by several stages, namely 1) Measurement of depth of peat soil and soil sampling, 2) Analysis of sample in laboratory and 3) Data processing. Determination of observation point is done by using transect system that is terrain technique. In each transect, soil observation and sampling were taken. Peat soil samples were taken with a half-cylindrical peat drill. Sampling is at every 50 cm interval of the depth of the peat layer. The depth of soil drilling is carried out from the surface of the peat until alluvial soils are found. At every 50 cm interval, soil sampling is taken to measure peat maturity, bulk density, ash content and carbon content. To know the level of organic material in peat material it is done by method of lost on Ignition (LOI). To calculate CO_2 emissions that are emitted due to peat decomposition process, this formula is used $CO_2 = C \times 3.67$ CO_2 = The amount of CO_2 gas from peat decomposition, C = the weight or amount of carbon lost during the decomposition process, 3.67 = constant to convert carbon to CO_2 .

3. Result and Discussion

The carbon stored in peat soils is determined by the thickness of the peat. The result of peat drilling at the study sites shows that the peat depth ranges from 600 to 800 cm. The process of peat soil formation lasted for a very long time. This correlates to the depth of peat soil that can reach tens of meters. The result of peat age measurement in Bengkalis Regency of Riau Province shows that peat age in Bengkalis Regency is 4,740-5,730 BC (Neuzil [3]) According to Widjaya Adhi [4] , peat depth 600 to 800 cm is grouped deep into very deep peats. (Harjowiegeno, [5]) . The degree of organic soil decomposition is indicated by the fiber content. Based on laboratory results, the peat fiber content in each peat depth varies. The peat fiber content in the fire area ranged from 38-75%, grouped into hemic peat (half-baked) according to Agus et al. [1] . Peat on the surface (top layer) is generally relatively more mature due to faster decomposition rate. The development of canals and forest fires will lead to soil compaction as well as decreased moisture content that impact on high bulk density values. The average bulk density in the fire area is 0,253 gram cm^{-3} with the value of 0,087-0,896 gram cm^{-3}. The value of bulk density in some depth of peat in fire area can be seen in figure 1.

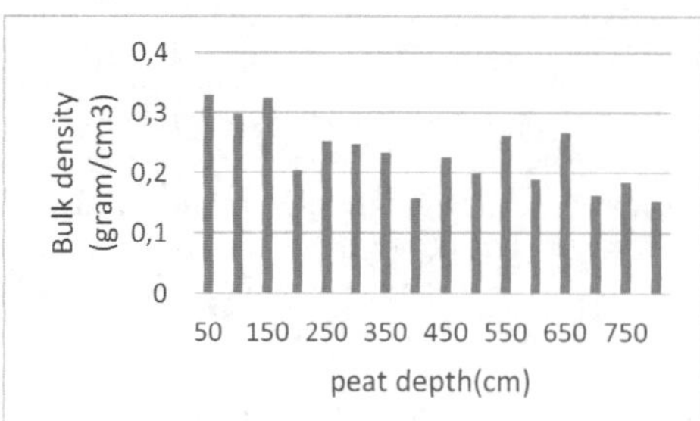

Figure 1. Bulk density on some depths of Peat

Figure 1 shows that the bulk density value on the peat surface is higher than the inside part. The differences in bulk density values are due to differences in peat maturity in each soil profile. Peat maturity is found on the surface of the peat. This occurs because the surface layers of peats both naturally and also due to human activities are able to accelerate the maturity of the peat.

The distribution of bulk density value is higher than Yulianti's research result [6] which ranges from 0.07-1.73 gram cm^{-3}. According to Regulation Number. 4 of 2001, the standard criteria for damage to the physical properties of peat causes the compaction of the soil and decrease the water content that affects the high value of the density. Soil with low bulk density values generally have high porosity, so the potential to absorb and channel the water becomes high, but if the bulk density is too low, the soil has a low bearing capacity (Nugroho et al. [7] ; Widjaja-Adhi [4]). Bulk density value is influenced by moisture content. According to Hanafiah [8] the higher the moisture content the lower the bulk density .

The level of ash in peat soil shows the amount of mineral content in the peat. In some locations generally the ash content in peat soils increases with the proximity of the peat layer to the substrate layer (Dariah, [9]. The average value of ash content on each peat in the study area was 2.24%. Ash content at each depth of peat layer can be seen in figure 2.

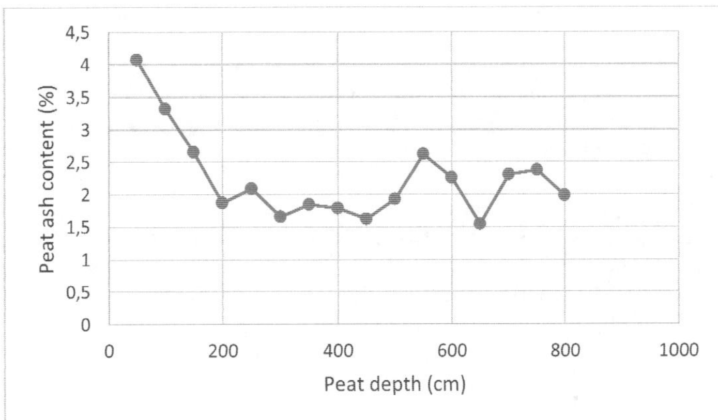

Figure 2. Peat ash content in some depth

The ash content on the peat surface layer is higher than the lower layer. The weight or ash content from the surface layer of peat soil to the bottom layer at forest fire location is 0.0057 grams cm^{-3}. The increasing of ash levels in forest fires is assumed to come from minerals stored in peat and burning biomass. Peat fires and plant biomass that grow on it cause the material to oxidize into gaseous form, especially CO_2 which is emitted into the atmosphere. The minerals contained in peat and plant biomass accumulate on the burning surface layer of the peat, thus increasing the ash or mineral content of the surface of the peat. The amount of carbon stored in peat soils is strongly influenced by the thickness of the peat. The distribution of carbon content per hectare in thickness / into peat can be seen in Figure 3.

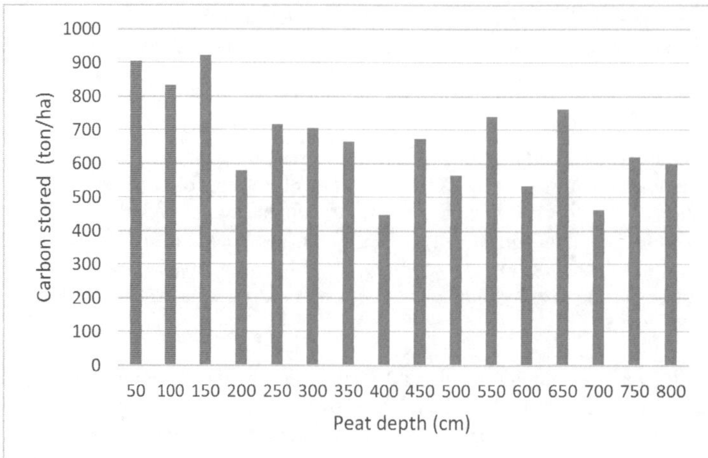

Figure 3. The storage of peat carbon in some depths of peats

The storage of peat carbon at study site with 8 meters peat thickness of 10723.69 tons / ha is equivalent to 1341.14 ton ha^{-1} m^{-1}. The value of carbon stocks is higher than Prayitno et al, [10] of 1675,361 -9055,922 tons ha^{-1}. Peat with a depth of 1 meter at the study site has a peat content of about 1739.95 tons ha^{-1}. According to Page et al. [2] peat with a depth of 1 meter has a carbon content of about 600 tons ha [1].

The carbon storage of peatlands is relatively high which causes these ecosystems to contribute greenhouse gas emissions, if carbon materials stored in peat form decompose or fire. The forest fire that occurred in 2014 has resulted in CO_2 emissions. Fires in peatlands can reach an average depth of 22 cm (between 0-42 cm), even in certain places it can reach a depth of 100 cm (Limin et al., [11]). It is estimated that the burning peat reaches a depth of 1 meter. In spite of the fact there was never equally even fire. Based on the calculation of the fire emission amount of 0-50 cm layer is 19.850.484 tCO2 and at layer 50-100 cm is 18.284.371 tCO2. So total emission of the depth of peat layer 0-100 cm with fire area 6000 hectares is estimated 38,134,855 tCO2.

4. Conclusion

The research location has a peat depth of 600-800 cm which is considered very deep peat. The peat fiber content ranges from 38 to 75, classified as hemic peat. The average bulk density is 0.253 gram cm^{-3} (0.087-0,896 gram cm^{-3}). The soil ash content is 2.24% and the stored peat carbon stock with 8 meter peat thickness is 10723,69 ton ha^{-1}. Forest fires are predicted to burn peat to a depth of 100 cm and produce CO_2 emissions of 6,355,809 tons ha^{-1}. The peat depth characteristics is the major factor in estimating carbon stocks in peatlands.

5. Aknowledgement

The researchers would like to thank KemenristekDikti who has assisted in financing the research, PT. Sekato Pratama Makmur which has provided facilities for researchers to conduct research in Hampar working area, Bengkalis Regency. The researchers also thank Research and Community Service and Forestry Department, Lancang Kuning University.

6. References

[1] Agus F, K. Hairiah dan A. Mulyani. 2011. *Pengukuran Cadangan Karbon Tanah Gambut*. Petunjuk Praktis. World

Agroforestry Centre-ICRAF, SEA Regional Office dan Balai Besar Penelitan dan Pengembangan Sumberdaya Lahan Basah (BBSDLP), Bogor, Indonesia, 58 p.

[2] Page, S. E., Siegert, F.., Rieley, J.O., Boehmn,. H.V., Jayak, A. and Limin, S. 2002. *The amount of carbon released from peat and forest fires in Indonesia during 1997*. Nature, vol 420, 2002.

[3] Neuzil, S.G. 1997. *Onset and Rate of Peat and Carbon Accumulation in Four Domed Ombrogenous Peat Deposits, Indonesia, in*: Biodiversity and Sustainability in Peatlands- Proceedings of International Symposium of Tropical Peatlands, palangkaraya, Indonesia, Edited by: rieley, J.o and page, S.e., Samara Publishing, Cardigan.

[4] Hardjowigeno, S. 1993. *Klasifikasi Tanah dan Pedogenesis*. Akademika Pressindo. Jakarta.

[5] Yulianti, N. 2009. *Cadangan Karbon Lahan Gambut dari Agroekosistem Kelapa Sawit PTPN IV Ajamu Kabupaten Labuhan Batu Sumatera Utara*. Thesis. Sekolah Pascasarjana. Institut Pertanian Bogor.

Detergent Disposal into Our Environmentand Its Impact on Marine Microbes

Effendi, I., S. Nedi [1], Ellizal [2], Nursyirwani[3], Feliatra [4], Fikar [5], Tanjung, R Pakpahan[6], Pratama[7]
University of Riau, Indonesia
effendy@gmail.com

Abstract. Detergents figure in an extensive array of industrial and home cleaning applications, released into the flow of wastewater coming from the home, can far-reaching environmental impacts. Microorganisms are crucial to nutrient recycling in ecosystems as they act as decomposers, pathogen, antibiotic producer, biodegradation of pollutants etc. The research is aimed to examine effect detergent disposal to bacterial population growth in marine environment both *in vitro* and *in situ* condition. Seawater samples were collected from Sungai Kayu Ara Village, and Dumai River estuary, Siak Regency and Dumai City, Riau Province. Experimental method with complete randomized design (RAL) 2 (two) factors; a detergent brand (a1: ATTACK, a2; RINSO and a3; SURF) and b concentration of detergent concentration with 5 (five) concentration level, b1 (0%) as control, b2 (0.3%), b3 (0.6%), b4 (0.9%) and b5 (1.2%) wass applied. The study showed that there was an effect of detergent addition, periode of exposure, and doses to the growth of bacterial population both *in vitro* and *in situ* conditions. The higher levels of detergent in the water column and the longer contamination duration, causing more and more depressed bacterial populations. It is suggested to run a further research on identification, and growth optimatioan of the species capable of degrading detergent.

Keywords **:** Deteregent, pollution, seawater, biodegradation, and environment.

1. Introduction

Detergents has been part of of our life nowadays, commonly available as powders or concentrated solutions. The chemical figure in an extensive array of industrial and home cleaning applications, including laundry and dishwasher detergents. Released into the flow of wastewater coming from the home, these detergents can have far-reaching environmental impacts.

Detergents work because they are amphiphilic: partly hydrophilic (polar) and partly hydrophobic (non-polar). Their dual nature facilitates the mixture of hydrophobic compounds (like oil and grease) with water. Because air is not hydrophilic, detergents are also foaming agents to varying degrees. The main component of the detergent is surface active agents or surfactants. The most commonly used surfactant contains LAS or linear alkylbenzen sulfonate. LAS is an anionic detergent that is classified as hard detergent. This becomes an important issue because the concentration of surfactants in detergents entering a waters can be toxic, affecting the life of the organism in the ocean waters.

Microorganisms or microbes are very diverse and include all bacteria, archaea and most protozoa. This group also contains some fungi, algae, and some micro-animals such as rotifers. Microorganisms are crucial to nutrient recycling in ecosystems as they act as decomposers. As some microorganisms can fix nitrogen, they are a vital part of the nitrogen cycle, and recent studies indicate that airborne microorganisms may play a role in precipitation and weather. There is growing interest in the role of marine microorganisms in biogeochemical processes, biotechnology, pollution and health. In recent years, many authors have focused on the great potential of marine microbes for use as prolific producers of bioactive substances, and they have exploited the vast marine microbial treasures for utilization as sources of drugs, antimicrobial agents, biodegradation of pollutants etc. This research was carried out to examine effect detergent disposal to bacterial population growth in marine environment both *in vitro* and *in situ* condition.

2. Method

This research was carried out from February to April 2017. Seawater samples were collected from the seawaters of Sungai Kayu Ara Village, Sungai Apit Subdistrict, Siak Regency and Dumai River estuary, Dumai City, Riau Province. The sample analysis was conducted at the Laboratory of Microbiology and Marine Chemistry Laboratory, University of Riau.

Experimental method with complete randomized design (RAL) 2 (two) factors; a detergent brand (a1: ATTACK, a2; RINSO and a3; SURF) and b concentration of detergent concentration with 5 (five) concentration level, b1 (0%) as control, b2 (0.3%), b3 (0.6%), b4 (0.9%) and b5 (1.2%) is applied. Each treatment was given 3 (three) repeated treatments and measured with 3 (three) replication measurements. While the survay method was done to see the correlation between detergent concentration with the population of heterophic bacteria in *in situ* condition in seawater. Water quality in ie. temperature, pH, salinity and dissolved oxygen with a view to directly observing the condition of the waters to support the growth of microorganisms and to avoid extreme conditions at the time of sampling.

In this study the effect of detergent brand with various concentrations was observed *in vitro* in the microcosms. The microcosms used is a 500 ml erlenmeyer flask filled with 250 ml of sea water. Each flask was poured with detergent with 0% concentration (control) 0,3, 0,6, 0,9 and 1,2% for each detergent brand. Each flask was wrapped with aluminum foil to minimize the effects of light on microbial growth. All microcosms are placed in a cold room with a temperature range of 5-10 ^{o}C. Seawater samples were taken aseptically by using a pipette with sampling time at 0, 5, 10, 15 and 20 day intervals.

Population growth of heterophic bacteria in seawater is counted by using a spreader plate method. The seawater (1 ml) sample was taken aseptically, diluted with physiologic solution (NaCl 0.9%) with dilution levels of 10^{-1}, 10^{-2}, 10^{-3}, 10^{-4} and 10^{-5}. Voleme of 0.1 ml from each dilution was then taken and poured into a petri dish that has contained nutrient agar medium. The sample was then immediately spreaded using a glass spreader. The inoculated petri dishes were then incubated at 18-22 °C for 48 hours. The number of bacterial cells is calculated based on the number of colony forming units in which the colony count sample is taken from a petri dish with a range of 30-300 colonies. The correlation between the concentration of detergent dissolved in seawater with the population of heterotrophic bacteria *in situ* conditions in seawater was obtained by taking samples of seawater at the estuary of the Dumai River. The samples were then processed the same as the previous procedures.

The calculation of dissolved surfactant (detergent) content in the waters of the Dumai River estuary is obtained through the following procedure. Seawater sampling is carried out under tidal conditions. Samples are taken as much as 300 ml on surface water with a depth of 30 cm. The sample is poured into the sample bottle and labeled, stored in an ice box, given ice and taken to the Marine Chemical Laboratory for calculation of dissolved surfactant levels. Surfactant concentration is measured according to SNI procedure (1992). The stages includes; making a 1000 ppm mains solution, preparing a standard solution of 10 ppm LAS, providing sash solution calibration curve configuration and surfactant value analysis. The surfactant content was obtained by using a simple linear regression with the equation Y = a + bx, where Y is the detergent concentration, x is the absorbance value and a and b are the constants.

The data obtained were analyzed statistically and presented in the form of tables and drawings, using SPSS and MS applications, Excel, which is used in the advanced test of Anova, (analysis of variance) and regression analysis.

3. Result

In this study, water quality parameters at the Sungai Kayuara Village are as follows; salinity (24 ppt), acidity (pH 5-6), temperature (28-30 ^{0}C) and dissolved oxygen (6.28 mg/ l). While at the the Dumai River estuary, water quality are as follows; temperature (30-32 ^{0}C), salinity (14-28 ppt), brightness (45-88 cm), current velocity (0.6 m / sec), and pH (7). From the data collected from both research sites, it can be seen that the condition of the waters is in normal condition and can support the growth of the local organism so that it can be used as sample for the research interest.

There was an effect of the addition of detergent (RINSO, SURF and ATTACK) with different doses to the growth of bacterial population in the sea water column. The pressure on the growth of heterotrophic bacteria population was observed from the beginning of observation (0 days), second observation (5 days) to the last observation on day 20. The most prominent effect on population growth resulted from the addition of SURF detergent with a concentration of 1.2%. Then followed by RINSO and ATTACK detergents at the same concentration. The higher the concentration of detergent added, the stronger the impact of the suppression of bacterial population growth resulted. The duration of contamination also affected growth pattern of bacterial population in seawater. In general, the longer the contamination occurs, the lower the total number of viable count (TVC) shown for each microscosm. More detailed data are presented in the following Table 1. The effects of adding detergents with different concentrations and length of contamination will be more clearly illustrated in the following Figure 1, 2 and 3.

Table 1.
Effect of concentration and length of contamination on the growth of sea water bacteria (Log TVC)

	Detergent concentration (%)				
Detergent Brand/ Day of Sampling	**0**	**0,3**	**0,6**	**0,9**	**1,2**
RINSO 0 day	6,85	6,83	6,82	6,83	6,82
RINSO 5 days	6,83	6,62	6,51	6,41	6,10
RINSO 10 days	6,81	6,41	6,39	6,11	5,76
RINSO 15 days	6,79	6,29	6,03	5,83	5,31
RINSO 20 days	6,55	6,06	5,74	5,30	5,06
SURF 0 day	7,99	7,44	7,27	7,12	7,26
SURF 5 days	7,77	7,52	6,12	6,11	6,11
SURF 10 days	7,73	7,04	6,04	6,05	6,01
SURF 15 days	6,73	6,78	6,10	5,78	6,00
SURF 20 days	6,51	6,21	5,36	5,27	5,03
ATTACK 0 day	7.72	7.77	7.48	7.81	7.82
ATTACK 5 days	7.54	7.47	7.22	7.15	6.98
ATTACK 10 days	6.44	6.25	6.19	6.17	6.08
ATTACK 15 days	6.34	6.26	6.23	6.11	5.99
ATTACK 20 days	6.36	6.31	6.23	6.18	5.46

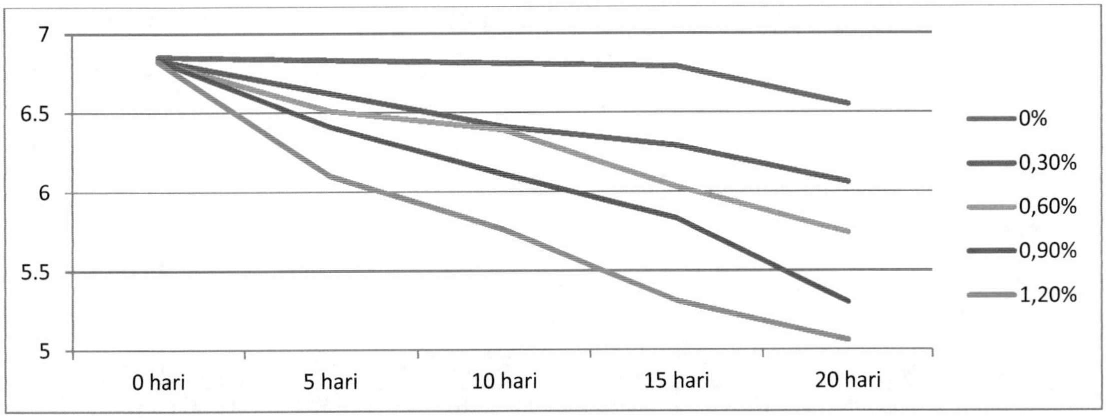

Figure 1. Effect of RINSO detergent on population growth of heterophic bacteria in seawater.

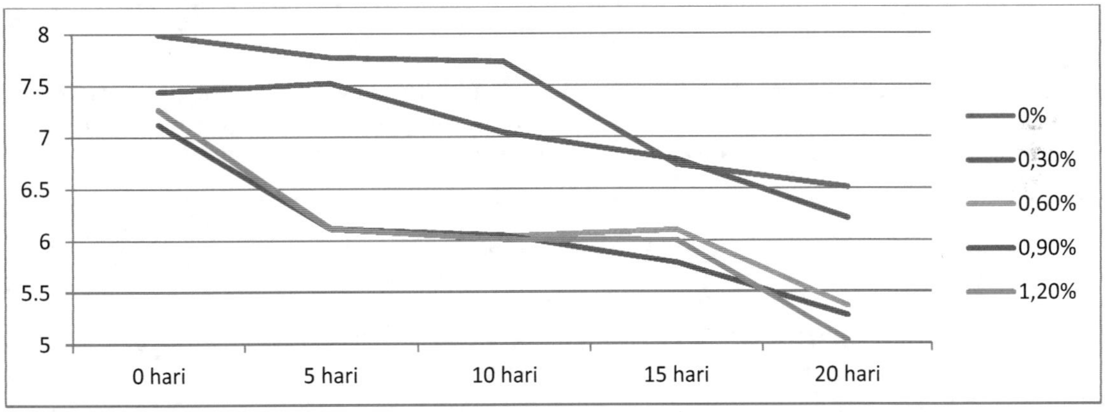

Figure 2. Effect of SURF detergent on population growth of heterophic bacteria in seawater.

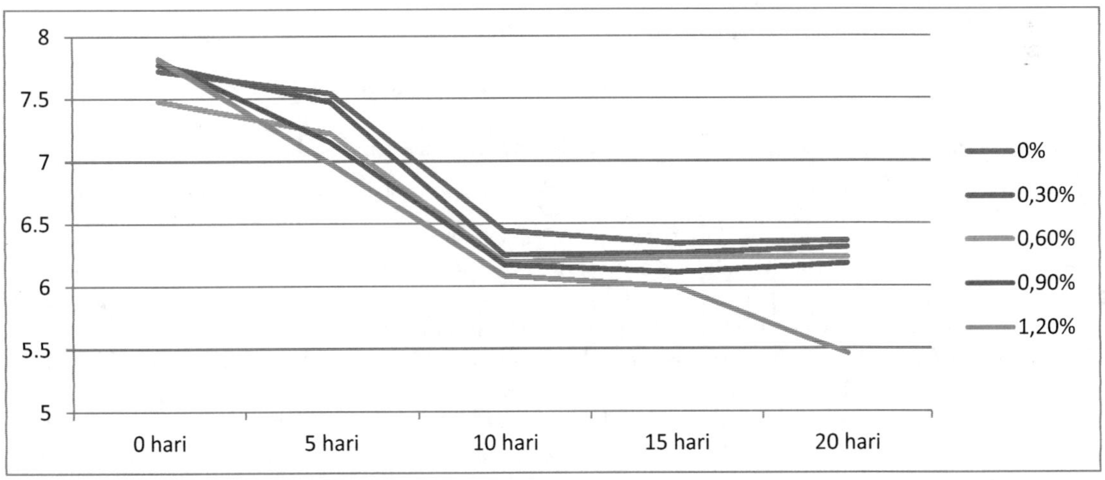

Figure 3. Effect of ATTACK detergent on population growth of heterophic bacteria in seawater.

Table 2.
Number of heterotrophic bacterial cells in seawater at each observation station.

No.	Station 1	Station 2	Station 3	Station 4	Station 5	Station 6
Sample 1	6,26	6,15	6,13	6,67	6,71	6,62
Sample 2	6,08	6,06	6,18	6,76	6,68	6,69
Sample 3	6,21	6,19	6,12	6,74	6,58	6,67
Rata-rata	6,19	6,13	6,15	6,73	6,66	6,66

Table 3.
Concentration of surfactant in seawater at each observation station.

No.	Station 1	Station 2	Station 3	Station 4	Station 5	Station 6
Sample 1	0,1987	0,3091	0,3043	0,7829	0,8599	0,7829
Sample 2	0,2078	0,3353	0,3711	0,8019	0,8723	0,8111
Sample 3	0,2011	0,2989	0,3794	0,8103	0,8291	0,8673
Rata-rata	0,2025	0,3144	0,3516	0,7984	0,8538	0,8204

The correlation analysis showed that the concentration of surfactant has a significant effect on the population of heterotrophic bacteria. These two factors have negatif correlation, the higher surfactant levels in each station cause the decrease bacterial cell number in sea water column (Figure 4).

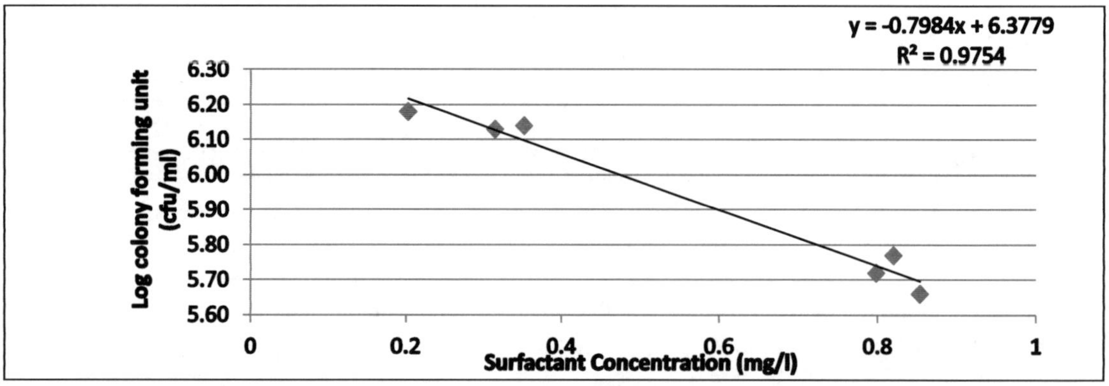

Figure 4. Correlation between concentration of surfactant with bacterial cell population in sea water.

4.Discussion

The results of this study showed that there was an effect of detergent addition (RINSO, SURF and ATTACK) with different brand, periode of exposure, and doses to the growth of bacterial population in the sea water column. Similarly, *in situ* conditions there was a negative correlation between surfactant concentration and bacterial populations in the waters of the Dumai River estuary. Negative effects of detergent or surfactant have been widely reported by researchers around the world. Scott, J. M. and M. Njones (2000) mentioned that detergent or surfactants which pass into the environment and the potential problems which can arise as a consequence. The chemicals can have poisonous effects in all types of aquatic life if they are present in sufficient quantities, and this includes the biodegradable detergents. All detergents destroy the external mucus layers that protect the fish from bacteria and parasites; plus they can cause severe damage to the gills. Most fish died when detergent concentrations approach 15 parts per million. Detergent concentrations as low as 5 ppm killed fish eggs. Surfactant detergents are implicated in decreasing the breeding ability of aquatic organisms.

Detergents are molecules that are amphiphilic in nature with a hydrophobic hydrocarbon tail and a hydrophilic head group. It is the hydrophile on the head group that defines the detergent as anionic, cationic, nonionic or amphoteric, all of which may be used in detergent formulations. Cationic in characteristic and cause damage to the outer membrane and promote their own intracellular uptake and entry (Russell 2002 and Holah *et al,* 2002). There are two kinds of detergents with different characteristics: phosphate detergents and surfactant detergents. Detergents that contain phosphates are highly caustic, and surfactant detergents are very toxic. The differences are that surfactant detergents are used to enhance the wetting, foaming, dispersing and emulsifying properties of detergents. Phosphate detergents are used in detergents to soften hard water and help suspend dirt in water (Siposova *et al,* 2017).

Sudiana (2004) explained that an increase in LAS concentrations that go beyond cell capacity (lysosomes) will cause disruption of cell metabolism. This lead to a decrease in LAS degradation and may even lead to cell death. Furthermore, the side effects of LAS and ABS presence that exceed the natural degradation capacity are polluted environments that can cause damage to ecosystem components (Bitton, 1983), and microbes (Niven et al., 1988). Further, Russell (2001) explained that surfactant as a detergent component can cause environmental problems. The stable foam formation in the river is highly undesirable because it blocks the transformation of the oxygen-mass from air to water. Hydrophilic constituents of toxic surfactants will endanger the survival of aquatic animals and bacteria in water. According to KEP / 51 / MENLH / 2004, the concentration of surfactants in waters is 1 mg / l. In this study the highest concentration of surfactant was located at station 5 (0.8538 mg / l) followed by station 6 (0.8204 mg / l) and 4 (0.7984 mg / l). Based on this, it can be seen that the waters of Dumai River estuary not yet contaminated seriously by surfactant so that it can still support the life of aquatic organisms.

Detergents also add another problem for aquatic life by lowering the surface tension of the water. Organic chemicals such as pesticides and phenols are then much more easily absorbed by the fish. A detergent concentration of only 2 ppm can cause fish to absorb double the amount of chemicals they would normally absorb, although that concentration itself is not high enough to affect fish directly. Phosphates in detergents can lead to freshwater algal blooms that releases toxins and deplete oxygen in waterways. When the algae decompose, they use up the oxygen available for aquatic life (Cohen and Keiser, 2017).

The main contributors to the toxicity of detergents were the sodium silicate solution and the surfactants-with the remainder of the components contributing very little to detergent toxicity (Lisa *et al*, 2017). The potential for acute aquatic toxic effects due to the release of secondary or tertiary sewage effluents containing the breakdown products of laundry detergents may frequently be low. However, untreated or primary treated effluents containing detergents may pose a problem. Bhat *et al* (2011) reported that 1% soap showed higher colony growth of *E. coli, Pseudomonas* and *Staphylococcus,* whereas 3% soap inhibited the growth of all bacteria except *E. coil* and *Micrococcus*. It was seen clearly that Gram positive bacteria were killed at low concentration of soaps than Gram negative bacteria. It is also reported that as the concentration of detergent or soap increases the intensity of inhibition also increases. Similar results were obtained in the present study (Bhat *et al*, 2011). The minimum inhibitory concentrations of detergents for *E. coli* are 3% and 2% respectively and for other bacteria minimum inhibition was observed at 2%.Surfactants are the most important ingredient in detergent products (up to 15-40% of total detergent formulations). This substance can activate the surface, because it tends to concentrate on the surface (interface), or substances that can raise and lower the surface tension. With surfactant there may be changes in surface tension accompanying the wetting process, stable foam power, stable emulsifying power (Scheibel, 2004).

Actually, the surfactants themselves show little toxicity their breakdown products, principally nonyl and octal phenols adsorb readily to suspended solids and are known to exhibit estrogen-like properties, possibly linked to a decreasing male sperm count and carcinogenic effects. While there is little serious risk to the environment from commonly used anionic surfactants, cationic surfactants are known to be much more toxic and at present there is a lack of data on the degradation of cationic and their fate in the environment (Cai and Hakkinen, 2014).

Detergent, however, nowadays has been well designed and in nature can be degraded. In this study we only looked at bacterial colonies in general, regardless of species. It is expected that the bacterial cells that survive inside the microcosms are bacterial species capable of decomposing the detergent. Number of 5-7 types of colony were noted in the research. Some microorganisms have a blessing of energy sources for their growth in the marine environment. Sources of nutrients derived from detergent compounds in a medium that can be utilized by heterotrophic bacteria for growth, where bacterial populations are able to utilize some compounds in the media as a source of carbon and energy for its growth (Chaturvedi and Kumar, 2011). Harijati (1994), reported that within 11 days certain bacteria can reduce the detergent level by 0.10%. It is expecpected the bacteria have used detergent composition in their foodstuff and the release of sulphonate group which then oxidized to sulfate. Heterotrophic bacteria themselves play a role in the waters as decomposers so that the growth of these bacteria in the water will affect the fertility of the waters in the marine environment, this is in accordance with Kurnia *et al*., (2016) stated that heterotrophic bacteria play an important role in aquatic systems due to the ability of metabolism activities.

Goodnow and Harrison (1972) studied for surfactant degradation among aerobic bacteria in peptone medium where such a degradation, if it occurs, will be gratuitous. Tallow-alkyl-sulfate (TAS), alkyl-ethoxylate-sulfate (AES), and linear-alkyl-benzene-sulfonate (LAS) were used. Fortyfive strains of 34 species in 19 genera degrade one or more of these detergent compounds. With some species, the surfactant inhibits degradation without inhibiting growth, whereas with one species slight degradation took place even at a toxic concentration of surfactant. Wide differences in ability to degrade surfactants may occur within a particular genus. For example, all three surfactants (TAS, AES and LAS) were degraded within 24 hr by *Acetobacter suboxydans*, whereas *A. peroxydans* had degraded no TAS in 72 hr and had degraded only 41 and 42%, respectively, of the available AES and LAS. Several species of *Azotobacter* were strong degraders, whereas *A. beijerinckii* was unable to withstand or appreciably degrade the three surfactants (Okpokwasili and Olisa (1991).

Schleheck *et al* (2000) found the bacterial community of the Proteobacteria class to dominate the bacterial community capable of degrading detergents from the marine water ecosystem. The ability of microbes, especially bacteria in using detergents as the main carbon source, indicates that bacteria play an important role in environmental bioremediation process of contaminated detergents waters (Kertesz et al., 1994). Several studies have shown that *Pseudomonas, Clostridium, Corynebacterium, Alcaligenes, Achromobacter, Bacillus* are able to degrade LAS. Amund *et al* (1997) isolated and identified SDS-degrading bacteria from different detergent contaminated ponds situated in Varanasi city, UP, India. Some selected isolates were identified on the basis of 16S rDNA sequencing. It was found that these isolates belonged to *Pseudomonas aeruginosa*, *Pseudomonas mendocina*, *Pseudomonas stutzeri*, *Pseudomonas alcaligenes, Pseudomonas pseudoalcaligenes, Pseudomonas putida* and *Pseudomonas otitidis* respectively (Chaturvedi and A. Kumar, 2011). Among these isolates *P. aeruginosa, P. putida and P. otitidis* have been previously shown to degrade and metabolize SDS.

The detergent-utilizing bacteria identified were mainly Gram-negative and of the following genera: *Vibrio, Klebsiella, Flavobacterium, Pseudomonas, Escherichia, Enterobacter, Proteus, Shigella and Citrobacter. Vibrio* was the most frequently encountered organism while *Proteus* was the rarest. The cyclohexanol degrading bacteria were identified as species of *Pseudomonas. Acinetobacter, Vibrio, Micrococcus and Flavobacterium. Pseudomonas* sp. had the highest growth potential on cyclohexanol. These organisms play important roles in reducing the pollutant loads (London *et al*, 2016).

5.Conclusion

The results of this study indicated that detergents, namely RINSO, SURF and ATTACK have an impact on the growth of bacterial population in the sea water. The higher levels of detergent in the water column and the longer contamination duration, causing more and more depressed bacterial populations, both *in vitro* and *in situ* conditions. This study only studied the number of bacterial cells globally. It is suggested further research on the species of microbes involved and their ability to degrade detergent or surfactant.

6.References

[1] Sudiana, I.M. 2004. *Peraan Komunits Mikroba Lumpur Aktif dalam Perombakan Detergen Alkyl Sulfonat Linear dan Benzena Alkyl Sulfonat*. Berkala Penelitian Hayati. LIPI

[2] Amund O.O.1, M.O Ilori, F.R Odetundun. Degradation of commercial detergent products by microbial populations of the Lagos lagoon. Folia Microbiol (Praha). 1997;42(4):353-6.

[3] APHA. American Public Health Association. 2005. *Standard Methods For the Examination of Water and Wastewater. Amer.* Publ. 17th Edition. New York Health Association.

[4] Kertesz MA, P Kolbener, H Stockinger, S Beil, and AM Cook, 1994. Desulfonation of linear alkylbenzenesulfonate surfactants and related compounds by bacteria. Appl. Environ. Microbiol. 60: 2296–303.

[5] Chaturvedi , V., and A. Kumar. 2011. *Diversity of culturable sodium dodecyl sulfate (SDS) degrading bacteria isolated from detergent contaminated ponds situated in Varanasi city,* India. International Biodeterioration & Biodegradation. Volume **65**, Issue 7, October 2011, Pages 961–971.

[6] Cohen, A.,and D. A. Keiser. 2017. *The effectiveness of incomplete and overlapping pollution regulation: Evidence from bans on phosphate in automatic dishwasher* detergent. Journal of Public Economics, Volume **150**, June 2017, Pages 53-74.

[7] Garno, Y.S. 2000. *Daya Tahan Beberapa Organisme Air Pada Pencemar Limbah Deterjen. Jurnal Teknologi Lingkungan.* **1(3)** : 212 – 218.

[8] Goodnow, R.A. and A.P. Harrison, Jr. 1972. *Bacterial Degradation of Detergent Compounds, Appl. Microbiol***24(4):** 555-560.

[9] Harijati, N., Suharjono dan T.H. Kurniati.1994. *Potensi Komunitas Bakteri Pemecah Deterjen Jenis Alkil Bensen Sulfonat (ABS) dan Linier Alkil Benzen Sulfonat (LAS).* Jurnal Universitas Brawijaya 6(2). Lembaga Penelitian Universitas Brawijaya. Malang.

[10] Holah, J.T., Taylor, J.H., Dawson, D.J. *and* Hall, K.E. *(*2002*) Biocide use in the food industry and the disinfectant resistance of persistent strains of* Listeria monocytogenes *and* Escherichia coli. J Appl Microbiol Symp Supp 92*,* 111S–120S.

[11] Hugo, W.B. *and* Russell, A.D. *(*2004*) Pharmaceutical Microbiology,* 7th edn. *Oxford: Blackwell Science ublications*

[12] Lisa J. White, Adam J. Taylor, Denver M. Faulk, Timothy J. Keane, Lindsey T. Saldin, Janet E. Reing, Ilea T. Swinehart, Neill J. Turner, Buddy D. Ratner, Stephen F . Badylak. 2017. *The impact of detergents on the tissue decellularization process*: A ToF-SIMS study. Acta Biomaterialia. Volume **50**, 1 March 2017, Pages 207–219.

[13] Okpokwasili, G.C. and A.O. Olisa. (1991). *River-water biodegradation of surfactants in liquid detergents and shampoos.*

[14] Bath, R. P., P.S. Prajna, V. P. Menezez and P. Shetty. 2011. *Antimicrobial Activities of Soap and Detergents.* ADVANCES IN BIORESEARCH. Volume **2**, Issue 2, December 2011: 52- 62

[15] Robert A. Goodnow and Arthur P. Harrison, Jr. 1972. *Bacterial Degradation of Detergent Compounds.* Appl Microbiol. 1972 Oct; 24(4): 555–560.

[16] Russell, A.D. *(*2001*). Mechanisms of bacterial insusceptibility to biocides.* Am J Infect Control 29*,* 259–261.

[17] Russell, A.D. *(*2002*). Antibiotic and Biocide resistance in bacteria: comments and conclusions.* J Appl Microbiol Symp Supp 92*,* 171S–173S.

IOP Conf. Series: Earth and Environmental Science **97** (2017) 012030 doi:10.1088/1755-1315/97/1/012030

[18] Scott, J. M. and M. Njones. 2000. *The biodegradation of surfactants in the environment.* Biochimica et Biophysica Acta (BBA) – Biomembranes. Volume 1508, Issues 1–2, 23 November 2000, Pages 235-251.

[19] Scheibel, J., 2004. *The Evolution of Anionic Surfactan Tehnology to Meet the Requirement of the Laundry Deterjent Industry.* Journal of Surfactan and Detergent. Vo**7**. No. 5.

[20] Siposova, K. , T. Kozar, A. Musatov. 2017. *Interaction of nonionic detergents with the specific sites of lysozyme amyloidogenic region − inhibition of amyloid fibrillization.* Colloids and Surfaces B: Biointerfaces. Volume **150**, 1 February 2017, Pages 445–455.

[21] Cai, Z., and P.J. Hakkinen. Detergent. 2014. *Reference Module in Biomedical Sciences.* Encyclopedia of Toxicology (Third Edition). 2014, Pages 10–13.

[22] London, A. S., C. Japutra, K., Planck, M. Lihon, and A. N. Nguyen. 2016. *A novel method to determine residual detergent in biological samples post endotoxin reduction treatment and evaluation of strategies for subsequent detergent removal. International Immunopharmacology*, Volume **37**, August 2016, Pages 16-22.

Effect of Binder on Combustion Quality on EFB Bio-briquettes

Nofriady Handra [1] and Hafni [2]
[1] Institut Teknologi Padang, Padang 25143 West Sumatera, Indonesia
Telp: +62 0751-7055202
E-mail: nof.hand11@gmail.com
[2] Institut Teknologi Padang, 25143 West Sumatera, Indonesia
Telp: +62 0751-7055202
E-mail: hafnimesin@gmail.com

Abstract: Energy demand in various sectors in Indonesia has increased in line with the rate of population growth and the national economy. Fulfillment of energy needs can be obtained from various energy sources such as fuel oil, solar, biomass, wind, water and others. So far, energy sources used in Indonesia are still using many non-renewable energy sources, such as fuel oil. The utilization of waste from empty palm oil bunches into bio-briquettes has helped the government in overcoming the problem of EFB waste. The availability of biomass has prompted researchers to utilize biomass waste that includes Agricultural and Forestry waste, to be processed into briquettes as an alternative energy substitute for fuel oil. This research aims to improve the utilization of waste of Palm Oil Bunches through the manufacture of bio-briquette as alternative fuel and determine the appropriate binder material for briquette making so as to produce optimal combustion value. The binders used for the manufacture of briquettes are pine sap and starch flour. The test result showed that the highest value of calorific was found in the mixture of 50% EFB composition with fibre size ± 1-5 mm with 50% pine resin which is 6331,7 cal/g. Meanwhile, lowest value on EFB ± with fibre size 5-10 mm composition EFB 60% and 40% starch flour binder that is 2295,7 cal/g. The results of a flame test study of several points that are known to turn on until it emits a flame for ± 30 seconds, it takes 22,2 minutes for the burnt-out briquette (to ashes). Based on visual observations that the fire colour of bio-briquette with finer fibre on the EFB composition 50% pine gum binder produces a bluish red fire colour. It is generally assumed that pine resin glues produce better fuel value compared to starch binder. Besides that, fibre particles size also affects the combustion quality produced.

Keywords: EFB, bio-briquette, calorific value, binder.

1. Introduction

Increasing environmental and energy dependency concerns have been the motivation for the increased use of bioenergy as a substitute for fossil energy in both space heating and electricity generation. The development of biomass has been an important issue for the past several decades and would remain to be attractive in the future due to its clean, renewable, and carbon–neutral properties [1]. Energy and fuels are the important links in the civilization and human development. The issues associated with the use of the fossil fuel, demand and supply gap, ever increasing prices, global warming and other environmental issues made the world to think for alternate sources of energy like solar, wind, ocean and biomass which are the only indigenous renewable energy sources capable of replacing large amount of solid, liquid and gaseous fossil fuel [2].

Reserves and production Petroleum fuels in Indonesia decrease 10% annually while the average oil consumption rate is up 6% per year [3]. Energy demand in various sectors in Indonesia has increased in line with the rate of population growth and the national economy. Fulfillment of energy needs can be obtained from various energy sources such as fuel oil, sun, biomass, wind, water, and others. So far, energy sources used in Indonesia are still using many non-renewable energy sources, such as fuel oil. This could trigger high subsidies to be issued by the government if world oil prices are experiencing a surge in prices as it is today that is almost reached US $ 100/barrel. According to data from British Petroleum Indonesia will experience a petroleum crisis in 2024 if no new oil reserves are found in large quantities. From the data is needed alternative energy that is renewable to reduce community dependence on petroleum. The solid waste biomass of the palm oil industry is one of the renewable alternative energy that has not been widely utilized in Indonesia. Industrial solid waste Oil palm can be used as alternative fuel in the form of briquettes [4], [7].

One of the steps that can be taken to overcome the problem is the utilization of alternative energy sources, especially renewable energy sources. The transfer of energy sources from oil to renewable energy sources is expected to reduce the level of dependence on petroleum, especially considering its abundant potential in Indonesia. One source of renewable energy that can be utilized is biomass. Biomass potential in 2004 was equivalent to 49.81 GW, but installed capacity was only 302.4 MW (www.esdm.go.id). Biomass is the result of photosynthesis of plants and their derivatives. The selection of biomass as fuel has more value because it is green energy. Burning biomass is considered not to disturb the environment because biomass can be replanted (renewable) and the combustion CO will be reabsorbed by the plant (zero emission). Biomass waste can be obtained from by products from a plantation industry, agricultural products, or industries that utilize raw materials derived from forests.

In Indonesia, coal is the main fuel in addition to diesel fuel which has been commonly used in many industries, in terms of economical coal is much more efficient than diesel oil, with the following comparison: diesel oil Rp 0.74/kilo calorie while coal only Rp 0.09/kilo calorie, (based on industrial diesel price of Rp 6.20/liter). In terms of quantity of coal including the most important fossil energy reserves for Indonesia. The amount is very abundant, reaching tens of billions of tons. This amount is actually enough to supply electrical energy needs for hundreds of years into the future. Unfortunately, Indonesia is unlikely to burn the coal out and turn it into electrical energy through the PLTU.

In addition to polluting the environment through CO2, SO2, NOx and CxHy pollutants this way is considered less efficient and less high added value, which is a special spotlight especially with the issue of global warming [5]. Oil palm is a plantation/industrial plant in the form of a straight tree from Palmae family. The main product of oil palm trees utilized is the fruit bunches that produce oil from the flesh and kernel (palm kernel). After doing the processing of oil palm, in the end leaving empty bunches of palm oil that is generally not processed by palm oil processing factory. The remaining empty bunches cause problems for the place and its disposal transport resulting in additional production costs for the processor. In the place of disposal is usually EFB burned, this also cause problems of environmental damage that is air pollution and odor.Biomass waste can be directly used as fuel, converted first into charcoal or hammered first into briquettes. The purpose of forging is to obtain better combustion quality and ease of use and handling. Optimization of adhesive and biomass waste on briquette production is an important factor in the production of briquettes The addition of adhesive on briquette production will increase the calorific value of the briquette due to the addition of carbon element. Waste of EFB can be seen in Figure 1.

Figure 1. Waste of empty fruit bunches after boiling process

Based on several problems in background discussions and see the potential of environmentalists, EFB waste must be managed wisely in order to demand the use of renewable alternative energy biomass, biomass utilization technology and produce environmentally friendly heat energy. The development of agricultural wastes for solid fuels is also triggered by the current condition, where fossil fuels which have been a mainstay for households or industries have increased prices. For household fuels such as kerosene has risen in price. Given the fact, the waste is very potential to be a solid fuel that is relatively cheaper and is expected to minimize environmental problems during the resulting waste handling. One alternative solution is to treat the agricultural waste into briquettes, a solid alternative fuel, cheap and meet the technical specifications, namely by the pressure on the mold.

2. Method

2.1 Raw Material

The results of the study will produce bio-briquette from palm oil waste. Palm wastes are obtained from the oil palm industry located in Taluak Kuantan - Riau Province. Waste obtained is after going through the process of enumeration with the engine so that the palm fibre are obtained that are rougher and still fresh.

Figure 2. (a). Empty Fruit Bunch (b). EFB fiber

2.2 Tools and material

Mold briquettes is one of the supporters of experiment. The mold of briquettes in use is formed by determining the mold classification. The process of this mold is one part pressed to get the density. Figure 3 is a cylindrical sample dimension with dimensions of 4 cm in length, Ø 5.5 cm and Ø 1.5 cm hole. The fiber are fed into the mold then pressurized to a pressure of 40 kg/cm^2. The volume of briquettes is determined by the equation:

$$\text{Volume} = \pi/4\ D^2 .\ t \qquad (1)$$

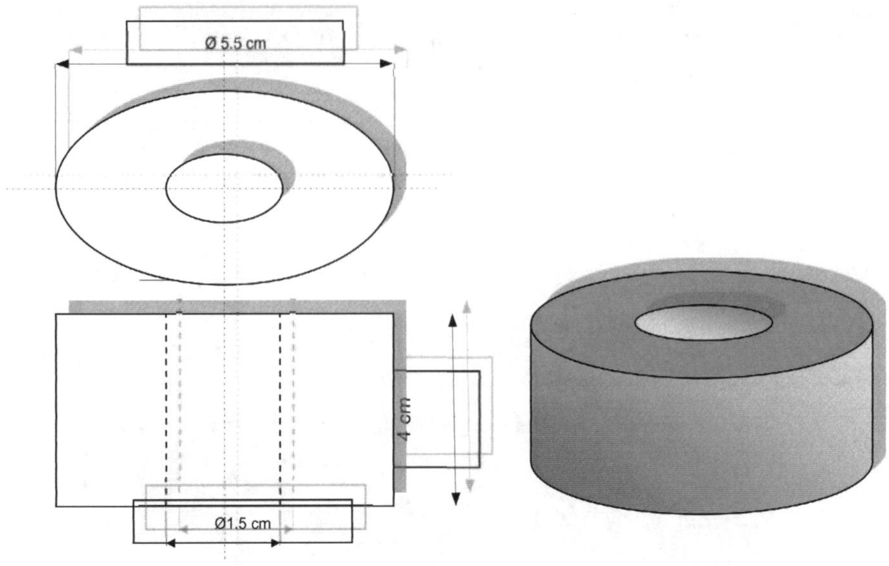

Figure 3. Dimensions of briquette

Figure 4. Molds of briquettes and samples

Density bio-briquettes in part on briquette technology processes produce bio-briquettes with densities above 1000 kg^{-3}, where briquettes will be immersed in water as quality testing. The highest physical density for ligno-cellulosic materials is between 1500 kg^{-3}. The highest pressure processes (such as mechanical pistons and compression of pellets or screw extruders make briquettes become compact from densities between 1200-1400 kgm^{-3}. Briquettes punched hydraulic pistons which make briquettes less dense, sometimes below 1000 kg^{-3}. Making briquettes to be solid is ineffective, as it will likely affect the resulting combustion properties [3]. In the above experiment, the density is determined by the equation :

$$D/\rho = B/V \quad (2)$$

Where: D/ρ = density (g/cm^3), B = Initial briquette mass (g) and v = volume (cm^2).

3. Results and Discussion

The availability of biomass has prompted researchers to utilize biomass waste that includes agricultural and forestry waste, to be processed into briquettes as an alternative energy substitute for fuel that is now increasingly scarce and its availability is dwindling. Briquettes made from oil palm empty bunches need to be utilized to help the government overcome the waste problem that has caused environmental pollution [8]. In the manufacture of briquettes, it should be known that the enumeration process must use the enumerator machine to make the resulting product more smooth and evenly, and there needs to be mixed variation between the raw material and the adhesive which aim to know the optimum and minimum calorific value. The mixed variations used consisted of 12 mixtures with EFB ± 1-5 mm and EFB ± 5-10 mm, ie :

1. (EFB ± 1-5 mm pine resin binder 30%, 40% and 50%)
2. (EFB ± 1-5 mm binder starch 30%, 40% and 50%)
3. (EFB ± 5-10 mm pine resin binder 30%, 40% and 50%)
4. (EFB ± 5-10 mm binder starch 30%, 40% and 50%)

From this variation can be known the value of effective caloric briquettes to be used as a reference in the manufacture of briquettes empty palm bunches using pine sap and starch. From the above mix it can find the optimum and minimum calorific value by conducting calorific value test. Testing the calorific value using oxygen bomb calorimeter tool, with this tool is known that the optimum calorific value is owned by the composition of 50% EFB 1-5 mm binder pine resin 50% that is equal to 6331,75 cal/g, and minimum calorific value is owned by EFB 5-10 mm 60% 40% kinder binder ie 2295,70 cal/g.

In addition to the heat value test, flame testing is also conducted to determine the performance/flame resistance. From the test results of flame to turn on until the flame arises in need time 30 seconds. For ash combustion briquette ash from beginning of EFB fine pine adhesive 50% by weight of briquettes before burning 81.01 g when finished burning to 7.34 g. Based on the graph of the picture 5 starch binder calorific value which has the highest calorific value is the fine fibers of pine resin glues with the binder composition of 40% (3137.83 cal/g.). Compared with crude fiber with pine resin binder of 50% (2891,54 cal/g.) binder composition, this is because the fine fibre at combustion inside the Unit Bomb Oxigen Calorimeter are perfect in comparison with crude fiber at the time of combustion in Unit Bomb Oxygen Calorimeter. Based on the graph, the highest heating value is owned by EFB ± 1-5 mm with the composition of 50% EFB 50% pine binder ie 6331,75 cal/g. Lowest value is owned by EFB ± 5-10 mm with the composition of EFB 60% binder starch flour 40% that is equal to 2295,70 cal/g.

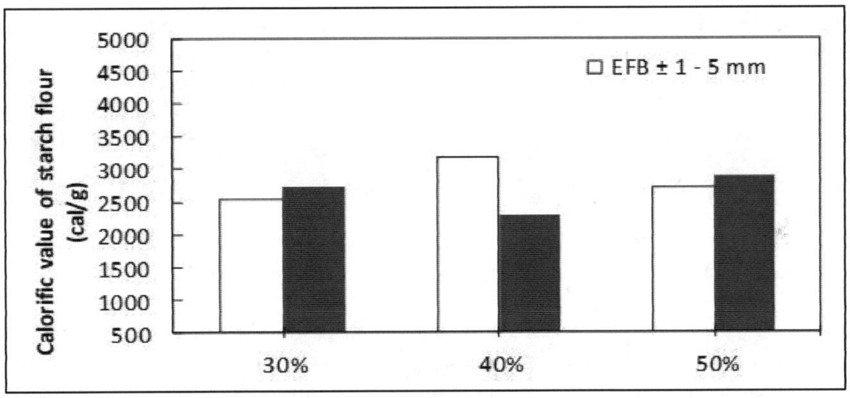

Figure 5. Graph calorific value of starch flour vs percentage of binders

Figure 6 is a graph of the relationship between the heating value of the pine resin glues and the percentage of each binder. Based on the graph of figure 6, the heating value of pine sap gum which has the highest heating value is a fine fibre pineapple binder with a 50% (6331.75 cal/g.) adhesive composition compared to crude fiber with pine resin binder 30% (5607.31 cal/g.) This is because the fine fibre at combustion in the unit of the oxygen calorimeter bomb are perfect in comparison with the crude fibre at the time of combustion inside the Bomb Oxigen Calorimeter unit. Based on the graph of fire ignition with pine resin glue shows longer burning value on fine fibres with a 50% (22.23 min) adhesive composition, compared with coarse fibre with a 50% (21.14 min) binder composition. The lower combustion value of briquettes is found in coarse fibre with a composition of 30% (16.32 min). This is due to the fibre factor used at the time of combustion.

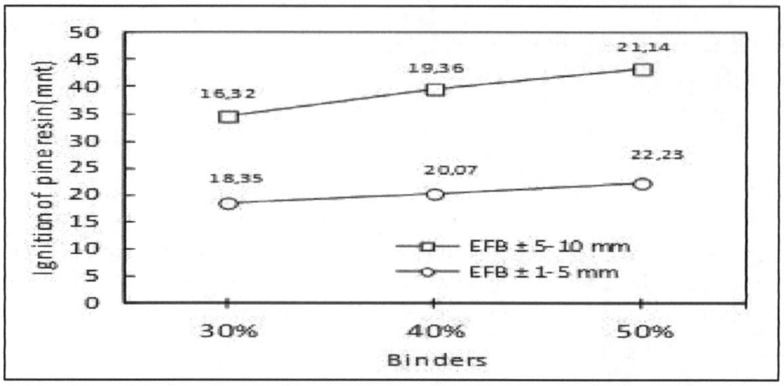

Figure 6. Graph of the relationship between the flame of pine resin glue and the percentage of each binders

Figure 7 shows a graph of flame binder starch with the percentage of each binder. Based on the graph of flame ignition on starch flour adhesive which has the longest burning value on coarse fiber with 50% (9.55 min) binder composition, while which has the lowest burning in fine fibers with 30% (1.27) binder composition, Fine fibres with 40% binder composition and 50% briquette combustion process have little difficulty in ignition.

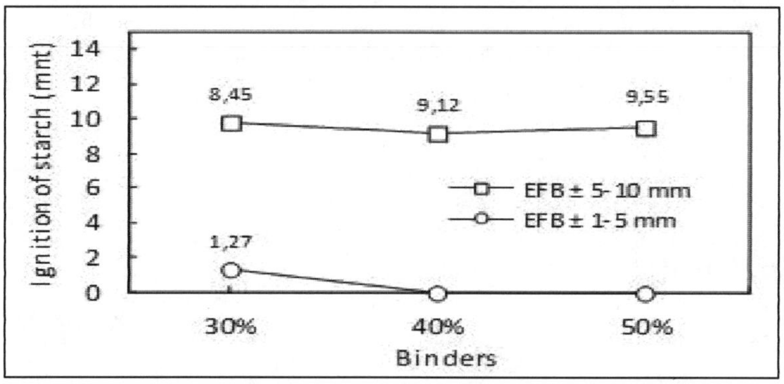

Figure 7. Graph of the relationship between flame starch glue binderwith the percentage of each binder

In general, briquettes change or increase in length when removed from the mold. This is due to the fibres having elasticity when removed from the mold, this greatly affects the binder and particle size of EFB fibers in each material [6]. The length of each briquette produced did not show any significant difference between one another.

Figure 8. Ignition of briquettes, (a). Pine sap binder with fine fibre 50%, (b). Flour binder starch fine fibre 50% (incomplete combustion)

4. Conclusion

Testing the heating value obtained the highest result is in mixed composition 50% EFB ± 1-5 mm binder pine resin 50% that is equal to 6331,75 cal/g. The lowest value is generated by EFB ± 5-10 mm with the composition of EFB 60% binder starch flour 40% that is equal to 2295,70 cal/g. The advantages of EFB use the perfect pine sap gluing, high heating value, while the shortcomings at the time of burning produce smoke. The advantages of EFB using starch binder that is not smoke, while the shortage of ignition initially difficult. The highest test result of carbon values was found in mixed EFB ± 1 - 5 mm 40% pine sap granules (40,27), the lowest carbon value was produced by EFB ± 1-5 mm 30% starch binder composition (16,42).

5. References

[1] Qiang Hu, Jingai Shao, Haiping Yang, *et al.*, 2015. *Effects of Binders on the Properties of Bio-Char Pellets, Applied Energy*, 157 pp; 508-516.

[2] N.S.L. Srivastava, S.L. Narnaware, J.P. Makwana, *et al*, 2014. *Investigating the Energy use of Vegetable Market Waste by Briquetting*,'' Renewable Energy, 68. pp; 270-275.

[3] Bambang. 2006. *Produksi bahan bakar minyak bumi di Indonesia.*

[4] A. Kurniawan. 2015. *Karakteristik pembakaran briket limbah industri kelapa sawit dengan variasi perekat dan temperatur dinding tungku 300°C, 400°C, dan 500°C menggunakan metode Heat Flux Constant* (HFC). Jurnal Penelitian. G.O. Young, "Synthetic structure," in Plastics, 2nd ed., vol. 3, J. Peters, Ed. New York: McGraw-Hill, 1964, pp. 15–64.

[5] E. Prasetyo Budiana *et al*, 2014. *Rekayasa heated die screw extruder untuk pembuatan binderless bio-briquette*. Simposium Nasional RAPI XIII. ISSN 1412-9612.E. P. Wigner, "Theory of traveling-wave optical laser," Phys. Rev., Vol. 134, pp. A635–A646.

[6] A. B. Nasrin, A. N. Ma, Y. M. Choo, M. Basri, W., L. Joseph, Michael, S., Lim, W.S., Mohamad, S., Rohaya, M. H. and Astimar, A. 2010. *Briquetting of empty fruit bunch fibre and palm shells using piston press technology*. MPOB Information Series. Malaysia: Malaysia Palm Oil Board.

[7] Nofriady. H and Hafni. 2017. *Effect of Particles Size on EFB Bio-briquettes of Calorific Value. JTM-ITP.* Vol. 7 No. 1 pp : 56-62.

[8] I. Muzi, Surahma Asti Mulasari, 2014. *The difference in concentration between Briquettes Adhesives Bioarang Empty Fruit Bunch Oil with Coconut Shell Briquette Bioarang against Time Boiling Water*, KESMAS, Vol.8, No.1.

[9] Khor, K.O. Lim and Z.A. Zainal. 2009. Characterization of Bio-oil : *A By product from slow pyrolysis of oil palm empty fruit bunches*. American Journal of Applied Science 6 (9):1647-1652.

[10] Bemgba Bevan Nyakuma, Anwar Johari and Arshad Ahmad. 2012. *Analysis of the pyrolytic fuel properties of empty fruit bunch briquettes*. Journal of Applied Sciences 12 (24): 2527-2533.

[11] M. Thabuot, T. Pagketanang and K. Panyacharoen, 2015. *Effect of applied pressure and binder proportion on the fuel properties of holey bio-briquettes*, Energy Procedia, 79, pp: 890-895.

[12] D. Andrijko, J. Grochowicz. 2007. *Effect of the moisture content on compression energy and strength characteristic of lupine briquettes*. Journal of Food Engineering, Vol. 83, Issue 1, Nov. pp: 116-120.

[13] Manoj Kumar. S, Gohil. P, and Nikita Sharma, M. Shell. 2015. *Biomass Briquette Production; A propagation of Non-Convention Technology and Future of Pollution Free Thermal Energy Sources*, AJER Vol. 04 Issue 02, pp; 44-50.

[14] C. Antwi-Boasiako, and B.B. Acheampong. 2016. *Strength Properties and Calori*fic Values of Sawdust-*briquettes as wood-residue Energy Generation Source from Tropical hardwoods of Different Dens*ities, Biomass and Bioenergy, 85 pp;144-152.

[15] Jitthep Prasityousil and Akarawit Muenjina, 2013. *Properties of solid fuel briquettes produced from rejected material of municipal waste composting*. Procedia Environmental Sciences, 17, pp: 603–610.

[16] Veronica Moset, Cristiane and Henrik Bjarne Møller, 2015. *Optimization of methane yield by using straw briquettes-in*fluence of additives and mold size. Industrial Crops and Products, 74. pp : 925-923.

[17] Teodora. D, V. Ros, F. Mariasiu, M. Deac., 2011. *The Influence of the Sawdust Briquettes Geometric Parameters on the Conversion Efficiency*'', Journal of Agricultural Science, 43 (1).

E-library Implementation in Library University of Riau

Yuhelmi[1]and Rismayeti[2]
[1]Universitas Lancang Kuning, Pekanbaru, 28265, Indonesia
Tel/Fax: +62761 72706
E-mail: yuhelmi@unilak.ac.id
[2]Universitas Lancang Kuning, Pekanbaru, 28265, Indonesia
Tel/Fax: +62761 53536
E-mail: rismayeti@unilak.ac.id

Abstract This research aims to see how the e-book implementation in Library University of Riau and the obstacle in its implementation. In the Globalization era, digital libraries should be developed or else it will decrease the readers' interest, with the recent advanced technology, digital libraries are one of the learning tools that can be used to finding an information through the internet access, hence digital libraries or commonly known as E-Library is really helping the students and academic community in finding information. The methods that used in this research is Observation, Interview, and Literature Study. The respondents in this research are the staff who involved in the process of digitization in Library University of Riau. The result of this research shows that implementation of e-library in Library University of Riau is already filled the user needs for now, although there is obstacle faced just like technical problems for example the internet connection speed and the technical problem to convert the format from Microsoft Word .doc to Adobe .pdf

Keyword: Digital libraries, E-library, Library University of Riau.

1.Introduction

1.1Background Issues

History, so that library has an important role. The recent technological development caused every institution ying to create the digital libraries (e-library). Digital libraries or E-Library using the recent information technology that connected to the internet, so that for those who want to read are able to do so through internet connected devices such as computer. The existence of E-library should be utilized because it could save and make the readers time more efficient because the reader itself don't need to physically be in the library, the reader could find article, journal, scientific works, and books through the internet connected devices.

1.2Research Purposes:

- Knowing the Process in the Digital libraries

- Knowing the Obstacle in Implementing the E-library.

1.3Benefits or Research:

- Provide convenience for individuals who need information
- Describe the Advantages of using E-Library

1.4Formulation of the Problems

- How to implement E-Library in the Library University of Riau
- The obstacle in its implementation

2. Literature Review

2.1 Definition of Library

According to the Library regulation in Indonesia (the law of the Republic of Indonesia number 43 year 2007) is explained that Library is institution who manage the collection of Papers, Printed works and/or Recorded Works professionally with Standard System to meet educational and research needs, preservation, information, and recreation site for the readers. According to Sulistyo-Basuki (Basuki, 1991) Library is a room or a building that can be used to store the books or other publications that arranged in particular arrangement to use by the readers, and not for sale. So, the conclusion is Library the place for storing the books and papers professionally to help the readers to get the information they need and not for sale.

2.2 Definition of Library Digital

According to Digital Libraries Federation, digital libraries has most of the collection in digital format and can be accessed through computer. The conclusion, digital libraries is the process of replacing printed libraries into digital libraries by converting written works into digital format or files that saved in the one of organization's database. Changes of Paradigm from Traditional Library to Digital libraries. When the development of library services (traditional) to the digital libraries (e-library), there will be a paradigm (Ishak, Dec. 2008).

Basically, every article that typed with the computer is the article that can be use to pusdokinfo. But the file that still in editable format such as Microsoft word's doc, or pagemaker pmg could be changed by certain parties, as a result the authenticity of the file itself is might be not valid. Therefore, the file need to be converted to secure format so that it is impossible to certain parties to change the content of the file. The most standard format for digital libraries implementation until today is still Adobe Acrobat PDF format (Februariyanti & Zuliarso, Juli 2012).

Sucahyo dan Ruldeviyani (Sucahyo & Ruldeviyani, 2007) said that there are three main elements that are needed in the development of information systems, Hardware, Software and Brainware (Human), The explanation for the hardware are:

- Central Web Server that provide web page request from the internet users.
- Database server, is the central of digital libraries, it means database server is the place where the collections of the digital libraries stored.
- FTP Server, provide a process where the users in Digital libraries are sending and receiving the file through the internet network
- Mail server, where the digital libraries providing every service related to e-mail

- Printer Server, where the printer is providing every service related to printing and managing the printing process
- Proxy server, is used to secure the digital libraries network
  With the existence of digital libraries, users can be conveniently finding the needed information. There are many convenient in digital libraries:
- Real time access, with the existence of digital libraries the user can access the library without physically going to the library, anywhere, anytime.
- Cost Effective, books digitalization or documents is much cheaper than buying new book or d ocuments.
- Security, in digital libraries its easier to preventing duplication or plagiarism.
  For Example: The PDF document format cannot be edited, the users only able to read it without editing the content
- Works Publication Globally, with the internet, every paper could be published globally with the efficient cost and time

2.3 Research methods and data collection

•Observation

According to Nasution (Sugiono : 2008 : 64) stated that observation is the basic of all knowledge, the researches is only using the valid data from observation.

•Interview

Interview is a meeting between two people to exchange information and idea through questions and answer process, so that it able to constructing the meaning of particular topic (Sugiyono, 2008 : 231)

•Literature Study

This research using many supporting information according to library study methods that collected from many source, such as collecting data and information from related books, online journal that related to E-Library topic.

According to collected information from many sources, it able to create discussion about E-Library implementation in Library University of Riau. E-Library discussion's conclusion will be made for the further development

3. Discussion

The library of the University of Riau was established in 1962 together with the establishment of the University of Riau. The University itself still owned by *Yayasan Universitas Riau* and only occupy 100 square meters. In 1980 the Library of University of Riau only has a 500 square meter building, located in the old campus location in *Jalan Patimura Nomor 9 Kota* Pekanbaru. Since 1992 the Library of University of Riau was moved to a new campus at *Kampus Bina Widya Panam Jalan HR Soebrantas KM 12.5 Simpang Baru, Pekanbaru* in line with the transfer of Rectorate office and some faculties. The Library of UR (University of Riau) has a building of four units each consisting of two floors with a total area of 4,000 square meters with a capacity of 250 seats, built on an area of 2 Ha.

The library currently serves more than 27,500 students, 1,400 lecturers and 654 employees. The Library of The University of Riau stores over 132,000 titles or 365,000 copies of the library collection, with an annual increase of about 8,500 copies. In addition the Library of University of Riau has an electronic collection consisting of journals and materials collection of local content. The number of e-journals subscribed to more than 5,000 titles in various disciplines is required primarily by the UR academic community. Both types of electronic journals collection can be accessed through www network: lib.unri.ac.id and www: ur-lib.com.

The Library of UR has been using the library automation system for two years using the *SliMS Cendana program* and its services have been developed complement of the SliMS program development, self-service application that has been applied for lending transactions, extension and return of books in an effort to improve the service. Application of this Self Service application since 30 August 2013.

The Library of UR gets the best in library accreditation in 2013, which is "A". The result of the Library UR accreditation assessment by National Library Assessors Team from 17 September to 18 December 2013 contained in Accreditation Certificate of National Library of Indonesia Number 19/1/ee/XII.2013

3.1 Implementation of E-Library in Library University of Riau

The form of digital libraries is closely related to the service availability. The activities in the manager's computer are Database creation, data entry, and electronic catalog and others activities to support the library administration. Whereas the activities in the user's computer is to provide the search services, and the activities to move the result of search activities. To Support the digital libraries there are many alternative such as :

- Files stored in the hard disk
- Every digital file can be stored in the hard disk, the consequences is the hard drive should have larger storage
- Every file in hard disk is linked to the database
- Standalone computer, or the computer who doesn't connected to the network, the users could just read the full article context from the database with only clicking to the full text icon
- File stored in the CD-ROM
- Every digital file can be stored in the CD- ROM,
- Every digital file can be stored in the CD-ROM, the consequences is there must be a field in the database or particular information stated that the particular article is stored in the CD-ROM with the particular number
- Standalone computer, or the computer who doesn't connected to the network, the users could just read the full article context from the database with only clicking to the full text icon
- Digital File Stored in the Database
- Every digital file stored in the sever hard disk.
- Every saved file in the hard disk linked to the database
- The computer must be connected to LAN (Local Area Network), so that there must be server and terminal computer

There are many processes that could be done, therefore in digitalization planning beside of identifying the collections type, it is also necessary to identify the existence of publication material. Whether particular publication is available in particular format such as Microsoft Word Format (.doc), pagemaker (.pmg) or any other formats. The Hardware that needed in addition to digitalization process are :

- Processing Computer.
- CD-ROM.
- CD writer drive.
- CD reader drive.

Digitalization could be done with 2 steps:

- Scanning the printed document and convert it to PDF File
- Transforming the editable documents to PDF file

According Allen and Retzlaff that Libraries are threatened because, in social terms, the internet might seem to render them less relevant. At the same time, the technologies of information brought to life in the internet make libraries so much more extensive that their relevance has never seemed more obvious.

According to the Library regulation in Indonesia (The law of the Republic of Indonesia number 43 year 2007) is explained that Library is institution who manage the collection of Papers, Printed works and/or Recorded Works professionally with Standard System to meet educational and research needs, preservation, information, and recreation site for the readers. Digital libraries or e-libraries providing service processes and manage their data using existing technologies. Broadly speaking, digital libraries indicate that generally owned collection of digital data and is accompanied by technology or certain features that will facilitate library visitors to find the data.

So far, digital libraries have grown rapidly, almost every local libraries and libraries owned by certain institutions such as the University have implemented the concept of e-library. One good example of implementing an e-library system is the University of Riau.

4. Obstacle in Implementing Digital Libraries

According to the Interview with Mr Firdaus Syam, Librarian who manage the digitalization of Thesis stated that the obstacle that commonly faced is Processing Time, where one thesis took a long time to convert Microsoft Word Format to PDF Format and giving the watermark, also there are CD's that cannot be opened (Interview on March 7th 2017). This statement also strengthened by Mrs. Head IT Service officer Mrs. Evi stated that the obstacle that usually faced is only Technical problem such as the Network Problem or the Processing time to convert Microsoft Word file to PDF (Interview on March 14th 2017).

The obstacle that common happened in digitalization processes are:

- The availability of the cost, the cost in building the technology infrastructure is high
- Copyrights problem, not every author allows their work to be digitalized, of course there are many author will be questioning about the compensation of their works
- The community who doesn't really know with technology, the technology development is really high in metropolitan area, however in the rural place people is not so familiar with computer and electronic devices, still there are many people in the rural place who does not know how to operate a computer.

5. Collection of E-Library in Library Technical Implementation Unit Of University Of Riau

- Article

Articles are complete factual articles of a certain length made for publication (through newspapers, magazines, newsletters, etc.) and aim at conveying ideas and facts that can convince, educate, and entertain readers.

- Journal

Journals are articles published periodically, an article commonly researched by colleagues.

- Papers

Where it is commonly used to illustrate, a scientific paper written for the completion of Associate Degree

- Skripsi (Scriptie)

Originally from Dutch Language, *Scriptie* is a commonly used word by Indonesian higher institution for Undergraduate. Which is usually used to illustrate a scientific paper written for thesis.

•Thesis
Thesis is scientific works to complete After Graduate degree.

• Dissertation
Thesis is scientific works to complete Doctoral degree.

6. Conclusion

Based on the discussion above, it can be concluded that:
Implementation of E-Learning in the Library in Library University of Riau has not achieved maximum results in the implementation because there are still technical obstacles such as the problem of network connection. Application of E-Library, Library University of Riau has many positive impact, so E-Library is already feasible to be used by the public. Initially the library applied at the University of Riau is only used by students of the University of Riau, but after the implementation of this E-Library, Library University of Riau can be famous in the community because it provides many online papers.

7.References

[1] Basuki, L. S.1991. *Pengantar Ilmu Peprpustakaan.* Indonesia: Gramedia Pustaka.
[2] Februariyanti, H., & Zuliarso, E. 2012. *Rancangan Bangun Sistem Perpustakaan.* untuk Jurnal Elektronik Jurnal Teknologi Informasi DINAMIK Vol.17, No.2, 126 – 127.
[3] Ishak. 2008. *Pengelolaan Perpustakaan Berbasis Teknologi Informasi.* Jurnal Studi Perpustakaan dan Informasi, Vol.4, No.2, 88.
[4] Lothar von Retzlaff. "*E-commerce for library promotion and sustainability: how library technicians can market themselves and their library's services online*". *The Australian Library Journal* 55. 2May 2006): p102(29).
[5] Ongowarsito, H. 2008. *Teknologi Informasi Untuk Perpustakaan: Digital libraries Sebagai Solusi Keterbatasan Informasi.* Jurnal Piranti Warta Vol.11 No.1, 117.
[6] Sucahyo, Y., & Ruldeviyani, Y. 2007. *PERPUSTAKAAN DIGITAL Perspektif Perpustakaan Perguruan Tinggi Indonesia*. Jakarta: Wiley.
[7] Supsiloani. 2006. *Perpustakaan Digital sebagai Wujud Penerapan Teknologi Informasi di Perguruan Tinggi.* Jurnal Studi Perpustakaan dan Informasi, Vol 2, No.1, 35.
[8] Sugiono.2008. *Metode Penelitan kuantitatif kualitatif dan R&D.* Bandung Alfabeta
[9] The Law of the Republic of Indonesia Number 43 year 2007

Post-Fire Peat Land Understory Plant in Rimba Panjang, Sumatera, Indonesia

L.N. Firdaus, Nursal, Sri Wulandari, Wan Syafi'i, and Yuslim Fauziah
Universitas Riau, Pekanbaru, Indonesia
E-mail: firdausln@lecturer.unri.ac.id

Abstract: The existence of understory plants during early post-fire succession is essential in term of natural post-fire ecological restoration. More than fifty percent of fire incidents in Riau, Sumatera, Indonesia occurred in shallow peat lands which have the huge impact on vegetation damage. This study aims to explore the understory plants species and diversity in post-fire peat land at Rimba Panjang, Kampar Regency, Sumatera, Indonesia. By using survey method, the observations were conducted on 150 plots which were distributed randomly over four locations based on the year after fire: 2009, 2014, 2015 and 2016. We found respectively 12, 14, 19 and 17 species at that sites with respective Shannon Wiener diversity index were 1.72, 2.00, 2.14 and 2.40. All the sites were dominated by Stenochlaena palustris (Burm.). Coverage percentage of understory vegetation were respectively 28.87%, 25.50%, 51.60% and 54.13%. Overall, we found 31 species of 17 familia. The result showed that the species composition, diversity index and coverage percentage of understory plant are likely to decrease in line with the length of time after the fire. Post peatland fires in Rimba Panjang are still having the characteristics of the peat swamp habitat which was dominated by Stenochlaena palustris (Burm.). Ecological restoration of that habitat is still possible, but it is necessary to consider technological and socio-economical aspects of local communities.

Keywords: Diversity index, Post-fire peat land, Understory plants, Vegetation structure, and Composition.

1. Introduction

Forest and land fires are a potential threat to sustainable development because they directly affect ecosystems. Land and forest fires in Riau not only have local impacts but also have attracted global attention, especially the effects of the smoke they generate. Riau Province has 4,221,000 ha of peat land [1] or 28.31% of the total peat area of Indonesia or about 65.54% of the peat area in Sumatra. During the period of June-October 2015, Riau's burning forest area reached 139 thousand hectares [2]. More than 50% of fire incidents in Riau occurred in peat more than 3 meters [3] which is mainly triggered by the breakdown of order hydrological peat [4].

Post-fire ecosystem restoration is dependent upon the regeneration capacity of plant species [5]. Plant communities with specific composition and structure are formed on burned areas [6]. Moser & Wohlgemuth [7] pointed out that the dominant species can act as 'switches' in vegetation development of forest regeneration determining whether or not can establish.

Soerianegara and Indrawan [8] state that the understory as one of the members of the forest community has an important role in forest ecosystems. The understory plants besides having the ecological function; it has also been used as food, medicinal plants, and an alternative energy source [9]. Lower plants have important ecological functions of degraded lands [10,11]. Lower plants need to be conserved because it has the value of environmental services and productive value [12].

Understory plants in a community are plants that live wild and naturally grow. The presence of lower plants in plantations than as a source of biodiversity also serve to protect the soil and soil organisms, helping to create a microclimate in the forest floor, keep the soil from erosion, and can maintain soil fertility [13]. Continuous surface erosion will dilute nutrients in the top soil layer so that the loss of soil fertility on a stand of plantation forest [14]. The existence of undergrowth on the forest floor can serve as a barrier blow rainwater and runoff so as to minimize the danger of erosion [15].

In addition, lower plants also often used as an indicator of soil fertility and litter producers in increasing soil fertility. In addition to ecological function, some lower plant species have been identified as plants that can be used as food, medicinal plants, alternative energy sources, mix / additional fodder, fertilizer, drinks and bedding for cattle [16].

This study aims to explore the understory plants species and diversity in post-fire peat land at Rimba Panjang, Kampar Regency, Sumatra, Indonesia.

2. Materials and Methods

The study was conducted in January-April 2017 on peatland and forest fire in Rimba Panjang Village, Kampar District, Riau Province (**Figure 1**). The research used survey method through direct observation in the field. Vegetation sampling was done by using purposive random sampling on 150 plots of 4 m^2 were distributed randomly at five different locations based on the length of time after which land and forest fires; 1 year (2016), 2 years (2015), 3 years (2014), 4 years (2013) and 8 years (2009) after the fire. Lower vegetation types and plant cover estimations of each species are recorded on each observation plot. The data were analyzed to determine the Importance Value of each species and the Index of diversity of the lower vegetation at each location according to [17]. As for the identification of species by using identification books [18-22].

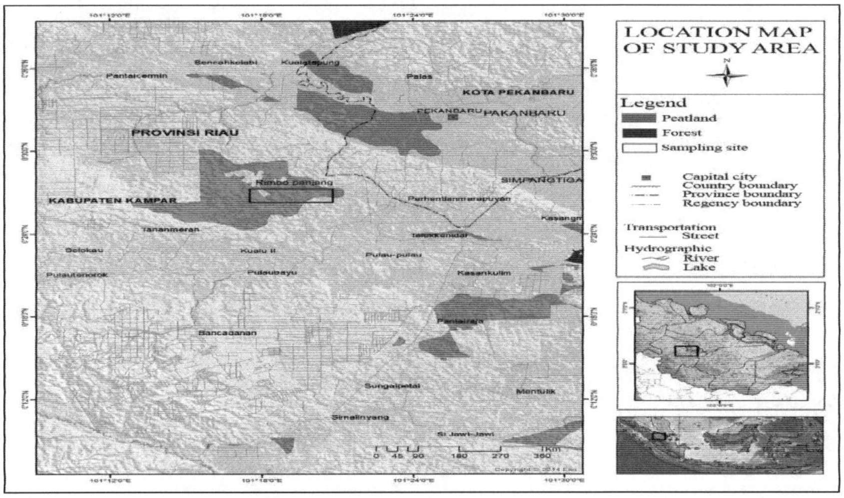

Figure 1. Location of research in the area of the former land of the village land and forest fires of Rimba Panjang-Kampar, Riau Province, Indonesia

3. Results and Discussion

The results of observation of the lower vegetation at 5 observation sites in Rimba Panjang Village showed different species composition and vegetation structure. Overall was found 31 species of understory vegetation from 21 Families (**Table 1**). Based on the composition of the constituent species have been found 5 types of ground vegetation tolerant to environmental changes post land and forest fires which are Stenochlaena palustris (Burm.) Bedd.), Carex phacota Spreng., Melastoma malabathricum L., Cyclosorus sp., Gleichenia linearis Burm., and Dicranopteris linearis Burm. which dominates the observation location.

The development of under vegetation for 8 years after land and forest fires in Rimba Panjang tends to be more fluctuating and dynamic. The number of vegetation types fluctuated from 17 species at the site 1 year after the fire to 12 species at 8 years after the fire. Observation of understory plant in the next 26-230 years after fire at Boreal forest of south east of Canada also showed that the variation of lower plants through a succession of changes is not constant [23]. Instead, Laughlin and Fule [24] reported that the total wealth of species and vegetation cover has not experienced the difference between the area burned and unburned in Grand Canyon National Park. Post-treatment was fire logging Also produced no significant effects on understory vegetation cover, diversity, or community composition 15 years after treatment [25]. Huisinga, et al. [26] states that the composition of the types of locations that burned much different from the location that is not on fire.

Vegetation cover tends to decrease from 54.13% at 1 year after the location of the fire became 28.87% in 8 years after the fire. Decrease in plant cover percentage may be related to light factor. In long-burned peatland, the presence of relatively high trees reduces the light reaching the soil surface so that the growth of understory is relatively slow. The Diversity Index (H ') tends to decrease from 2.40 at the 1-year post-fire location to 1.72 at 8 years after the fire (**Figure 2**). The diversity and abundance of lower plants tend to increase in long-burning sites. Probably this is caused by the opening of the forest canopy to dominate by certain types.

Table 1. Comparison of species Important Value in post-fire peatland, of Rimba Panjang, Kampar District, Riau Province

No	Species	Familia	Year of post-fire				
			2016	**2015**	**2014**	**2013**	**2009**
1	*Stenochlaena palustris* (Burm.) Bedd.	Blechnaceae	**36.44**	**77.53**	**86.68**	**36.06**	**84.48**
2	*Carex phacota* Spreng.	Cyperaceae	**29.67**	**7.81**	13.77	**29.35**	**46.72**
3	*Dianella ensifolia* (L.) DC.	Liliaceae	4.02	-	-	3.95	2.90
4	*Melastoma malabathricum* L.	Melastomataceae	**29.50**	**8.58**	**15.13**	**29.08**	**29.57**
5	*Imperata cylndrica* L.	Poaceae	**24.28**	5.49	-	**23.95**	-
6	*Uncaria* sp	Rubiaceae	-	3.12	-	-	-
7	*Cleome rutidosperma*	Capparaceae	-	**8.20**	3.04	1.74	-
8	*Amaranthus spinosus*	Amaranthaceae	3.83	-	-	3.77	-
9	*Physalis angulata* L.	Solanaceae	1.78	-	-	1.74	-
10	*Physalis minima* L.	Solanaceae	1.96	-	-	2.85	-
11	*Columella discolor*	Vitaceae	-	6.24	-	-	-
12	*Cyclosorus* sp	Thelypteridaceae	**24.51**	**43.00**	**16.05**	**24.24**	**8.35**
13	*Breynia racemosa*	Euphorbiaceae	-	3.55	**19.95**	-	**7.78**
14	*Coelorachis glandulosa*	Poaceae	-	-	-	-	4.64
15	*Gleichenia linearis* (Burm.) Clarke.	Gleicheniaceae	6.93	6.69	**14.79**	6.80	4.06
16	*Dicranopteris linearis* (Burm.)	Gleicheniaceae	6.60	4.32	4.74	6.49	3.48
17	*Nepenthes gracillaria*	Nepenthaceae		4.19	7.39	-	2.90
18	*Cyperus rotundus* L.	Cyperaceae	9.05	2.47	-	8.90	2.56
19	*Ipomoea obscura* L.	Convolvulaceae	-	-	-	-	2.56
20	*Uncaria* sp.	Rubiacee	6.10	-	-	6.00	-
21	*Dicranopteris linearis* (Burm.)	Gleicheniaceae	5.46	-	3.43	5.38	-
22	*Nepenthes ampullaria*	Nepenthaceae	5.34	1.08	-	5.25	-
23	*Leersia hexandra* Swartz.	Poaceae	2.27	4.32	-	2.23	-
24	*Paspalum conjugatum* Berg.	Poaceae	2.27	-	-	2.23	-
25	*Zingiber* sp.	Zingiberaceae	-	-	6.13	-	-
26	*Arundina* sp.	Orchidaceae	-	1.08	4.08	-	-
27	*Piper* sp.	Piperaceae	-	-	3.43	-	-
28	*Zingiber* sp.	Zingiberaceae	-	-	1.39	-	-

29	*Nephrolepis biserrata* Sw. Schott.	Nephrolepidaceae	-	6.30	-	-	-
30	*Cyperus iria*	Cyperaceae	-	3.74	-	-	-
31	*Smilax* sp.	Smilacaceae	-	2.28	-	-	-

Our study has found six importance major vegetation types in 5 locations observation post land and forest fires (Figure 3). *Stenochaena palustris* is a type of fern that grows climbing or creeping, smooth trunk, branching and roots along the stem attached to the supporting tree or on the ground. Spread over tropical Asia, Polynesia and Australia. Found near freshwater, brackish water, tidal forest, in sandy soil and especially along the river banks and water sources. The rod is very strong and not easily broken so it can be used as a strap. The young leaves are eaten as vegetable [27].

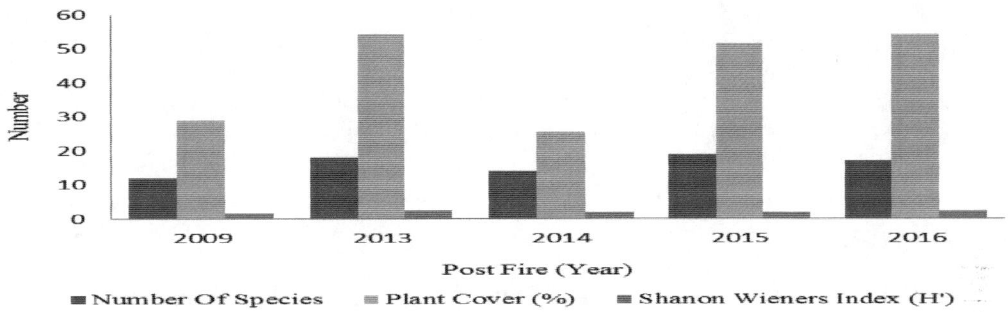

Figure 2. Comparison of species, vegetation cover and vegetation diversity index at five locations under observation

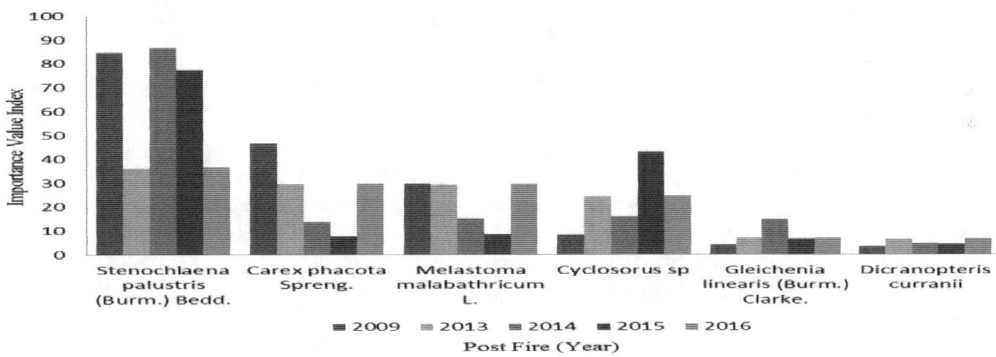

Figure 3. Comparison of Importance six major vegetation types in 5 locations observation post land and forest fires

Carex phacota Spreng is a herb that grows erect, 0,08-1,5m, triangular rods, sharp, long leaves wake up lines with sharp edges, inflorescence in panicles. In Java grow growing at an altitude of 1500-2500 meters above sea level in the wetlands [20] Other types namely *Carex baccans* grown in mountainous areas, especially along the edge of 1100-2400 meters above sea level in the forests and shrubs that shade. Other types of *Carex remota* found in damp places and marsh illuminated in full sunlight [27]. Melastoma species a shrub, upright, high reaches 4 m. In Java, they grows at an altitude of 10-1650 above sea level especially in the field and open areas that are not too arid and sandy. The young shoots and fruit can be eaten [27].

4. Conclusion

The species number, vegetation cover and Vegetation Diversity Index fluctuate and tend to decrease from the first year until the eighth year after the fire. We found respectively 12 (2009), 14 (2014), 19 (2015) and 17 Species (2016) at that sites respectively. Shannon Wiener diversity index were 1.72, 2.00, 2.14 and 2.40. All the sites were dominated by *Stenochlaena palustris* Burm.. Coverage percentage of understory vegetation were respectively 28.87%, 25.50%, 51.60% and 54.13%. Overall, we found 31 species of 17 familia. The result showed that the species composition, diversity index and coverage percentage of understory plant are likely to decrease in line with the length of time after the fire. Post peatland fires in Rimba Panjang are still having the characteristics of the peat swamp habitat.

The types of ground vegetation that can survive and are considered tolerant to changing environmental conditions is *Stenochlaena palustris* Burm., *Carex phacota* Spreng., *Melastoma malabathricum* L., *Cyclosorus* sp., *Gleichenia linearis* Burm. *Dicranopteris linearis* Burm. *Stenochaena palustris*, *Carex phacota* and *Melastoma malabathricum* tend to become dominant and achieve stable for 8 years after the fire. Ecological restoration of that habitat is still possible, but it is necessary to consider technological and socio-economical aspects of local communities.

5.Acknowledgment

This publication is part of the Professorship Research Grant of the University of Riau in Fiscal Year 2017, under contract number: 909 /UN.19.5.1.3/PP/2017, May 5, 2017. Special thanks to Ridho Abdul Gani and Afitrian Amran for the help during our research. The author also expressed his gratitude to Rudi Ependi (member of Fire Care Community) as field guide during the research. Special thanks to Mohd. Yunus at the Center for Environmental Studies of the University of Riau for technical assistance illustration of the Study Location Map.

6.References

[1] T.A. Jones. 2002 "Writing a good paper," *IEEE Trans. on General Writing*, Vol. 1, no. 2, pp.1-10.

[2] Lane and` Lange. 1993. *Writing Clearly: an Editing Guide.* USA: Heinle Publishers.

[3] Budi Wardhana. 2016. BRG's Roadmap for Peatland Restoration. *CBD & FAO Workshop: "Forest Ecosystem Restoration" Bangkok, 27 June 2016*

[4] World Bank. 2016. The Cost of Fire: An Economic Analysis of Indonesia's 2015 Fire Crisis. Indonesia Sustainable Landscapes Knowledge Note previously appeared in the *Indonesia Economic Quarterly* (IEQ), World Bank, Jakarta.

[5] Badan Litbang Pertanian. 2013. *Atlas arahan pengelolaan lahan gambut terdegradasi di pulau Sumatera. Recommendation for sustainable management of degraded peatland map. Skala 1:250.000*. Jakarta: ICCTF-BAPPENAS-Badan Penelitian dan Pengembangan Pertanian Kementerian Pertanian, Jakarta.

[6] Carmenta, R., Zabala, A., and Phelps, J. 2015. Indonesian peatland fires: Perceptions of solutions. Center for International Forestry Research (CIFOR), November 2015.

[7] Moreira, F., Arianoutsou, M., Vallejo, V.R., de las Heras, J., Corona, P., Xanthopoulos, G., Fernandes, P and Papageorgiou, K. 2012. Setting the Scene for Post-Fire Management. *In* F. Moreira et al. (eds.), *Post-Fire Management and Restoration of Southern European Forests, Managing Forest Ecosystems 24*, pp. 1-19. Springer Dordrecht Heidelberg London New York

[8] Radanova, S.S. 2014. *ECOLOGIA BALKANICA* 5 (*Special Edition*), 55-60.

[9] Moser, B and Wohlgemuth T. 2006. Which plant species dominate early post-fire vegetation in the Central Alps, and why? V International Conference on Forest Fire Research.

[10] Soerianegara. I dan Indrawan. 1988. *Ekologi Hutan Indonesia.* Laboratorium Ekologi Hutan. Fakultas Kehutanan. Institut Pertanian Bogor, Bogor.

[11] Hilwan I, Mulyana D, Pananjung G.W. 2013. *Jurnal Silvikultur Tropika* 4 (1), 6-10

[12] Katovai, E., Burley, A.L. and Margaret, M.M. 2012. *Biological Conservation* 145, 214–224.

[13] Nursal, Sri Wulandari, and Firdaus L.N., 2017. *Understory Plant Pada Tegakan Karet Pasca Revegetasi Lahan Bekas Tambang Bauksit.* Makalah hasil penelitian disampaikan pada SEMIRATA BKS-MIPA PTN WILAYAH BARAT. Jambi, 12-14 Mei 2017.

[14] Jarwaningsih. 2013. *Jurnal Penelitian Hutan Tanaman* 10 (2), 85-98.

[15] Iwan Hilwan, Dadan Mulyana, dan Weda Gelar Pananjung. (2013). *Jurnal Silvikultur Tropika* 4 (1), 6-10.
[16] Yadi Setiadi. 2015. Reklamasi dan Reboisasi Lahan Pasca Tambang. *Prosiding Seminar Hasil-Hasil Penelitian*, 27 November 2013, pp 63-88.
[17] Pinno, B.D, Sherr, I., Errington, R.C and Shea, K. 2016. Islands – Soil Patches And Plant Community Dynamics On A New Oil Sands Reclamation Design. Paper to be presented at the 2016 National Meeting of the American Society of Mining and Reclamation, Spokane, WA, *Reclaiming the West*, June 4-9, 2016.
[18] Sri Suharti. 2015. Pemanfaatan tumbuhan bawah di zona pemanfaatan Taman Nasional Gunung Merapi oleh masyarakat sekitar hutan. *Pros Sem Nas Masy Biodiv Indon.* 1 (6), 1411-1415.
[19] Dombois, D.M. and Ellenberg, H. 1974, *Aims and Methods of Vegetation Ecology*, John Wiley & Sons, New York.
[20] Backer, C.A., R.C. Bakhuizen Van Den Brink. 1963. *Flora of Java (Spermatophyta only) Vol I*, N.V.P. Noordhoff - Gronibgen – The Netherlands.
[21] Backer, C.A., R.C. Bakhuizen Van Den Brink. 1965. *Flora of Java (Spermatophyta only). Vol II*, N.V.P. Noordhoff - Gronibgen – The Netherlands.
[22] Backer, C.A., R.C. Bakhuizen Van Den Brink. 1968. *Flora of Java (Spermatophyta only). Vol III,* N.V.P. Noordhoff - Gronibgen – The Netherlands.
[23] Van Steenis, C.G.G.J., 2006. *Flora: Untuk sekolah di Indonesia,cetakan ke-11,* PT. Pradnya Paramita, Jakarta
[24] Piggot, A.G. and C.J. Piggot., 1988, *Ferns of Malaysia in Colour,* Tropical Press SDN, BHD, Kuala Lumpur.
[25] De Grandpre, L., Daniel, G., and Bergeron, Y. 1993. *Journal of Vegetation Science* 4, 803-810.
[26] Laughlin, D.C. and Fule, P.T. 2008. *Can. J. For. Res.* 38: 133–142. doi:10.1139/X07-118.
[27] Peterson, D.W and Dodson, E.K. 2016. *Forest Ecology and Management* 370, 56-64.
[28] Huisinga, K.D., Laughlin. D.C., Fule´, P.Z., Springer, J.D, and McGlone, C.M. 2005. *Journal of the Torrey Botanical Society* 132(4), 590–601.
[29] Heyne, K., 1987: *Tumbuhan Berguna Indonesia, jilid-1, cetakan ke-1*, Badan Litbang Kehutanan, Jakarta

Mechanical Properties of Composite Waste Material Based Styrofoam, Baggase and Eggshell Powder for Application of Drone Frames

Mastariyanto Perdana [1], Prastiawan [2], Syafrul Hadi [3]
Institut Teknologi Padang, Padang, 25143, Indonesia
E-mail: mastariyanto.perdana@gmail.com

Abstract: The garbage issue becomes a very serious problem at the moment. Much research has been done to make waste into useful materials. One of the utilization of waste is as the basic material of composite material that can be applied in the field of engineering. Some of the wastes generated are styrofoam, bagasse and eggshell. Styrofoam, bagasse and eggshell can be applied to a composite material. Styrofoam serves as a composite binder material while the bagasse and eggshells serve as a reinforcement. Volume fraction between styrofoam, bagasse and eggshell are 80%:10%:10%, 70%:15%:15%, 60%:20%:20%, and 50%:25%:25%. The aims of research are determine the mechanical properties of composite material based waste materials from styrofoam, bagasse and eggshell. Mechanical properties tested in this study are bending strength and toughness of composite materials. The results showed bending strength of composite for each volume fraction of 80%:10%:10%, 70%:15%:15%, 60%:20%:20%, and 50%:25%:25% are 5.07 MPa, 8.45 MPa, 8.68 MPa, and 11.01 MPa, respectively. Toughness of composite materials for each volume fraction of 80%:10%:10%, 70%:15%:15%, 60%:20%:20%, and 50%:25%:25% are 0.33 J/mm^2, 0.42 J/mm^2, 0.75 J/mm^2, and 0.75 J/mm^2, respectively. Composite materials based on waste materials from styrofoam, bagasse and eggshell can be used as an alternative material for drone frames.

Keywords: *Styrofoam, baggase, eggshell, waste material, bending strength, toughness, drone*

1. Introduction

The current waste problem has become a very serious problem for people especially in big cities, not only in Indonesia, but also around the world. Production of waste along increases with population growth, changes in consumption patterns, and people's lifestyles. They have increased the number of waste piles, species, and diversity of garbage characteristics. At the present time many studies concentrate on using waste as a usage or known as waste recycling. Utilization of waste in the field of engineering has not been much applied.

Some of the garbage that can be utilized are styrofoam, bagasse and eggshell. The garbage if processed can be used as material used in field of engineering. One of them is an environmentally friendly hybrid composite material.

Bagasse fiber has been widely used as a basic material of environmentally friendly composites. Research has been done, mechanical properties and properties of the fiber-based composite bagasse [1]. But research that focuses on bagasse with micro size (bagasse powder) is still not widely investigated. To improve the better mechanical properties of sugarcane-based composite can be done by converting the bagasse fiber into the bagasse powder. This is caused by the increasing number of bonds between powder of bagasse with matrix will cause the bonding power between bagasse also higher. it will have an effect on increasing the mechanical properties of composite. Bagasse size modification has also been done, bagasse powder usage will increase the strength of the composite compared to bagasse fiber [2].

Eggshell is a waste material that is available quite a lot and at this time has not been utilized. The eggshells are containing calcium carbonate (CaCO3) compounds of about 94% [3]. Therefore, the abundant use of eggshells as raw materials for composite materials is a fairly relevant undertaking, in order to improve mechanical properties of composite. Where calcium carbonate has a lightweight, strong and rigid properties that can be used as hybrid composites linked with a powder of bagasse to replace calcium carbonate derived from limestone. The use of hybrid composites based of bagasse and egg shell powder can also indirectly reduce the environmental burden.

Matrix material for binder of environmentally friendly composite can be utilized from waste of styrofoam. Generally, styrofoam is used for packing of food. Styrofoam is made from expanded polystyrene (EPS). According to data from the Central Bureau of Statistics, the import of styrofoam in 2006 reached 3,472,667 tonnes with a value of US $ 7,938,106. As a result, the waste is increasing. Type of plastic packaging of polystyrene often creates environmental problems because it is difficult to decompose biologically. The use of styrofoam as a binder on composites has been investigated. The use of styrofoam for binder on sandwich composites based on sisal fibers and banana stems makes the rise of rigidity of the composite material [4]. Other studies also explain that the use of styrofoam as the core matrix of natural fiber based sandwich composite (sisal fiber) provides a fluctuating sound damping effect but can be applied to soundproofing materials [5].

This study was conducted to find alternative composite materials to replace the construction of unmanned drones. The composites for this existing drone matrix mostly use composites made of resin and fiberglass or carbon fiber mixtures, in essence these composites include composites that are not environmentally friendly, since these composites if they are no longer used and are discharged to nature Will the process of decomposition for very long even hundreds of years of course things like this very pollute the environment.

The use of environmentally friendly hybrid composites based on styrofoam as binders and reinforcement from powdered bagasse and eggshell powder will be able to reduce the environmental burden. Environmental problems can be resolved and this composite material can be applied for engineering material, especially material for the frame/ body of drones.

The purpose of this research is to know mechanical properties (bending strength and impact toughness) of environmentally friendly composite material based on waste material from Styrofoam, bagasse and eggshell so that can be applied as environmentally friendly materials in the field of engineering later, especially for drone's frame.

2. Literature Review

The composite is a combination of two or more materials into a material consisting of a reinforcement and a binder (matrix) to obtain a better character than the constituent material. Composite materials can be divided into three types when viewed from the type of matrix used. The three types of composites are (1) metal matrix composites, (2) polymer matrix composite and (3) ceramic matrix composite.

The polymer composite-based natural fiber are superior in their mechanical and physical properties [6]. Natural fibers generally have low density (low density), low cost fiber, high specific properties and most importantly have biodegradable properties [7, 8].

In the last decade, researchers have done a lot of research on composite materials that can decompose in nature quickly. One of his researches is the utilization of natural fibers for composite materials. Composite natural fiber or natural fiber composite is widely developed in the automotive field as a substitute of fiberglass composite fiber [6, 9].

Recent research explains that people have turned to micro-sized or nano-sized materials, so are on the composite material. Shear strength and friction resistance of micro-carbon composites and nano-carbon composite are higher than that of macro-sized carbon fiber composites [10]. The use of macro fiber has been replaced with the use of micro-sized fibers. It also includes the use of natural fibers for composite materials. Natural fibers have been widely transformed into micro size or also called powder/natural fiber particles. The purpose of changing the size of the macro-sized fibers into micro-sized fibers is to obtain a better bonding power so that it will impact on better mechanical properties [2, 11, 12, 13].

3. Method

In this study, materials used were styrofoam, bagasse and eggshell. Styrofoam serves as binder material for composite while the bagasse and eggshells serve as a reinforcement. The styrofoam used was unused styrofoam. Styrofoam is dissolved in an acetone solution to make styrofoam turn into a liquid. After that added additive/compatibilizer on fluid styrofoam. This additive serves to make the styrofoam liquid more homogeneous. In making this binder, the compatibilizer used is acetate anhydride. Due to the brittle nature of styrofoam, it is necessary to add a Dioctyl Phthalate (DPO) stabillizer that serves to bend [14].
Bagasses were taken in lawang, West Sumatera. Eggshells were taken in Payakumbuh, West Sumatera. Bagasse and eggshells which still contain impurities cleaned using clean water. After cleaned, then bagasse and egggshell were drained and dried without direct sunlight. After bagasse and eggshells was dried, then the process of blending up bagasse and eggshells into powder with size 74 microns.

Prepare a square-shaped mold with dimensions of 170 mm x 170 mm x 5 mm (bending and impact test) and which have been waxed throughout the surface. The wax is intended to keep the composite from sticking if removed from the mold. Combine styrofoam that has been diluted with the powder of bagasse and eggshell powder, then stirred using mixer for 10 minutes until mixed evenly. Mixing (mixing) is done at room temperature. Pour mixture of styrofoam, powder of bagasse and eggshell into mold. The volume fraction between hybrid fiber and styrofoam matrix is 80:20, 70:30, 60:40 and 50:50. Volume fraction between binder and reinforment 80%:20% (80% styrofoam, 10% bagasse powder and 10% eggshell powder), Volume fraction between binder and reinforment 70%:30% (70% styrofoam, 15% bagasse powder and 15% eggshell powder), Volume fraction between binder and reinforment 60%:40% (60% styrofoam matrix, 20% bagasse powder and 20% eggshell powder), and Volume fraction between binder and reinforment 50%:50% (50% styrofoam matrix, 25% bagasse powder and 25% eggshell powder).

After the mold is full, there is a compressing molding process on hybrid composite. Then wait for the composite to dry evenly. After this hybrid composite is made (figure 1), then after that will be made bending and impact test specimens. Bending test specimens and impact tests were prepared based on ASTM-D790 and ASTM D5942-96.

Figure 1. Panel of composite material based on waste material from styrofoam, bagasse and eggshell

The type of bending test to be performed is three points bending using ASTM-D790 standard. Bending test specimen according to ASTM-D790. Testing bending using universal machine testing merk simadzu UH-300 kNI type in Mechanical Engineering Laboratory – Institut Teknologi Padang. On impact test, methods used was charpy flatwise impact (figure 4). The impact specimens were prepared according to ASTM D 5942-96.

4. Result and Discussions

4.1 Bending Strength

Figure 2 shows a graph of bending test results of a hybrid composite. On the graph shows that the highest bending strength is found in the volume fraction between styrofoam, bagasse powder and eggshell powder at 50:25:25 of 11.01 MPa. The lowest bending strength was found in the volume fraction between styrofoam, bagasse powder and eggshell powder at 80:10:10 of 5.07 MPa. This indicates that the increasing composition of bagasse powder and eggshell powder on the composite or the reduction in the amount of the composition of styrofoam makes the hybrid composite bending strength even higher. Bagasse powder and eggshell powder make composite properties relatively strength, elastic and rigid. While styrofoam tends to make composites become lighter.

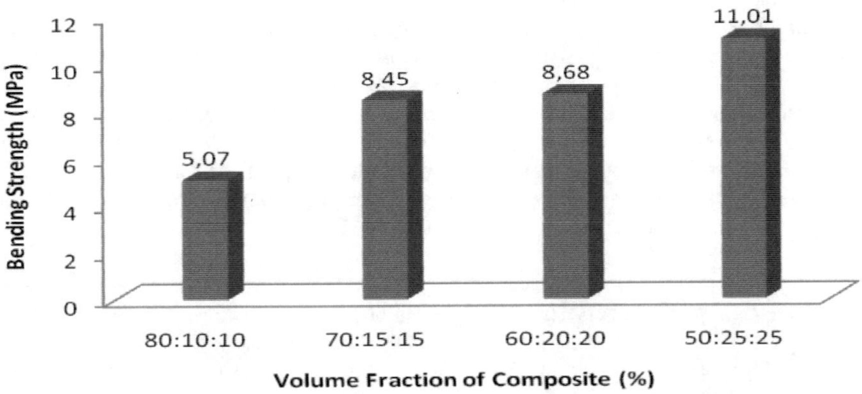

Figure 2. Bending stength of composite based waste material

Bending strength of composite will increase with addition of bagasse powder and eggshell powder. This is due to the increasing composition of bagasse powder and eggshell powder making the bond between styrofoam, bagasse powder and eggshell getting better. Perfect interfacial bonding between binder (styrofoam) and reinforment (bagasse and eggshell) is indicated by at no more air buble in the composite. The more air bubble in the composite will cause decrease of composite strength [15].

(a)

(b)

(c)

(d)

Figure 3. Fracture surface of bending test specimens. (a) 80:10:10, (b) 70:15:15, (c) 60:20:20, (d) 50:25:25

Figure 3 shows the fracture surface of bending test specimen of the hybrid composite for each volume fraction variation. From the observation of the fracture surface of the composite, shows the fracture surface of the composite with a volume fraction of 80:10:10, the composite having the most bubble (figure 3.a). And on composites with a 50:25:25 volume fraction having the least bubble (Figure 3.d). The more bubble in the composite make composite strength decrease. This is because the air bubble makes no bond (interfacial bonding) between the binder and reinforcement. The results show that with the addition of the composition of the reinforcement (bagasse powder and egg shell powder) and the reduction of the binder composition (styrofoam) makes the composite better in terms of interfacial bonding between the matrix (styrofoam) and the reinforcement (bagasse powder and egg shell) Characterized by the number of air bubble produced on the composite.

Volume fraction of the powder/particles affects the strength of the hybrid composite, wherein the composite with volume fraction 50% of particle (bagasse powder + eggshell powder) provides the highest bending strength compared with the volume fractions of 40% and 30% and 20% of the particles. At volume fraction 50% of particles show a bond between the binder and reinforcement higher. This is caused because the reinforcement has the discretion to fill the empty area between the binders [16]. The bonds between the particles are better and fill the empty space that is not filled by the matrix, so the composite has a high strength and rigid. It's because the smaller size of the material the more it will bind to the matrix. It is in accordance with the reported [10] that a reinforcement having a particle size (micro-sized) provides better on mechanical properties than fiber-shape.

The results of bending tests was obtained, hybrid composites based on waste materials from styrofoam, bagasse powder and eggshell powder can be applied to the drone body. This is due bending strength of hybrid composite is higher than the maximum bending force for the drones calculated through the Catia software. Where the results of calculations using the Catia software, the bending strength for the drones frame is 3.04 MPa [17]

4.2 Impact Toughness

In Impact tests on hybrid composites styrofoam-based bagasse and eggshell powder was obtained as shown in Figure 4.

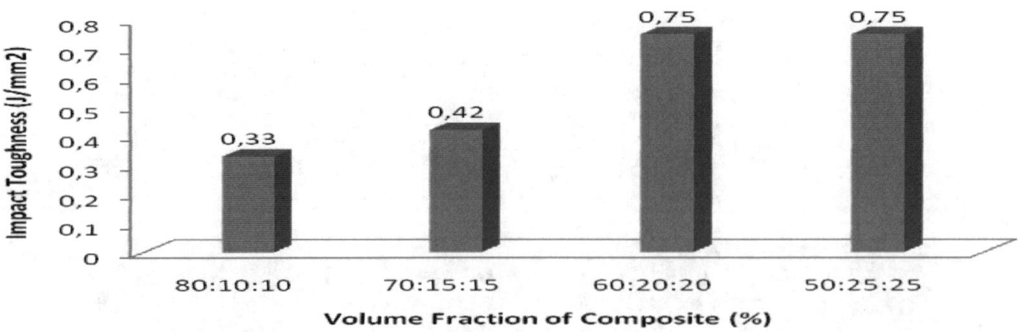

Figure 4. Impact toughness of composite based waste material

Figure 4 shows a graph of Impact test results of a hybrid composite. On the graph shows that the highest impact toughness is found in the volume fraction between styrofoam, bagasse powder and eggshell powder at 60:20:20 and 50:25:25 of 0.75 J/mm^2. The lowest impact toughness was found in the volume fraction between styrofoam, bagasse powder and eggshell powder at 80:10:10 of 0.33 J/mm^2. This indicates that the increasing composition of bagasse powder and eggshell powder on the composite or the reduction in the amount of the composition of styrofoam makes the hybrid composite impact toughness even higher too. Result of impact test, also similarity with bending test result. Bagasse powder and eggshell powder make composite properties relatively strength, elastic and rigid. While styrofoam tends to make composites become lighter.

(a)

(b)

(c)

(d)

Figure 5. Fracture surface of impact test specimens. (a) 80:10:10, (b) 70:15:15, (c) 60:20:20, (d) 50:25:25

Figure 5 shows the fracture surface of impact test specimen of the hybrid composite for each volume fraction variation. Similar with bending test result, from the observation of the fracture surface on impact test specimens, shows the fracture surface of the composite with a volume fraction of 80:10:10, the composite having the most bubble (figure 5.a). And on composites with a 50:25:25 volume fraction having the least bubble (Figure 5.d). The more bubble in the composite make composite impact toughness decrease too. This is because the air bubble makes no bond (interfacial bonding) between the binder and reinforcement. The results show that with the addition of the composition of the reinforcement (bagasse powder and egg shell powder) and the reduction of the binder composition (styrofoam) makes the composite better in terms of interfacial bonding between the matrix (styrofoam) and the reinforcement (bagasse and eggshell powder).

Impact testing of composite based waste material from styrofoam, bagasse powder and eggshell is still below from composite based polyester and rockwool fiber. From impact testing, the highest impact toughness is owned by 4% rockwool composition that is equal to 3.7 J/mm^2, while the composition of 12% rockwool has impact toughness of 2.4 J/mm^2, and the composition of 8% rockwool has the lowest impact toughness is 2.2 J/mm^2 [18].

5. Conclusion

The results showed bending strength of composite for each volume fraction of 80%:10%:10%, 70%:15%:15%, 60%:20%:20%, and 50%:25%:25% are 5.07 MPa, 8.45 MPa, 8.68 MPa, and 11.01 MPa, respectively. Toughness of composite materials for each volume fraction of 80%:10%:10%, 70%:15%:15%, 60%:20%:20%, and 50%:25%:25% are 0.33 J/mm^2, 0.42 J/mm^2, 0.75 J/mm^2, and 0.75 J/mm^2, respectively. it can be concluded that the highest bending strength and impact tougness of environmentally friendly composite material based on waste material from Styrofoam, bagasse and eggshell bases based on bagasse on volume fraction 50:25:25 was 11.01 MPa and 0.75 J/mm^2, respectively. Composite materials based on waste materials from styrofoam, bagasse and eggshell can be used as an alternative material for drone frames.

References

[1] Shabiri, Akhmad Nadji, Rizky Salaam Ritonga, and M Hendra S Ginting. 2014. "*Pengaruh Rasio Epoksi/Ampas Tebu Dan Perlakuan Alkali Pada Ampas Tebu Terhadap Kekuatan*." *Jurnal Teknik Kimia - Universitas Sumatera Utara* 3 (3): 28–31

[2] Oladele, Isiaka Oluwole. 2014. "*Effect of Bagasse Fibre Reinforcement on the Mechanical Properties of Polyester Composites*." *The Journal of the Association of Professional Engineers of Trinidad and Tobago* 42 (1): 12–15

[3] Yahya, Muhammad, Yelmida Aziz, and Zultiniar. 2016. "*Sintesis Hidroksiapatit Dari Precipitated Calcium Carbonate (PCC) Kulit Telur Ayam Melalui Proses Hidrotermal*." *Jurnal FTENIK* 3 (1): 1–8. doi:10.6052/j.issn.1000-4750.2014.04.S008

[4] Sinarep, and Agus Dwi Catur. 2012. "*Pengaruh Tebal Styrofoam Core Terhadap Berat Jenis Dan Kekuatan Tekan Komposit Sandwich Matrik Poliester Diperkuat Serat Sisal Dan Serat Pohon Pisang*." *Jurnal Dinamika Teknik Mesin* 2 (1): 1–5

[5] Sinarep, Agus Dwi Catur, and M Hafidzul. 2014. "*Redaman Suara Pada Komposit Sandwich Polyester*." *Jurnal Dinamika Teknik Mesin* 4 (1): 30–37

[6] Sanjay, M R, G R Arpitha, L Laxmana Naik, K Gopalakrishna, and B Yogesha. 2016. "*Applications of Natural Fibers and Its Composites : An Overview*." *Journal Natural Resources* 7: 108–14

[7] Saheb, D Nabi, and J P Jog. 2015. "*Natural Fiber Polymer Composites : A Review*." *Journal of Advances in Polymer Technology* 18 (4): 351–63. doi:10.1002/(SICI)1098-2329(199924)18.

[8] Bongarde, U. S, and V .D Shinde. 2014. "*Review on Natural Fiber Reinforcement Polymer Composites*." *International Journal of Engineering Science and Innovative Technology (IJESIT)* 3 (2): 431–36.

[9] Holbery, J, and D Houston. 2006. "*Natural-Fibre-Reinforced Polymer Composites in Automotive Applications*." *Jom* 58 (11): 80–86.

[10] Liu, Ning, Jianzhang Wang, Jie Yang, Gaofeng Han, and Fengyuan Yan. 2015. "*Effects of Nano-Sized and Micro-Sized Carbon Fibers on the Interlaminar Shear Strength and Tribological Properties of High Strength Glass Fabric / Phenolic Laminate in Water Environment*." *COMPOSITES PART B* 68. Elsevier Ltd: 92–99. doi:10.1016/j.compositesb.2014.08.035

[11] Iñiguez, C G, C J J Bernal, M W Ramírez, and N J Villalvazo. 2014. "*Recycling Agave Bagasse of the Tequila Industry*." *Advances in Chemical Engineering and Science* 04 (02): 135–42. doi:10.4236/aces.2014.42016.

[12] Jahjono, Arif T. 2011. "*The Influence of Wood Powder Adding on the Porosity of Composite Material*," no. February: 2–3.

[13] Ramli, Ridzuan, Maksudur Rahman Khan, Najmul Kabir Chowdhury, Mohammad Dalour, Rohaya Mohamed, Astimar Abdul Aziz, Zawawi Ibrahim, and Nahrul Hayawin Zainal. 2013. "*Development of Cu Nanoparticle Loaded Oil Palm Fibre Reinforced Nanocomposite*." *Journal Advances in Nanoparticle* 2 (November): 358–65

[14] Nurhajati, Dwi Wahini, and Ihda Novia Indrajati. 2011. "*Kualitas Komposit Serbuk Sabut Kelapa Dengan Matrik Sampah*." *Jurnal Riset Industri* V (2): 143–51

[15] Perdana, Mastariyanto, and Romi Perdana Yulsardi. 2015. "*Pengaruh Fraksi Volume Penguat Terhadap Kekuatan Lentur Green Composite Untuk Aplikasi Pada Bodi Kendaraan*." *Jurnal Iptek Terapan, Kopertis Wilayah X* 3: 71–77.

[16] Siswanto dan Kuncoro, D., 2011. *Pengaruh Fraksi Volume dan Ukuran Partikel Komposit Polyester Resin BerpenguatPartikel Genting Terhadap Kekuatan Tarik dan Kekuatan Bending. Jurnal Politeknosains*. Vol. X, No. 2, 91 – 99

[17] Sayuti, Syahril, and Muhammad Hakim Soffan. 2016. "*Perancangan Dan Pembuatan Rangka Drone Yang Ekonomis Dan Praktis*." In *Prosiding Seminar Nasional Aplikasi Sains & Teknologi (SNAST)*, 485–90

[18] Putra, F., K. 2014. *Analisa Pengaruh Variasi Komposisi Terhadap Kekuatan Tarik Statik dan Impak Komposit Berpenguat Serat Rockwoll Pada Pesawat Tanpa Awak*. Thesis. Universitas Sumatera Utara.

IOP Conf. Series: Earth and Environmental Science **97** (2017) 012035 doi:10.1088/1755-1315/97/1/012035

Application Profile Matching Method for Employees Online Recruitment

Sunarti[1] and Rangga Rahmadian Y[2] and Yulvia Nora Marlim[3]
STIKOM Pelita Indonesia, Pekanbaru, 28127, Indonesia
Phone: +6276124418, Fax: +6276135508
E-mail: sunarti@lecturer.pelitaindonesia.ac.id

Abstract: Employees is one of the determinant factors of company's success. Thus, reliable human resources are needed to support the survival of the company. This research takes case study at PT. Asuransi Bina Dana Arta, Tbk Pekanbaru Branch. Employee recruitment system at PT. Asuransi Bina Dana Arta, Tbk Pekanbaru Branch still uses manual system as seen in application letter files file so it needs long time to determine accepted and rejected the application. For that it needs to built a system or application that allows companies in determining employees who accepted or rejected easily. Pofile Matching Method is a process of competency assessment that is done by comparing the value of written, psychological and interview test between one applicationt with other. PT. Asuransi Bina Dana Arta, Tbk Pekanbaru branch set the percentage to calculate NCF (Core Factor Value) by 60% and NSF (Secondary Factor Value) by 40%, and set the percentage to calculate the total value of written test by 40%, the total value of psycho test by 30%, and the total value of interview 30%. The final result of this study is to determine the rank or ranking of each applicant based on the greater value which, the greater that score of final result of an application get, the greater the chance of the applicant occupy a position or vacancy. Online Recruitment application uses profile matching method can help employee selection process and employee acceptance decisions quickly. This system can be viewed by directors or owners anywhere because it is online and used for other company branch

Keywords: *Online Recruitment,* Employee, *Profile Matching Method, NCF (Core Factor Value), NSF (Secondary Factor Value)*

1. Introduction

Recruitment is a process that generates a number of qualified applicants for employment in a company or organization [7]. Meanwhile, according to other experts mentioned that recruitment is the process of finding, finding, inviting, and setting a number of people, both from within and outside the company as a candidate with certain characteristics as specified in the planning of Human Resources [10]. Recruitment deals with activities that affect the number and type of applicants, whether the applicant then receives the job offered [4]. Selection is the process of collecting and evaluating current and accurate information about individuals who are getting a job offer [2]. PT Asuransi Bina Dana Arta, Tbk (ABDA) is still using manual system in its human resources processing that is on data processing of new employee recruitment. For example, the collection of application files that have been sent applicants who will be selected in the form of documents that will certainly take a long time to select the file one by one manually by the HRD, and will the accumulation of data applicants. The selection process of employee recruitment has not yet been used the appropriate method so that the company gets employees that are not in accordance with the required.

Therefore, this company wants to get qualified employees and in accordance with the field and its ability then, the company must do the selection. The decision-making process for employee recruitment takes a long time for selection. For that reason, it is necessary to design computerized employee acceptance applications, as well as to provide a means to facilitate applicants to send files through upload to the system. Building a decision support system for selecting candidates using profile matching method to improve effectiveness and efficiency as well as accurate in processing test results.

2. Literature Review

Successful companies consider their employees as a key resource necessary for the survival of the company. Companies can work only if they are able to ensure sufficient human resources. Human resources are the heart of the company, as they have the ability, skills and talents. Hence the need for an employee recruitment system. [6]. Recruitment is the process of finding, finding and attracting potential employees to be employed in and by organizations. Recruitment is also a series of activities seeking and attracting job applicants with the motivation, skills, expertise, and knowledge needed to cover the shortcomings identified in personnel planning. The purpose of recruitment is to meet as many offers as possible from prospective employees so that the organization has a greater chance to make choices against prospective applicants who are considered to meet the organization's qualification standards. [9]. Recruitment is the process of finding, finding and attracting capable applicants to be employed within and by an organization. The recruitment process begins when the applicant seeks and ends when applicants apply. In carrying out the recruitment task the labor-seekers base their activities on predetermined human resource planning. It should be emphasized that recruitment activity cannot be based on human resource planning because in the plan has been established various requirements that must be fulfilled by people want to work in organization concerned. [11]

“Recruitment is the process of searching for prospective employees and stimulating them to apply for job in the organizational”. (Withdrawal is the process of searching and binding of prospective employees who are able to work within the organization). [3]. According to Law no. 13 year 2003 article 1, the workforce is any person capable of performing work both inside and outside the employment relationship, in order to produce goods or services to meet the needs of the community. Furthermore, Article 4 states that an employee is a permanent worker who works under the orders of others and receives compensation and guarantees.The employee is the seller of services (mind and energy) and gets compensation that the amount has been determined first. In this case, employees are obliged and bound to do the work given and are entitled to compensation by agreement [3]. Selection is an activity of selecting and determining applicants who are accepted or refused to become company employees. Selection is based on certain specifications of each company concerned. [3] From the definition can be concluded that the selection is to obtain the most appropriate employees in the quality and quantity of candidates to be withdrawn. And can be added back that employee selection is very important if it turns out the employees perform well in accordance with the expected.

The theoretical lens used in this study is the Unified Theory of Acceptance and Use of Technology (UTAUT), extended to include the impact of the recruiters' managerial position and educational level on technology adoption. So essentially must be adjusted positions to be given to prospective employees with the level of education and ability [1].There are various tests that are used to select employees who apply for a company. One is a psychological test. Such tests are the primary means of measuring the mind and are often used by employers as part of the selection process, to help them obtain accurate assessments of whether an individual is capable of performing the required work and whether one's character matches the job. Profile Matching is a research method that can be used in decision support system, competency appraisal process is done by comparing between one profile value with some profile of other competence value, so that can know the result of difference of requirement of competence needed, the difference of competence is called gap, where The smaller gap has a higher value

According to Kusrini [5] Method profile matching is a frequently used method as a decision-making mechanism by assuming that there is an ideal level of predictor variable that must be met by the subject under study, rather than the minimum level to be met or skipped. In the process of profile matching outline is a process of comparing the actual data value of a profile to be assessed with the expected profile value, so that it can be known the difference of competence (also called gap), the smaller the resulting gap the greater the value. Once the importance of human resources management, especially the process of employee acceptance led to many people doing research on employee acceptance system, among others, [9] the research results are: Competency GAP Mapping Calculations Based on Aspects.

3. Profile Matching Method

The gap in question is the difference between the vacancy profile (position) with the applicant's profile. For the calculation formula gap can be seen below:

Gap = Profile of Applicants- Position Profile

Before determining the Gap of each aspect, first PT. Asuransi Bina Dana Arta, Tbk Pekanbaru Branch determines the scoring range as a Gap calculation reference, As in the following table:

Table 1. Assessment Range

Assessment Range	Category	Value
X<50	Very less	1
50≤x<60	Less	2
60≤x<70	Enough	3
70≤x<85	Good	4
X≥85	Very good	5

Range assessment can be adjusted to the needs of the organization concerned, then the next step is to explain each aspect so that gap (difference) in accordance with the formula GAP. After obtaining each GAP of each prospective worker, each profile of the candidate labor (applicants) given the weight of the value in accordance with the benchmark value in the weight table GAP value as can be seen in table 2

Table 2. Table Weight GAP Value

Number	Difference (Gap)	Value Weight	Information
1	0	6	Not GAP (competence as required)
2	1	5,5	The individual competence is excess of 1 level / level
3	-1	5	Individual competence is less than 1 level / level
4	2	4,5	Competence of individual excess 2 levels / level
5	-2	4	Individual competence is less than 2 levels / level
6	3	3,5	The individual competence is 3 levels/ level
7	-3	3	Individual competence is less than 3 levels / level
8	4	2,5	Competence of individual excess 4 levels / level
9	-4	2	Individual competence is less 4 levels / level
10	5	1,5	The individual competence is 5 levels / level
11	-5	1	Individual competence is less 5 levels / level

So, each candidate labor (applicants) will have the weight of the gap value in accordance with table 2.

IOP Conf. Series: Earth and Environmental Science **97** (2017) 012035 doi:10.1088/1755-1315/97/1/012035

4. Counting and Grouping Core Factor and Secondary Factor

After determining the weight of the gap values for all aspects in the same way, each aspect is subdivided into two groups namely the Core factor (primary factor) and Secondary factor (supporting factor)

a. Core Factor (Main factor)

Core factor is the aspect (competence) that is most prominent or most needed by a position that is expected to produce optimal performance. The core factor calculation can be shown in the following formula [5] :

$$NCF = \frac{\Sigma IC}{\Sigma NC}$$

Information:

NCF : The average value of core factor
NC : Total number of core factor values (aspect 1, aspect 2, aspect 3, etc.)
IC : Number of core factor items

b. Secondary factor (Supporting factors)

Secondary factor is the items other than the aspect of the core factor (supporting factor). Calculation of secondary factor can be shown in the following formula [5]:

$$NSF = \frac{\Sigma NS}{\Sigma IS}$$

Information:

NSF : Average value of secondary factor
NS : Total number of secondary factors (aspect 1, aspect 2, aspect 3, etc.)
IS : Number of items secondary factor

c. Calculating the Total Value of Each Aspect

From the calculation of these aspects then calculated the total value based on the percentage of core factor and secondary factor which is expected to affect the performance of each profile. The calculations can be seen in the following [5] :

$$Nilai\ Total = 60\%\ NCF + 40\%\ NSF$$

Information:

NCF : The average value of core factor
NSF : Average value of secondary factor

d. Counting the Final Results (Ranking)

The final result of the profile matching process is the ranking of the candidate who can be a candidate for the workforce who can fill a certain position. Ranking refers to the results of certain calculations, the calculation can be shown in the following formula [5]:

$$Hasil\ Akhir = (x)\%\ N1 + (x)\%N2 + (x)\%N3 + (x)\%N4 + \cdots$$

Information:

N1 : Total aspect value 1
N2 : Total aspect value 2
N3 : Total aspect value 3
N4 : Total aspect value 4

(x)% : The percent value of the final result formula (100% total)

After each applicant gets the final result, it can be determined the rank or rank of each prospective worker (applicant) based on the greater the final value, the greater it will be to obtain the position or position, and vice versa.

IOP Conf. Series: Earth and Environmental Science **97** (2017) 012035 doi:10.1088/1755-1315/97/1/012035

5. Results and Discussion

Explanation of Matching Profile calculation process to be more clear, will be exposed through an example. Research process of employee acceptance selection done at PT. Insurance Bina Dana Arta, Tbk Pekanbaru Branch. Before doing the calculation by using Profile Matching, here is the table of applicant values of each test that has been done by PT. Insurance Bina Dana Arta, Tbk by taking 5 sample applicants, where each test is a variable (criteria) in the process of calculating profile matching and each variable has aspects (sub criteria).

PT. Asuransi Bina Dana Arta, Tbk Pekanbaru branch set the percentage to calculate NCF of 60% and NSF by 40%. And set the percentage to calculate the total write test score of 40%, the total psycho test score of 30%, and the total value of the interview 30%.

a). Define criteria and sub-criteria

Table 3. Criteria and sub criteria

Criteria	sub criteria	Type	Standard Value
Writing test	Field Competencies	Core factor	85
	General knowledge	Secondary Fact	75
Psychology test	Confidence	Core factor	85
	Responsible	Core factor	75
	Achievement motivation	Core factor	80
	Creativity	Core factor	75
	Initiative	Secondary Factor	70
	Accuracy	Core factor	80
	Speed	Secondary Factor	75
	Organizational capability	Secondary Factor	70
	Imagination	Secondary Factor	85
	Cooperation	Core Factor	75
	Leadership	Secondary Factor	80
	Optimistic	Core Factor	75
	Work attitude	Core Factor	80
Interview test	Honesty	Secondary Factor	75
	Knowledge of work	Core Factor	80
	Work experience	Secondary Factor	70
	Character	Core Factor	80

Each applicant will be tested according to predetermined criteria and sub criteria. the value obtained will be processed and compared with the value of other applicants. the highest applicant's value to be accepted as an employee. The are 2 steps in process of competency gap mapping

b). Collection of gaps from each aspect

For the collection of gaps that occur in each aspect have different calculations. Here the researchers cite the written test that is:

Writing test: In this case will be calculated gap between the applicant profile and vacancy profile with the variable (criteria) is a written test that has 2 aspects (subcriteria), can be seen in table 4.

Table 4. Test Write for grouping gaps

Sub Criteria	Applicant's Number and Applicant's Value					Profile Lowongan
	ID01	ID02	ID03	ID04	ID05	
01	75=4	65=3	60=3	70=4	75=4	85=5
02	70=4	75=4	60=3	70=4	70=4	75=4
Gap						
01	-1	-2	-2	-1	-1	Gap
02	0	0	-1	0	0	

c). Determination of Gap Value Weight

The following is a written test table on competency gap mapping

Table 5. Test Write Competitiveness Gap mapping results

Sub Criteria	Applicant's Number and Applicant's Value				
	ID01	ID02	ID03	ID04	ID05
01	-1	-2	-2	-1	-1
02	0	0	-1	0	0

Table 6 Table of Tests Write Results Weight Gap Value

Sub Criteria	Applicant's Number and Applicant's Value				
	ID01	ID02	ID03	ID04	ID05
01	5	4	4	4	5
02	6	6	5	5	6

Calculation and grouping of Core Factor and Secondary Factor

Table 7. Core Factor Value Written Test

Sub Criteria	Applicant's Number and Applicant's Value				
	ID01	ID02	ID03	ID04	ID05
01	5	4	4	5	5
02	6	6	5	6	6
CF	5	4	4	5	5

Calculation of Total Value of Each Aspect (Value Factor). The calculation of the total score on the written test criteria is as follows:

Table 8. Table Total Tests of Write Tests

Gap Value Weight	Applicant's Number and Applicant's Value				
	ID01	ID02	ID03	ID04	ID05
NCF	5	4	4	5	5
NSF	6	6	5	6	6
Total Value	5,4	4,8	4,4	5,4	5,4

Final Value Calculation (Ranking). The calculation of the final result or ranking is the final stage of the calculation of the profile matching process. The final result can be a ranking of candidates that can be used as labor. The calculation of the final result (ranking) can use the formula:

Table 9 Final Value After Ranking Table

Gap Value Weight	Applicant's Number and Applicant's Value				
	ID01	ID02	ID03	ID04	ID05
NT	5,4	4,8	4,4	5,4	5,4
NP	5,62	5,53	5,46	4,99	5,54
NW	5,85	5,7	6,00	6,00	5,4
Final score					
Nilai Akhir	5,6	5,29	5,20	5,46	5,44

After obtaining the final result as in table 9, it can be determined rank or rank of each applicant based on the greater the value of the end result, the greater the chance of an applicant to occupy a position or vacancy.

Table 10. Rank of Each Applicant

Applicant Number	NT	NP	NW	The final result
ID01	5,4	5,62	5,85	5,60
ID04	4,4	4,99	6,00	5,46
ID05	5,4	5,54	5,4	5,42
ID02	4,8	5,53	5,7	5,29
ID03	4,4	5,46	6,00	5,20

The output generated from the decision support system using the Profile Matching method is the ranking of each applicant according to the registration wave. Based on table 10 there is one of the five applicants that have the highest score, then the ID01 applicant number is the applicant with the highest score. Then Applicants with ID01 are recommended by the system to become employees.

6. Conclusions

Based on the result of the research, it can be concluded; firstly, with a new employee recruitment system to support the decision by using Profile Matching method so it does not require long time in new employee selection. Secondly, the E-Recruitment information system provides convenience to the HRD section in the process of receiving new employees, where in the case of sending the application files have been through the website. So, there is no pile of application files. Thirdly, the decision support system that has been built can help the leadership to determine the applicants who are accepted in the company in accordance with the vacancy he proposed. Fourthly, Web-based E-Recruitment application can support the implementation of online tests conducted at PT. Insurance Bina Dana Arta, Tbk Pekanbaru Branch. We recommend adding the appropriate criteria to each required section in each division

7. References

[1] El, M., El, A., Segers, J., & Pais, I. 2016. *Computers in Human Behavior Technology adoption in employee recruitment: The case of social media in Central and Eastern Europe*. Computers in Human Behavior, 57, 240–249. http://doi.org/10.1016/j.chb.2015.12.043
[2] Gatewood, RD dan H.S. Field. 2001. *Human Resource Selection*. Thomson Learning.
[3] Hasibuan, Melayu SP. 2005. *Manajemen Sumber Daya Manusia*. Bumi Aksara. Jakarta.
[4] Ivancevich, John M. 2001. *Human Resource Management*. New York: Mc. Grow – Hill Companies.
[5] Kusrini. 2007. *Konsep Dan Aplikasi Sistem Pendukung Keputusan*. Andi Offset. Yogyakarta.
[6] Lorincová, S. 2015. *The Improvement of the Effectiveness in the Recruitment Process in the Slovak Public Administration*, 34(15), 382–389. http://doi.org/10.1016/S2212-5671(15)01644-5
[7] Mathis. Robert I, Jackson John H. 2001. *Manajemen Sumber Daya Manusia*. Jakarta: Salemba Empat.
[8] Puspitasari, Luckyana. 2013. *Penerapan Metode Profile Matching Dalam Sisem Pendukung Keputusan Penilaian Kinerja Karyawan (Studi Kasus: PT. Perkebunan Nusantara III Medan)*. Medan.
[9] Rachmawati, Ike Kusdyah. 2007. *Manajemen Sumber Daya Manusia*. Yogyakarta
[10] Samsudin, S. 2006. *Manajemen Sumber Daya manusia*. Bandung: CV. Pustaka Setia.
[11] Siagian, Sondang P. 2003. *Manajemen Sumber Daya Manusia*. Ed. I. Jakarta: Bumi Aksara.

Thermal Cyclic Resistance Polyester Resin Composites Reinforce Fiber Nut Shell

Hendriwan Fahmi
Institut Teknologi Padang, Padang, 25146, Indonesia
Telp/fax : +62751443317
Email : hendriwan.basyaruddin@gmail.com

Abstract: The purpose of study is to determine the effect of fiber length and thermal cyclic of the bending strength of polyester resin composite reinforced by fibers nut shell. The materials used in this study is a nut shell fibers with fiber length of 1 cm, 2 cm and 3 cm and polyester resin with composition 70-30%wt. Fiber nut shell treated soaking in NaOH 30% for 30 minutes, then rinse with clean water so that the fiber free of alkali and then dried. Furthermore, the composite is heated in an oven to a temperature of 100^0C for 1 hour and then cooled in the open with a variety of thermal cyclic 30, 40, and 50 times. Bending properties of composites known through the testing process using a three-point bending test equipment universal testing machine. The test results show that the bending strength bending highest in fiber length of 3 cm with 30 treatment cycles of thermal to the value of 53.325 MPa, while the lowest occurred in bending strength fiber length of 1 cm with no cycles of thermal treatment to the value of 30.675 MPa.

Keywords: Composites, fiber nut shell, Long Fiber, Thermal Cyclic, BendingTest

1. Introduction

The prospect of developing natural fiber-based composite in Indonesia is very good, since the majority of natural fiber-producing plants can be cultivated in the country, such as cotton, kapok, flax, jute, hemp, sisal, coconut, palm, banana,. The development of natural fiber-fiber composite technology is also in line with government policy to explore local potential genius. This will certainly increase the empowerment of renewable local natural resources (Boimau, 2015). Technological developments turned out to cause the emergence of new problems, namely environmental problems that can interfere with human life. One of the environmental problems that began to feel today is the destruction of the environment caused by the amount of material that can not be destroyed by nature, therefore more environmentally friendly replacement materials are needed. Natural fiber materials include environmentally friendly materials, these materials can be broken down by nature, the fiber has a tendency to suck water, the fiber can be decomposed by nature under certain conditions by bacteria / fungi. In addition to environmentally friendly natural fiber materials have various advantages of cheap price. A study of HDPE waste composites with natural fibers has been done Thimothy and Baillie (2007), the effect of agave fiber extract method and HDPE waste as roof ceilings indicates that fibers with previous heat treatment will increase their strength compared to non-treated fibers previously.

Research on long-term heat cycles between fiberglass and epoxy with a temperature cycle of 27 ° C and a peak temperature of 50 ° C, with repetitions of 6, 12, 24, 36, and 48 retained within 5 minutes indicates that the length of time at high temperatures Will result in damage to the interfacial bond between the amplifier and the matrix resulting in a decrease in composite strength (Papanicolaou and Xepapadaki 2009). The mechanical strength of the HDPE matrix with volume fractions of 5, 10, 20, 30, and 40% of clay particles and rubber tires pressed at 10 MPa at 190ºC for 10 min indicates a decrease in mechanical strength (Avila and Fajula 2009).

Fiber nut shell is one of alternative natural fiber materials in the manufacture of composites scientifically utilization was continuously developed. This areca nut begins by the user because in addition to easy to obtain, cheap, can reduce environmental pollution (biodegradability) so that this composite can overcome environmental problems, and do not endanger health. The development of fiber nut shell as a composite material is very understandable given the availability of natural fiber raw materials, Indonesia has abundant raw materials. K.Boimau and T.D. Cunha (2015), conducted research on the influence of fiber length of gewang leaves to composite properties showed that composite with 70-30% v composition with 3 cm fiber length has bending strength of 45.35 MPa, length 2 cm 39,7 Mpa and length 1 Cm of 35.1 Mpa. This result shows that the longer the fiber the higher the bending power. This study aims to determine the effect of fiber length and the effect of thermal cyclic produced on bending strength with random fiber orientation.
A composite is a material formed from a combination of two or more materials, in which the mechanical properties of the constituent material vary. Due to different forming characteristics, it will produce new material that is composite which has mechanical properties and the combined characteristics of the material forming. The matrix in the composite functions as a fiber binding material into a unit of structure. Protect from external damage, forward or move external load on the sliding plane between fiber and matrix, so that matrix and fiber are interconnected (Schwartz, 1992). The fabrication of fiber composites requires a strong surface bond between the fiber and the matrix. In addition, the matrix must also have chemical compatibility so that unwanted reactions do not occur on the contact surface between the two. To select the matrix should be considered its properties, among others, resistance to heat, poor weather resistance, and resistance to shocks that are usually a consideration in the selection of matrix materials. Polymer materials that are widely used as a matrix material in composites are two kinds: thermoplastic and thermoset (Schwartz, 1992).

According to Gibson R.F (1994), matrices in composite structures can be derived from polymer, metal and ceramic materials. In general, the matrix has the following functions:

- Bind fibers into one unified structure.
- Protects fibers from damage due to environmental conditions.
- Transfer and distribute foil loads.
- Contribute several traits such as stiffness, strength, and
- physical resistance.
- Below are the conditions that must be met as matrix materials for printing of composite materials (Surdia, 1985):
- The resin used should have a low viscosity, in accordance with the reinforcing and permeable materials.
- Can be measured at room temperature in optimal time.
- Has a small shrinkage in preservation.
- It has good adhesiveness with reinforcing materials.
- It has good properties of the preserved material.

In general the matrix is divided into two groups namely (Feldman and Hartomo, 1995):

- Thermosets are materials that can not melt or soft when heated because the molecules are cross-linked so that the material can not be recycled back. For example epoxyesin, polyester, urea formaldehyde.
  Phonolformaldehyde, melamine formaldehyde and others.

- hermoplastic is a material that can become soft again when heated and hardened when cooled so that the formation can be done repeatedly because it has a linear structure. Examples of thermoplastics: PVC (poly vinyl cloride), FE (Polyethylene), PP (polypropylene) nylon 66, polyamide poiasetal and others.

Composite amplifiers generally have less ductile properties but are more rigid and stronger. The main function of the amplifier is to support the strength of the composite, so that the high strength of the composite depends greatly on the amplifier used, since the voltage applied to the composite is initially received by the matrix to the amplifier, so that the amplifier will hold the load to the maximum load. Therefore the amplifier must have higher tensile stress and modulus of elasticity than the composite matrix. The polyester resin is a matrix of composites. These resins are also included in thermoset resins. In the polyester thermoset the liquid resin is converted into a hard and brittle solid formed by the chemical crosslinking which forms a very strong polymer chain. According to Mubarak (2006) thermoset resins do not melt due to heating. At the time of printing, this resin does not need to be applied pressure, because when it is liquid it has a relatively low viscosity, hardened at room temperature with the use of a catalyst without producing gas (unlike other thermoset resins).

In general, the polyester resin is strong to the acid except the oxidizing acid, but has a low resistance to the base. If the resin is inserted into boiling water for 300 hours it will break and crack. Widely polyester is used in the form of composite materials.

The mechanical properties of the epoxy proposed by F.smith (1993) can be explained based on Table 1:

Table 1. Mechanical properties of epoxy resin

Properties	
Tensile Strength	40-90 MPa
Modulus of elasticity	2,0-4,4 GPa
Bending Strength	60-160 MPa
Impact	10,6-21,2 J m
Density	1,10-1,46 gr/cm^3

Many betel nut utilization is still limited. Nuts that are still often used so far are the leaves, stems and seeds. Meanwhile, the coir part of the nut is wasted with no use or used further. Given the very limited use of fiber nut shell, there needs to be innovation for the use of fibers nut shell such as the manufacture of natural composites. Fiber nut is one of the alternative fiber materials in the manufacture of scientific composites, the utilization is still developed because nut shell nutrients in addition to easy to obtain, cheap, can reduce environmental pollution so that the composite is able to overcome environmental problems, and not harm the health.

Development of betel fiber as a composite material is very understandable given the availability of raw materials of natural fiber Indonesia which has abundant raw materials. The composites with natural fiber reinforcement have advantages such as specific strength and high modulus, low density, low price, abundant in many countries, lower pollution emissions.

The content in betel nut fiber is:

a. Cellulose content 70,2%
b. Water 10,92%
c. Ash 6.02%

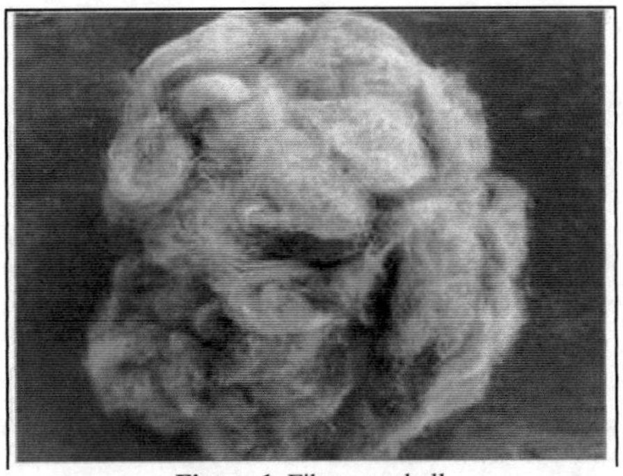

Figure 1. Fiber nut shell

2. Methodology

Fiber Preparation

Betel nut fibers are taken only fine, the skin and fiber are rough. The selected betel nut is dry fiber. After the betel nut fibers are sufficient for the manufacture of specimens, a specimen will be made using polyester resin, the position of the areca nut is made randomly or randomly. Preparation stage of betel nut fiber is:

2.1 Preparation of betel nut

The areca fruit is an old and ripe fruit, to indicate the areca nut that has been able to take this usually from the discoloration of the skin (from green to reddish yellow), There is also a fruit that has been old or ripe this will fall from itself Stem nut.

2.2 Separation of the skin and the contents of areca nut

After the fruit is taken, it will be separated between the contents and the skin. There are two ways of separating the skin and the contents of areca nut. The first ripe fruit is directly split and separated with aids, for example: a screwdriver or a knife. Both nuts are dried in the sun until the skin is dry, just separated.

2.3 Drying of betel nut skin

For betel nut fiber that is used is, fiber that comes from the skin of fruit that has been dried. Ripe areca nut will be dried first.

2.4 Separation of betel nut fiber

The dry skin consists of 3 parts:

- Outer skin, black or gray. Being on the outside.
- Fine fiber, yellow. Is a constituent after the outer skin of betel nut, located between the outer shell and coarse nut shell fiber.
- The coarse nutrient shell, yellow, is shaped like a coarse grain and larger than fine fibers of areca nut. Coarse fibers directly blend with the shell of betel nut

Separation between these fine fibers in a simple and manual way. The outer shell and coarse shell fiber are removed, leaving only fine fibers, the results obtained in the form of a cotton-like yarn.

2.5 The formation of dimension (length) fiber of betel nut

Cutting will be done after the fine fibers are separated on the skin of betel nut fibers, the lengths made in this case are 1 cm, 2 cm and 3 cm.

2.6 Soaking with NaOH solution
After the formation of the length dimensions of betel nut fiber, it will be immersed in NaOH solution, the areca nut is treated by immersion in 30% NaOH solution for 30 minutes. The soaked betel nut is then cleaned with water and dried.

Specimen manufacture
After immersion in the NaOH solution and resin preparation is complete, the next step is to perform the specimen, the fiber orientation in this case is random, the composite panel making begins by preparing a square mold with dimensions of 300 mm x 170 mm x 5 mm (for bending test) That have been waxed throughout the surface. Then the fiber nut is positioned randomly / randomly in the mold. After that the polyester resin and hardener mixed with a ratio of 30 : 1 until evenly distributed, then poured on molds that have been filled with areca nut. After that the mold is left to harden and dry in room temperature for ± 24 hours. All these composite manufacturing processes are carried out by the same method for each fiber length. The bending test specimen complies with ASTM-D790.

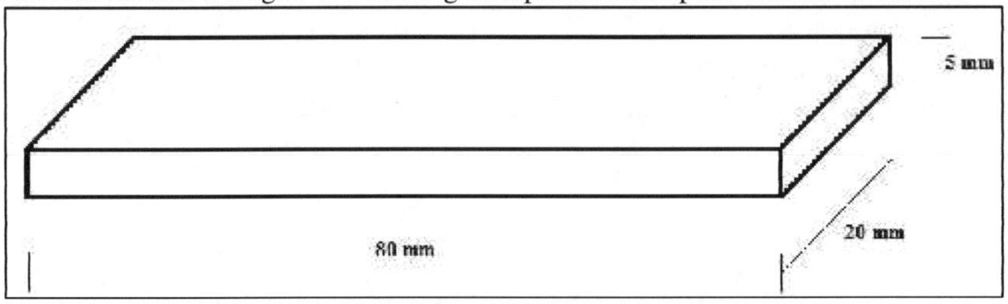

Figure 2. Dimensional specimen bending test

Figure 3. Specimen

Thermal cyclic
After the specimen is ready to be formed, the thermal cycle test is 30, 40 and 50 times at 100^{0}C and held for 60 min. Further testing is done bending.

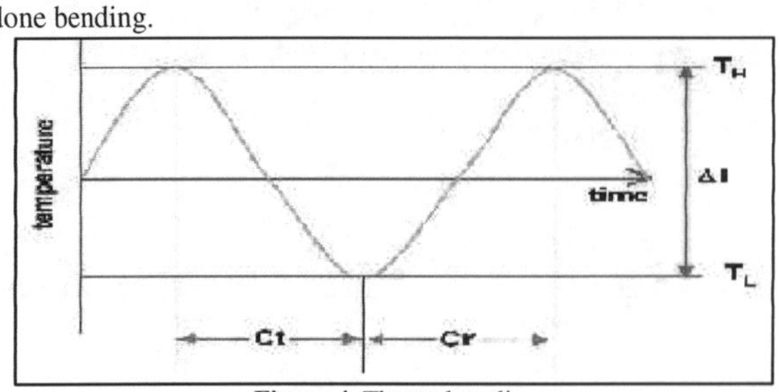

Figure 4. Thermal cyclic

3. Results and Discussion

The tested composite was a polyester resin composite with a betel nut fiber reinforcement, with a composition of 70: 30 wt. The fiber length used in this study was 1 cm, 2 cm and 3 cm and used thermal cyclic variations of 30, 40 and 50 cycles at 100 ° C and held for 60 minutes. Furthermore, the bending test with the method of Three Point Bending.

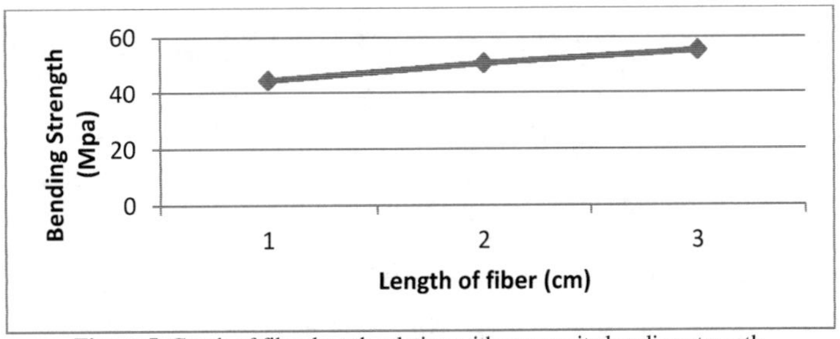

Figure 5. Graph of fiber length relation with composite bending strength.

From the picture above got the highest strength value is 55.2 MPa at 3cm fiber length and the lowest value obtained at 1 cm fiber length is 44,63 MPa. In general, the manufacture of composites is aimed at combining properties possessed by the resin as a matrix and fiber as a reinforcement, therefore the incorporation of these two things will result in the composite material being formed to be stronger. This fiber will eventually retain most of the forces acting on the composite, when the test is carried out the initial voltage will be received by the matrix (resin) and forwarded to the fiber so as to withstand the maximum load.

Thermal cyclic

From the thermal cycling test the results obtained as in table 2
Table 2. Thermal Cyclic testing data

Thermal Cyclic	Length of fiber (cm)		
	1	2	3
0	44,63	50,7	55,2
30	35,93	49,05	53,3
40	30,9	43,5	53,13
50	30,68	37,65	48

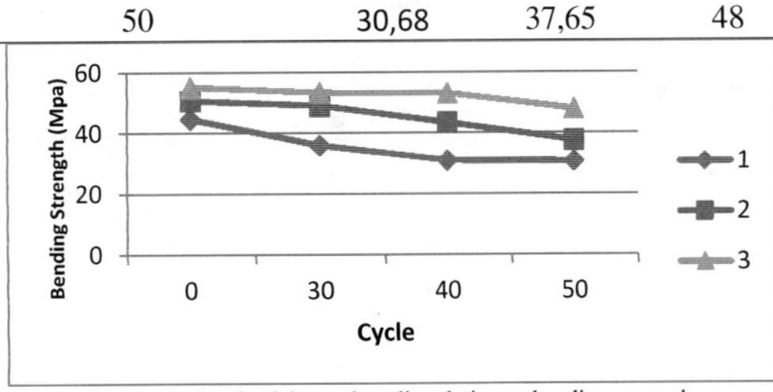

Figure 6. Graph of thermal cyclic relation to bending strength.

From the picture above, the bending strength decreases with the increase of thermal cycle. Thermal cycling treatment results in lower tensile strength as the larger cycle is given to the composite properties. This is because during the process of thermal cycling, there is a shift of the interface between the matrix and the amplifier, so that will cause a gap in this composite is the beginning of the formation of cracks. The time cycle shifts occur because of the expansion coefficient between different matrices and fibers (Avila and Fajula 2009), so that over time the fibers will be detached from the matrix causing the tensile strength of the reduced composites. The decrease in bending strength occurs due to the occurrence of a gap caused by the release of the bond between the matrix and the areca nut which causes the area of the specimen to accept the smaller the force. The influence of thermal cyclic treatment on composites using epoxy matrices, debonding failure of the matrix due to the influence of thermal fatigue, research (Surdia 2010) of time and high temperatures can be the cause of decreasing polymer strength, polymers in a short time at higher temperatures Give effect damage.

4. Conclusion

The highest bending strength in epoxy resin matrix composite reinforced with betel leather fiber with a fiber length of 3 cm is 55.2 MPa and the lowest value at 1 cm fiber length is 44.63 MPa. In testing the thermal cycle in getting the value of bending strength decreases with increasing number of heating cycles due to the occurrence of a gap caused by the release of the bond between the matrix and pinang fiber which causes the area of the specimen to accept the smaller the force.

5. References

[1] A.J., Hartomo., 1992. *Polimer dan Perekat.* Yogyakarta

[2] Boimau, Kristomus., dan Cunha, T.Da., 2015. *Pengaruh Panjang Serat Terhadap Sifat Bending Komposit PolyesterBerpenguat Serat Daun Gewang,* Universitas Nusa Cendana, Kupang.

[3] Bate, Yanuarius., Maliwemu, E., dan Pell, Y.M., 2015. *Pengaruh Panjang Serat dan Fraksi Volume Serhadap Sifat Bending Pada Komposit Widuri Polyester*, Universitas Nusa Cendana, Kupang.

[4] Saferi Rozi., Dahlan Hendery., dan Bur Mulyadi., 2015. *Pengaruh Thermal Shock Terhadap Kekuatan Tarik dan Bending Komposit Resin Berpenguat Serat Rami,* Universitas Andalas, Padang.

[5] Bahtiar, M., Sutjahjono, H dan Dwilaksana, D S., 2014. *Analisa Fraksi Volume Filler Terhadap Sifat Mekanik Komposit Laminat Matriks Polyester Berpenguat Serat Sisal*, Universitas Jember, Jember.

[6] Onal, L., Karaduman, Y. 2009. *Mechanical Characterization of Carpet Waste Natural Fiber- Reinforced Polymer Composites,* Departement of tektile Engineering, Erciyes University , Kayseri, Turkey.

[7] Papanicolaou, G.C. Xepapadaki, G.D. Tagaris. 2009.*Effect of Thermal Shock Cycling on the Creep Behavior of Glass-Epoxy Composite,* Composites Material Group, Department of Mechanical and Aeronoutical engineering, University of Patras, Greece

[8] Sudargo P.H., Suhardoko., dan Baroto B.T., 2015. *Pengaruh Fraksi Volume dan Panjang Serat Terhadap Sifat Bending Komposit Polyester yang diperkuat serat Limbah Gedebog Pisang*, ATWS, Surakarta.

[9] Sabuin, A., Boimau, K., dan Adoe, D.G., 2015. *Pengaruh Temperatur Pengovenan terhadap Sifat Mekanik Komposit Hibrid Polyester Berpenguat Serat Glass dan Serat Daun Gewang*, Universitas Nusa Cendana, Kupang.

[10] Schwartz MM., 1984. *Material Komposit Handbook,* McGraw-Hill, New York, Amerika Serikat.

[11] Schwartz MM., 1992. *Material Komposit Handbook : 2^{nd}ed.* McGraw-Hill Inc, New Jersey.

[12] Gibson, R.F., 1994. *Principles of Composite Material Mechanics.* McGraw-Hill Inc, St. Louis

IOP Conf. Series: Earth and Environmental Science **97** (2017) 012037 doi:10.1088/1755-1315/97/1/012037

The Influence of Calcium Carbonate Composition and Activated Carbon in Pack Carburizing Low Carbon Steel Process in The Review of Hardness and Micro Structure

Hafni[1], Syafrul Hadi[2] and Edison[3]
Mechanical Engineering Department, Padang Institute of Technology, Padang, Indonesia
E-mail: hafnimesin@gmail.com

Abstract. Carburizing is a way of hardening the surface by heating the metal (steel) above the critical temperature in an environment containing carbon. Steel at a temperature of the critical temperature of affinity to carbon. Carbon is absorbed into the metal form a solid solution of carbon-iron and the outer layer has high carbon content. When the composition of the activator and the activated charcoal is right, it will perfect the carbon atoms to diffuse into the test material to low carbon steels. Thick layer of carbon Depending on the time and temperature are used. Pack carburizing process in this study, using 1 kg of solid carbon derived from coconut shell charcoal with a variation of 20 %, 10 % and 5 % calcium carbonate activator, burner temperature of 950 ^{0}C, holding time 4 hours. The test material is low carbon steel has 9 pieces. Each composition has three specimens. Furnace used in this study is a pack carburizing furnace which has a designed burner box with a volume of 1000 x 600 x 400 (mm^3) of coal-fired. Equipped with a circulation of oxygen from the blower 2 inches and has a wall of refractory bricks. From the variation of composition $CaCO_3$, microstructure formed on the specimen with 20% $CaCO_3$, better diffusion of carbon into the carbon steel, it is seen by the form marten site structure after quenching, and this indicates that there has been an increase of or adding carbon to in the specimen. This led to the formation of marten site specimen into hard surfaces, where the average value of hardness at one point side (side edge) 31.7 HRC

Keywords: *the composition of activated carbon and calcium carbonate, burner temperature, holding time.*

1. Introduction

Because of soft, pliant and easy to manufacture, steel is widely used as an ingredient in the manufacture of a product. To obtain a hard nature of the surface and remains soft at the core is carried out surface hardening process (face hardening), so that the product can be used with its design purpose. Steel with content low carbon(below 0.3% C), can be carbon , specifically for the S-35C steel with 0.20 % to 0.35% C, which have less good properties for gilded but can disseminator [1] Process this surface hardening is influenced by the number of levels of carbon contained in the steel. Carbon steel is a mixture of iron and carbon and sulfur (S), phosphorus (P), silicon (Si) and manganese (Mn) [5]. Properties of carbon steel depend on the carbon content, and therefore carbon steel can be grouped based on the levels of carbon [2]:

- The extra low carbon steel, carbon content more than 0.08%
- Low Carbon Steel (Low Carbon Steel), the carbon content of 0.08 – 0.35%.

- Carbon Steel Medium (Medium Carbon Steel), the carbon content of 0.35 - 0.5%.
- High Carbon Steel (High Carbon Steel) 0.55 -1.7% carbon content.

According to Kuswanto [3], the quality low-carbon steel can be improved, particularly for improved than not afford hardened be able to be hardened. This can be done by way do carburizing process. Where one of the method is to use a solid carbon media or pack carburizing. The depth of the carbon atoms also diffuse sufficiently successful for the benefit of the technique is ± 1000 μm to do this carburizing process, required a stove burner that is usually installed permanently. The furnace is operated with fuel coal or charcoal, so it should be shunned from flammable materials. Due to limited land, it is necessary to design a stove that can be moved (mobile) so can be operated the safe area and when not in use can be placed in that which does not take up large areas. In this study, the authors designed and made the pack carburizing furnaces mobile[4.5], as well as doing test, carburizing products practice student products, in order to obtain properties better is hard, wear-resistant, tough and resilient through heat treatment.

2. Material

One effort to improve the quality of metal mainly hardness is the process carbonization, According to research conducted by Kuswanto [2] which examined the treatment *pack carburizing* on low carbon steel as a material alternative for the cutter, where the process of *pack carburizing* in the kitchen space is gradually warming underway, the first phase of 200 °C for 1 hour, second phase of 500 °C for 1 hour and 700 °C for 1 hour, the last in carburizing temperature of 900 °C for 2 hours. Further cooling is done slowly, where the kitchen is turned off and waits until it goes down at a temperature of 350 °C. After reaching that temperature, the kitchen door was opened to remove the carburizing box. Outdoors kitchen carburizing box lid is opened, all specimens out to be cooled in the open. The process of heating and cooling of the specimen can be seen in the image below.

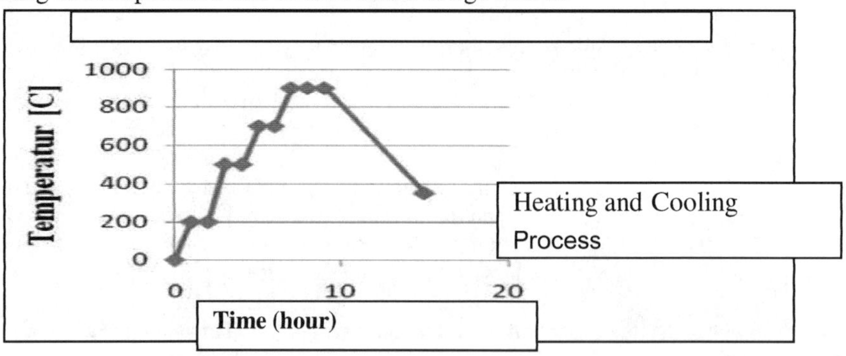

Figure 1. Initial data test material:

Table 1. Vickers hardness test data micro initial test material

Specimen	Hardness	No	Diagonal 1 d1 (mm)	Diagonal 2 d2 (mm)	Diagonal rata-rata d (mm)	P (Kg)	Hardness (Kg/mm²)
Bahan baku	Mikro Vickers	1	0,062	0,062	0,06200	0,3	144,69
		2	0,062	0,062	0,06200	0,3	144,69
		3	0,0615	0,062	0,06175	0,3	145,87
		4	0,062	0,061	0,06150	0,3	147,06
		5	0,061	0,061	0,06100	0,3	149,48

Table 2. Mickro vickers hardness data after *carburizing pack*

Spesimen	Hardness	No	Diagonal 1 d1 (mm)	Diagonal 2 d 2 (mm)	Diagonal rata-rata d (mm)	P (Kg)	Hardness Value
A	mikro Vickers	1	0,05	0,05	0,05	0,3	222,48
		2	0,051	0,0505	0,05075	0,3	215,95
		3	0,056	0,0565	0,05625	0,3	175,79
		4	0,059	0,059	0,05900	0,3	159,78
		5	0,061	0,0615	0,06125	0,3	148,26

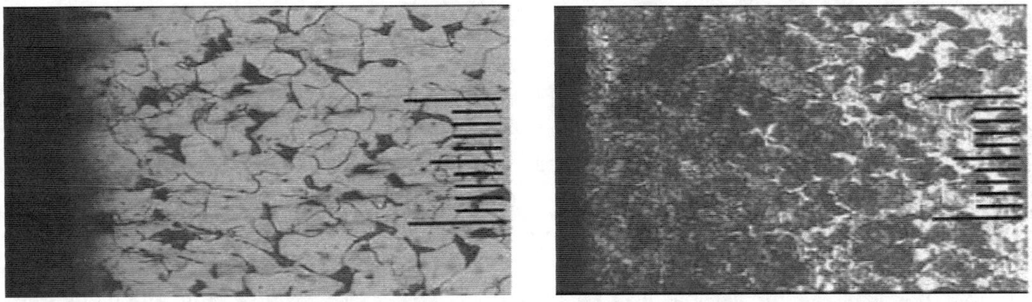

Figure 2. Low carbon steel micro Structure A before the process of carburizing, B After carburizing process

Test results concluded. Vicker hardness prices rose by 26%. and the depth of the carbon atoms also diffuse sufficiently successful for the benefit of the technique is ± 1000 μm [2].

Figure 3. Furnace *pack carburizing*

3. Method

Research process conducted through the stages of preparation of the furnace, active carbon manufacturing process, the manufacturing process the test material, the *pack* carburizing, quenching process, the testing process to test the material to see the magnitude of the diffusion of carbon atoms on the surface of the test material (carbon steel low) [5,6].

3.1. Materials Research

Materials used in this study are:

- Low carbon steels as a material that hardened surface
- Media carburizing, Charcoal active selected shell coconut and activators selected for each of calcium Carbonate ($CaCO_3$).
- Heating Temperature is 950 ^oC
- Holding time for 4 hours.

Instrument used for the process of carburizing this is designed furnace with coal fuel

Pack carburizing process was made a furnace with shape:

1) The furnace is made up of steel frame
2) Material furnace room made of refractory bricks
3) Air is blown by the blower
4) Material from the coal

Furnace is designed mobile [5].

Carburizing process, Coconut shell is very good in use as activated charcoal for media carburizing [8], large granular activated charcoal that is in use 30 mesh and then mixed with calcium carbonate ($CaCO_3$) with the composition of the three variations of composition:

a. Variations 1-1 kg of activated carbon, 20 % $CaCO_3$
b. Variation 2-1 kg of activated carbon 10 % $CaCO_3$
c. Variation 3-1 kg of activated carbon 5% $CaCO_3$

After mix homogeny calcium carbonate and activated carbon for each composition, and then input into 3 pieces of steel box 4 cm, a steel box with one variation. Then add 5 test material of low carbon steel for each box, set each test material by 2 cm, input again a mixture of calcium carbonate and activated carbon material covering the test, after which each is in a steel box lid. Tick each frog based composition.

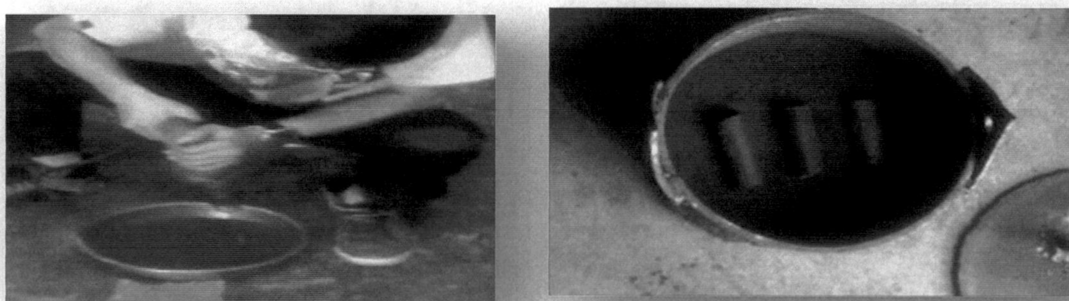

A B

Figure 4. (a) The mixing process carburizing media, (B) the laying of the test material in a steel box.

After combustion in the furnace chamber carburizing pack is perfect, the input box in the room stove. Then steel lid, heat control heat at a temperature of 950 ^0C [9. 10], and hold for 4 hours. Holding time 4 hours have a higher hardness values for all media quenching [6.7.8] The choice of temperature is based on the chemical composition of carbon steel is used, namely 0.082 %C. Atom diffusion process will occur at a temperature of approximately 0.5 *melting point. Base on* Fe-C phase diagram, it is known that carbon steel has a *melting point of* ± 1600 ^oC [3]. After carburizing, carbon steel in *quenching* simultaneously into room temperature water to obtain a hard coating on its surface. Metallography process to view the microstructure of the test material after the process of carburizing and hardness test to determine the hardness on the surface material testing.

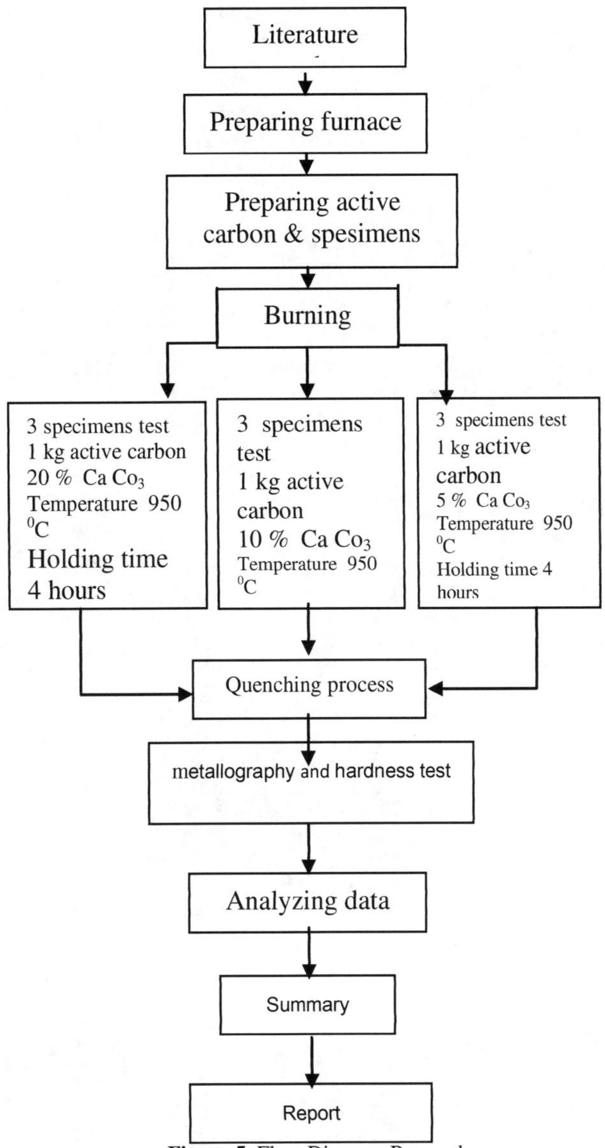

Figure 5. Flow Diagram Research

4. Results and Discussion

4.1 Micro structure

Test material that is low carbon steel test material in the form of bars with diameter 16 mm length of 5 cm, In this test there are 2 kinds of test materials; (1) The test materials without carburizing treatment. The test material with *carburizing* treatment, the test material after the process of *carburizing* process is carried out *quenching* with water as the cooling medium. After test material quenching there in the waste 1 cm, which examined the microstructure is a length of 4 cm. it aims to see the extent to which carbon of activated carbon diffuses into the steel with *activator* calcium carbonate ($CaCO_3$). Amount of *activator* the will affect the diffusion of carbon into the steel. Photo microscope of each test material to 200 x magnifications, the microstructure taken at the edges and the middle side of the test material, the data of data is as follows:

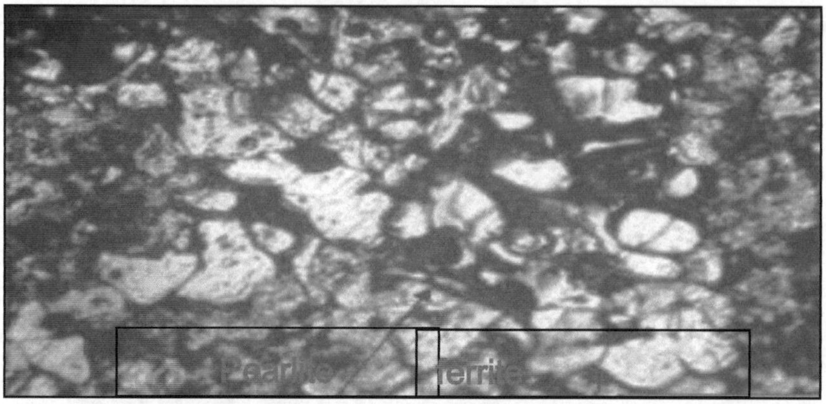

Figure 6. The microstructure without treatment

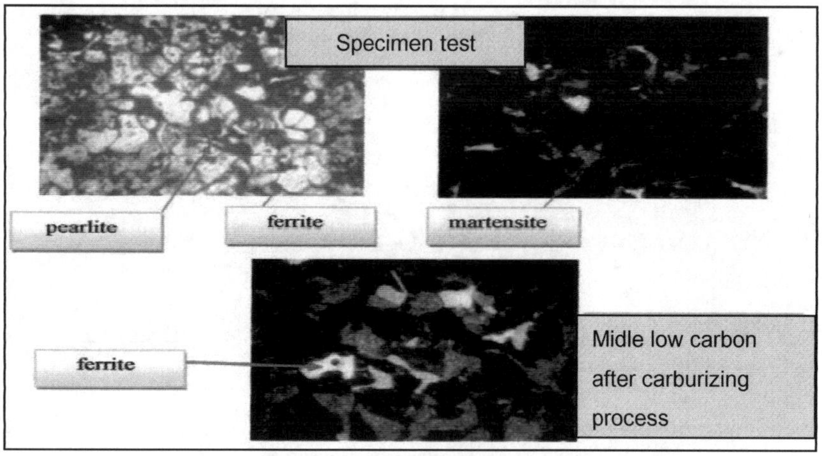

Figure 7. The activated carbon composition of 1kg with 20 % Ca Co_3

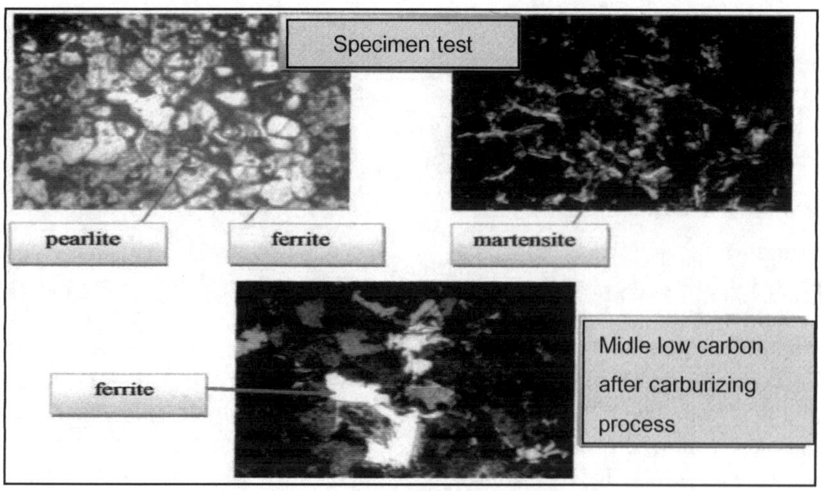

Figure 8. The activated carbon composition of 1kg with 10 % Ca Co_3

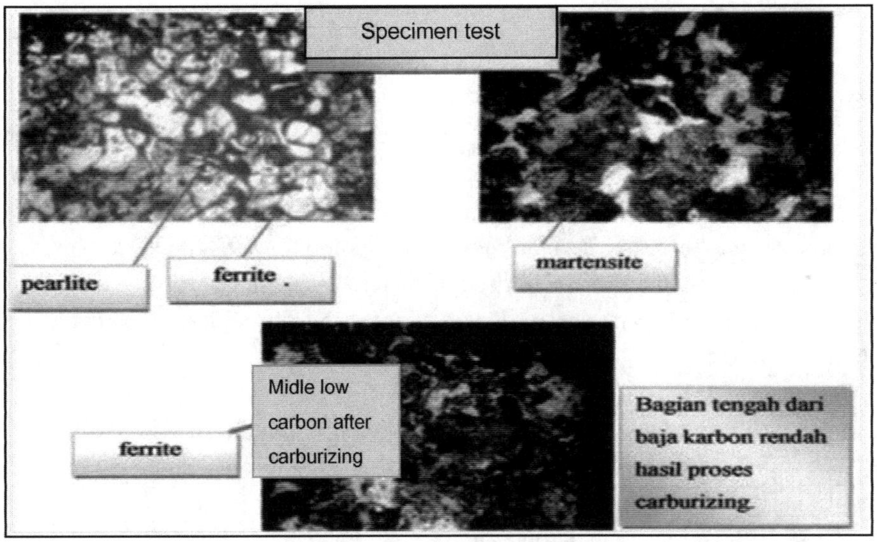

Figure 9. The activated carbon composition of 1kg with 5 % Ca Co_3

4.2 Hardness

Hardness defined as the ability of a material to withstand the load indentation or penetration (emphasis added). Violence is one of the mechanical properties *(mechanical properties)*of a material. Test equipment used is a Rockwell hardness tester. From the hardness testing performed on untreated carburizing steel and steel with carburizing treatment obtained average hardness as shown in Figure 10.

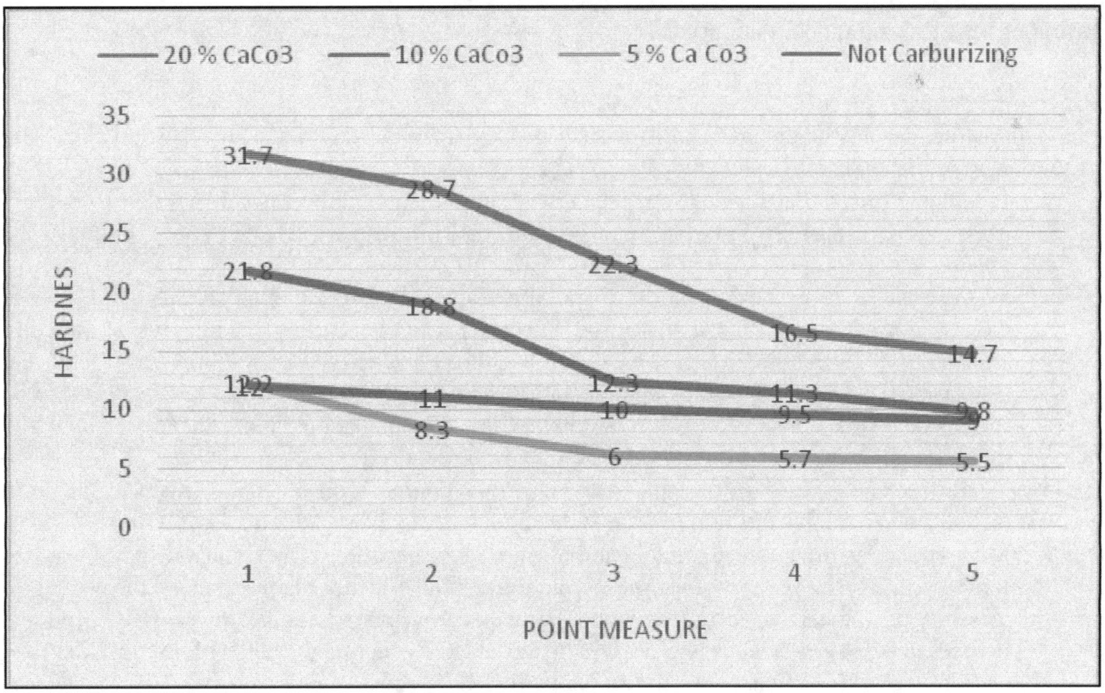

Figure 10. Graph HRC hardness test material

5. Discussion

From microstructure formed on low carbon steel test material without carburizing treatment (Figure 6) looks ferrite (white) and pearlite (dark), ferrite was shaped coarse grains, of value hardness on the edge 1 (Figure 10) 12 HRC, showed carbon steel this is a soft and relatively low carbon steel, microstructure formed on the material composition test with 20% $CaCO_3$, better diffusion of carbon into the test material carbon steel, case this can be seen with martensitic structure after quenching process, this indicates that there has been an increase or addition of carbon into the test material. This martensitic formation causes the surface of the test material becomes hard. Hardness average at one point (the edge) 31.7 HRC, then hardness showed a decrease to the side of the center (point 5) 14.7 HRC as in Figure 10.

The composition of 10% $CaCO_3$ showed a decrease of violence in point 1 (side edge) with violence an average of 21.8 HRC (Figure 10) and formed in the form binate micro structure. While the composition of the 5% $CaCO_3$, the hardness value at point 1 (side edge) equal to the values to violence on the test material without treatment carburizing is equal to 12 HRC (picture 10), and the hardness at the center side is 5.5 HRC is lower than the hardness value of materials carbon steel test without carburizing treatment by 9 HRC. The decline in violence in the composition of 10% and 5% $CaCO_3$ is because of decreased diffusion of carbon into the test material carbon steel during carburizing process, decrease carbon diffusion is caused lower percentage of $CaCO_3$ in the activated carbon.

From Figure 10, can be seen in the test material with a composition of 5% $CaCO_3$, a decline in violence from the side edge (point 1) to the side of the center (point 5), hardness is lower than the test material without treatment carburizing, this was due to warming test material, while at the carbon fusion of activated carbon does not occur.

6. Conclusions

The microstructure and hardness testing on steel process results carburizing and steel without the process of carburizing, it can be concluded:

- From microstructure formed on the test material low carbon steel without treatment carburizing form of ferrite (white) and pearlite (dark), ferrite was shaped coarse grain, of the value of hardness at the edge of the side 1 to 12 HRC.
- Microstructure formed on the test material with a composition with 20% $CaCO_3$, better diffusion of carbon into the test material carbon steel, it is seen to form martensitic structure after the quenching, this indicates that there has been an increase or addition carbon into the test material , The formation of martensitic This causes the surface of the test material to be loud, where violence was the average of her on the side of a single point (the edge) 31.7 HRC then the hardness shows a decrease to the side of the center (point 5) 14.7 HRC
- Microstructure formed on the test material with a composition of 10% $CaCO_3$ in the form binate, a decline in violence in point 1 (side edge) with an average hardness of 21.8 HRC.
- Microstructure formed on the test material with a composition of 5% $CaCO_3$ in the form of ferrite pearlite fine grains of hardness value at point 1 (side edge) equal to the values to violence on the test material without treatment carburizing is equal to 12 HRC, and the hardness at the center side is 5.5 HRC lower than the hardness value of the test material carbon steel without treatment carburizing by 9 HRC
- $CaCO_3$, an ingredient *energizer* in the *pack carburizing,* the fewer the number of materials *energizer* then the hard carbon of activated carbon diffuses into the steel

7. References

[1] Nevada (2011). *Analysis hardness of low carbon steel (S35C) With the influence of holding time through a process of pack carburizing With the use of shells as a catalyst*. Volume **8** Number 2, 927-935 Journal of Technology.

[2] G. Budinski, Budinski K. (1999), *Engineering Materials Properties and Selection*, 6th edition, Prentice Hall International, Inc., New Jersey, USA.

[3] B. Kuswanto, (2010). *Pack carburizing Treatment in Low Carbon Steel Materials as Alternative to Cut Knife on Application of Appropriate Technology*, Proceedings of the National Seminar on Science and Technology, 2010. Faculty of Engineering, University of Wahid Hasyim Semarang.

[4] Samsudi Raharjo. (2007). *Analysis of Agricultural Products Using Tools Pack Furnace Carburizing With Conventional Furnaces. Traction*. Vol. **5**. No. 1, Jurnal.unimus.ac.id 12.

[5] Hafni. (2013). *Design furnace. Pack carburizing* Vol. **VII**, No.40, Sept 2013 Journal of Science Tower LPPMUMSB.

[6] WattimenaWME (2014) *Effect of holding time and quenching to violence carbon steel st 37 in the process of carburizing Pack Using charcoal nutmeg (Myristica fagrans)* Volume **11** Number 1, 1163-1171. Technology Journal.

[7] Hafni (2015), "*The Effect of Hold Time Process Pack carburizing In Low Carbon Steel by Using calcium carbonate and Coconut Shell Charcoal*, The Review Of Violence," Journal of *Mechanical Engineering*, Vol **5** ISSN 2089-4880. Institute of Technology Padang.

[8] Mulyadi (2010). *Study of Changes in Carbon fusion of violence and as a result of the process kuacing on Dental Materials Removal Power Thresther.*Vol **7**. No. 1, ISSN 1829-8958.Journal of Mechanical.

[9] Aramide FO Engineering. (2010), *Pack carburization of Mild Steel, using Pulverized Bone as Carburizer: Optimizing Process Parameters*. ISSN 1583-1078. Leonardo Electronic Journal of Practices and Technologies.

[10] Hazizi Azri Bin Ahmad Sabri (10) *Experimental study of pack carburizing of carbon Steel. Bachelor of Engineering University Malaysia Pahang*.

Acknowledgments

The author would like to thank the Ministry of Research and technology and Higher Education that has helped in support financially, and the academic community of the Institute of technology of Padang in the process of implementing this study.

Analysis of Factors Affecting the Success of Onions Development Program in Kampar Regency

AMALIA AND ASGAMI PUTRI
Universsitas Lancang Kuning, Pekanbaru, 28265, Indonesia
Amaliamasjkur@unilak.ac.id, asgami@unilak.ac.id

Abstract. The purpose of this study is to analyze the factors influencing the success of the onion plant development program in Kampar regency. The research method used was the applied survey method using interview technique and observation or direct supervision on the location of the object. The briefing of the interviews as well as the accuracy of collecting the required data was guided by the structured questionnaires. Determination technique of location / region sampling was done purposively based on the potency and capacity of commodity development. While the respondents were taken by cluster purvosive sampling method in order to classify the samples in accordance with the purpose of the study, determined by as many as 100 people taken from members of the farmer group. Analytical technique used is by using Logic Regression Analysis to determine the factors that influence the success of the program seen from the characteristics of farmers. From the results of this study it can be concluded that the factors influencing the success of onion development program in Kampar regency were a age (X1), education (X2), income (X3), ethnicity (X4), jobs (X5) And family responsibility (X6) could be made as follows:

$$\text{Log Y (P/1-p)} = -1.778 + X1^{0.021} + X2^{0.028} - X3^{0.213} + X4^{1.986} + X5^{2.930} - X6^{0.455}$$

From the above equation, it can be explained that the attributes that are positively related are X1 (age), X2 (education), X4 (ethnicity) and X5 (jobs) while the negative correlates are X3 (income) and X6 (family responsibility). From the logical regression result it can be seen that the significant value <0,05, then the independent variable influenced the dependent variable, so that when viewed from the table in the equation it was found that factors affecting the success rate of red onion development program in Kampar regency were X2 (education), X4 (ethnicity), X5 (jobs), and X6 (family responsibility).

Keywords: Development program, onion, factors affecting

1. Introduction

The various policies and development programs in Kampar regency, especially in the field of Agriculture and Food Security as outlined in the Strategic Plan (Renstra) are clearly designed by considering the amount, quality and distribution of potential resources. This situation is increasingly important in the era of regional autonomy which on the one hand provides opportunities and considerable flexibility for regions in designing the implementation of regional development area. But on the other side is a challenge in itself, as a result of the region should be able to grow creativity, especially in the form of concrete efforts in accelerating the progress of regional development.

The Government of Kampar Regency in efforts to improve the welfare of its people through the horticulture sub-sector has done the development of red onion cultivation, where red onion cultivation has never been developed in Kampar District, because most of the land in Kampar Regency is

peatland. Related to red onion production, the Ministry of Agriculture has conducted various development and research to meet the needs of red onion production domestically including in the development of onion cultivation in peatlands. The onion planting in peatlands is a groundbreaking program to broaden the development of peatland specific to red onion inclusions, including in off-season conditions (high rainfall). In addition to looking at the territory of Kampar Regency that has the potential of lowland and wide peatlands.
Based on the findings of science and technology on the development of onion peatland agriculture, Kampar Regency government seizes opportunities to improve the welfare of its people through the breakthrough of the development of onion cultivation in Kampar Regency. Until now, the production of onions in Kampar Regency is only for the purposes of the region only. But will slowly expand this commodity market to the District to neighboring provinces. Kampar regency government also targets to become the center of red onion crops in Sumatra in 2016. One of the largest land is located in Geringging Village Kampar Kiri District which has 13 hectares of onion land.

According to Titiek Purbiati (2012), in the cultivation of onion in peatland needs to adopt the research and modification so that according to its agro-ecosystem which include peat incision, soil processing, fertilizer application, weed cleaning, pH neutralizing and paving dams. The application of technology certainly requires a large cost, for that the government also helps the cost of development with APBD worth Rp.6 billion and APBN worth Rp.3 billion, while the ministry of agriculture will realize assistance Rp.2.5 billion for land in the village of Bima, Kampar Village and Nganjuk Village. Based on the research result of Yasid, H, and co. (2015) that in Geringging River Village Kampar Kiri Subdistrict Kampar Regency, the commodity of red onion has the potential to be the superior commodity in the area with the note of the need for continuous effort in bracing obstacles that still plagues such as related to infrastructure, cultivation technology, and farmer skill which should be continuously improved.

The number of farmers who cultivate red onion crops as a whole should be a benchmark that can provide significant value for the income of the community, but in line with the government program until now there has been no facts in the form of figures that will be the comparison of how much income is obtained by the community With the availability of existing land to produce red onion crops.Based on the detailed description, the research team is interested in conducting research entitled "Analysis of Factors Affecting the Success of Onion Development Program in Kampar Regency". The purpose of this study is to analyze what factors influence the success of the red onion plant development program in Kampar Regency.

2. Research Method

This research was conducted in March 2016 until September 2016. The type of data used in this study is Primary and Secondary Data. The research method used is the applied survey method using interview technique and observation or direct supervision on the location of the object of development of onions. The briefing of the interviews as well as the accuracy of collecting the required data is guided by a structured questionnaire.

Determining the technique of location / region sampling is done purposively based on the potency and carrying capacity of commodity development. While the respondent sample taken by cluster purvosive sampling method to classify the sample in accordance with the purpose of the study, is determined as 100 people taken from members of farmer groups who receive the assistance of APBD and APBN each 50 farmers taken with probability sampling as a businessman of onion production.

The analysis used to determine the factors that influence the success of the program seen from the characteristics of farmers is logical regression analysis. Logistic regression is part of the regression analysis used when the dependent variable (response) is a dichotomous variable. Dichotomous

variables usually consist of only two values, representing the occurrence or absence of an event usually labeled 0 or 1 (Hosmer, Lemeshow, 1989).
Logistic regression models can be established by looking at the estimation values of parameters in Variables in The Equation. Regression model that is formed based on parameter estimation value in Variables in The Equation is as follows.

Log (P / 1 – p) = β0 + β1X1 + β2X2 + ... + βkXk

Hypothesis testing is done by comparing the level of significance (sig) with the error rate (α) = 5%.

3. Result and Discussion

Factors tested against the success of the development program of red onion crop in Kampar is age (X1), education (2), income (X3), ethnicity (X4), jobs (X5), family responsibility (X6). The probability variable in this case consists of two alternative options: "Success" is coded (1) and "Not successful" is coded (0). After analyzing the results obtained as follows:

Tabel 1. Classification Table*

Observed	Predicted		
	Success		Persentage Correct
	Not Successful	Successful	
Step 1 Successful Not Successful	51	8	86.4
	9	32	78.0
Overall Percentage			83.0

a. The cur value is .500

From table 1 it is known that there are 59 respondents from Not Successful categories and among them there are 8 respondents who have the possibility to succeed and 51 respondents totally failed. So it is possible for this program to work if it is improved. From then there are 41 respondents who succeed, but it turns out there are 9 respondents who have a chance to fail, but these 9 respondents are trying so they will still succeed.

Attributes that become factors influencing the success of onion development program in Kampar regency are age, education, income, ethnicity, jobs, family responsibility. The equations are formed as follows:
Log Y (P/1-p)= -1.778 $+X1^{0.021} + X2^{0.028} - X3^{0.213} + X4^{1.986} + X5^{2.930} - X6^{0.455}$

From the results of the above equation can be explained what attributes are significant. Positively related are Age (X1), e(X2), enhnicity (X4) and jobs (X5), while Negative related are income (X3) and family responsibility (X6).

From the equation above it is known that if there is an age increase of respondents as much as 1 year then there is an opportunity for the respondents success rate in developing red onion will increase by 2.1%. This means that the chances of unsuccessful respondents will decrease with age, this is because the higher the age of respondents the tendency of respondents to develop onion will be more serious by following the advice of agricultural extension in planting onions.

If there is an increase in education level of 1 year then there is a chance of successful respondents in developing onion cultivation increased by 2.8%. This means that the higher the respondent's education, the easier it will be to absorb information and be more aggressive in finding information for success in developing red onion.

If the level of consumer income increased by Rp.1, then there is an increased opportunity for unsuccessful respondents as much as 2.13%. This is due to the higher consumer income than the respondent's only makes the onion development program just as a side income so that respondents tend to not be serious in managing the cultivation of onions.

From the above equation is known that if the Javanese is increased by 1 person then there is a chance for successful respondents in developing the red onion of 19.86%. This is because the Javanese ethnic has a better work ethic characteristics in onion cultivation.

From the equation is known that if the farmer's jobs increases then there is a chance for successful respondents in cultivating onions as much as 29.3% This is because the jobs of farmers to make respondents have experience in cultivating plants so that the success rate will be higher than other types of jobs.

From the above equation is known that if the responsibility of the respondent's family increased by one person, then there is a chance for respondents to not succeed in cultivating onion by 4.5%. This is due to the increasing number of responsibility in the families of respondents, the tendency to use government assistance is not in accordance with the recommendation because it diverted its use for consumptive costs, where the greater the responsibility of the family then the higher the daily needs. So there is a tendency of respondents to cover the cost of daily consumption by using government assistance funds.

4. Conclusion

From the results of this study, which affect the success rate of red onion development program in Kampar regency is X2 (education), X4 (etnicity), X5 (jobs), and X6 (family responsibility). The most positive influence is the factor X5 that is equal to 29.3%, this is because the jobs of farmers to make respondent have experience in cultivating plants so that the success rate will be higher tahn other types of jobs.

REFERENCES

[1] Amalia, Hamdan Yasid dan Asgami Putri. 2015. *Analisis Komoditi Unggulan Bawang Merah di Kecamatan Kampar Kiri Kabupaten Kampar.* Jurnal Ilmiah Pertanian Vo. 10 No.2 Agustus 2015. Pekanbaru

[2] Badan Litbang Pertanian. 2006. *Prospek dan Arah Pengembangan Agribisnis Bawang Merah.* http://www.litbang.deptan.go.id/special/publikasi.

[3] *BPTP Sultra (Balai Pengkajian Teknologi Pertanian Sulawesi Tenggara).* 2009. Teknologi Budidaya Bawang Merah di Lahan Kering. http://sultra.litbang.deptan.go.id.

[4] Ditjen PHP (Direktorat Jendral Pengolahan dan Pemasaran Hasil Pertanian). 2006. *Road Map Pascapanen, Pengolahan dan Pemasaran Hasil Bawang Merah.* http://agribisnis.deptan.go.id.

[5] Ginting, H. 2002. *Pengaruh Daya Serap Teknologi Pterhadap Usaha Tani yang dilakukan oleh Petani. Laporan Hasil Peneltian Universitas Sumatra Utara Fakultas Pertanian.*

[6] Iriani, E. 2013. *Prospek Pengembangan Inovasi Teknologi Bawang Merah di Lahan Optimal (Lahan Pasir) dalam Upaya Peningatan Pendapatan Petani*. Jurnal Litbang Propinsi Jawa Tengah Vol.11 No.2 Desember 2013.

[7] Sudjana. 2001. *Metode Statistika, Edisi Revisi Cet 6*. Bandung:Tarsito.

IOP Conf. Series: Earth and Environmental Science **97** (2017) 012039 doi:10.1088/1755-1315/97/1/012039

Mass Media's Perspective on the Challenges of Information and Communication Technology (ICT) in Indonesia

Junaidi
Universitas Lancang Kuning, Pekanbaru, 28265, Indonesia
junaidi@unilak.ac.id

Abstract. This study investigated the challenges of Information and Communication Technology (ICT) in Indonesia based on mass media's perspective. The challenges of ICT on mass media's perspective are very complex. Some of these challenges include the consideration of the advantages and disadvantages of ICT, threats to the unity of the country, cultural problems, regulation, business competition among operators, Pokemon Go booming, the existence of social media which triggers hostility between people and the increasing rate of the interconnection cost. Among the main reasons to this phenomenon is a need of massive investment to build the infrastructures. This study used Robert N. Entman framing model to analyze the headlines of Majalah ICT (an ICT magazine).

Keywords: *Challenge, ICT, Mass Media, Framing.*

1. Introduction

Information and Communication Technology (ICT) has developed rapidly in Indonesia and it influences many aspects of life. ICT market in Indonesia is predicted to increase by 12.2% on CAGR in 2016 – 2020. Based on BMI research, the comunication Asia and Broadcast Asia events were very successful, shoved this business to take the advantages from mass media and digital broadcasting market development through Communic in Indonesia. Some technology vendors such as IBM, Siemens, Huawei and Indosat have come into the market with some Smart Cities solutions. The business also supports a national campaign on '1000 Digital Startup' by 2020, which is initiated by the Indonesian Communication and Information Ministry and local start-up incubator company Kibar.

Warsito (2008:70) stated that the development of ICT today has reached the third phase. The first phase emerged in the form of agricultural technology. The second phase emerged in the form of industrial technology. Now, the third phase is signified by electronic and information technology development. As a matter of fact, it is inevitable that electronic and information technology has influenced many aspects of life (Olivia, 2016).

The trend and development of ICT in Indonesia is well reflected on the report of ICT Index released by the International Telecommunication Union (ITU) in late 2015 (ICT Development Index, 2015). ITU, the UN section concerning with telecommunications and information technology, has released the world ICT Development Index (IDI) in the progress report of the Information Society. In this index, Indonesia is on the 108th position in which its IDI value is only 3.94. This position is below other ASEAN countries such as Singapore getting 8.08 point and it is on the 19th position in the world. Malaysia is far above Indonesia in which it stays on the 64th position. Other ASEAN countries, such as Brunei Darussalam occupies the 71st rank, Thailand stays on the 74th level, Philippines gets the 98th position, and Vietnam at the 102nd position (4.28 points). Compared to its position in 2010, Indonesia moves up 1 point (from 109th to 108th position). At that time, Indonesian IDI value was 3.11. Meanwhile, for the regional position, Indonesia is ranked 18th. The highest IDI score in Asia-Pacific is achieved by Korea followed by Hong Kong, Japan, Australia, New Zealand and Singapore. For the global level, Korea is on the first rank in which it gets 8.93 points followed by Denmark, Iceland, the UK and Luxembourg.

Thus, the researcher generated the research question for this study based on the above mentioned facts. Therefore, the research question is 'What are the challenges of Information and Communication Technology (ICT) in Indonesia based on mass media's perspective? A framing approach analysis is on the news in ICT' magazine. Based on this research question, the purpose of the study was formed to identify the challenges of ICT in Indonesia based on mass media's perspective.

2. Method

This is a descriptive qualitative research which applied framing analysis method and constructivism paradigm (Moelong, 2007). Constructivism paradigm believes that there is no objective reality, because it is created through construction process and particular views (Creswell, 1998; Lowenberg, 1993; Schwandit, 1994). This research sought to describe and summarize various conditions, situations or phenomena of social reality in a society taken as the object of the research, and to lift the reality up to surface as a characteristic, a sign or a description of particular condition, situation or phenomenon. Thus, qualitative descriptive format is more appropriate to be used to study the issues that require in-depth analysis such as the study on consumer behaviors, media effects, and the implementation of a policy (Berger, 1967; Bungin 2008). This format commonly uses sentences that match to the theories in data presentation.

Table 1: The unit of study

No	Edition of the Magazine	News
1	Majalah ICT No.46-Year IV-June 2016	Telecommunication Operator Business Fierce War
		Indonesia Cyber Council Vacillation
		The Launch of Brisat and Lapan-IPB Satellite
		The Head of the Indonesian Police's Evaluation on Social Media
2	Majalah ICT No.47-Year IV-August 2016	The Arrest of Application-based Taxis
		The commotion of Pokemon Go
3	Majalah ICT No.48-Year IV-September 2016	The Intermittent Interconnection
		Hunting Down Google Tax
		Jack Ma as the E-Commerce supervisor

2.1 Framing

Robert N. Entman (1993) is one of the experts putting the foundation for framing analysis intended to study the content of the media. The concept of framing was written in an article for the Journal of Political Communication and other writings applying the concept in a case study of media coverage. The concept of framing by Entman is used to describe the selection process and highlights certain aspects of reality (Eriyanto, 2001).

According to Entman (1993), framing in the news is done in four different ways. The first is problems definition. This element is regarded as the master of the frame. The second is causes diagnosis. This element is used to frame those regarded as actors in an event. The causes in this case could be 'what' or 'who'. The third is moral judgment making. This framing element is used to justify or provide an argument on the problems that are already defined. The fourth is treatment recommendation. This element is used to justify the definitions.

3. Discussion and Result

3.1Telecommunication Operator Business Fierce War

The price war between *Indosat Ooredoo* and *Telkomsel* is taken as the Problem Definition (PD) from a headline entitled 'Telecommunication Operator Business Fierce War' in the ICT No.46-Year IV-June 2016 edition magazine. The price war reached the culminating point when *Indosat Ooredoo* challenged the most respected telecommunication operator in Indonesia, *Telkomsel*. In fact, *Indosat* attacked *Telkomsel* publicly within its products promotion.

The price war would never happen without any reasons and triggers. This competition was begun from a picture which became viral in social media. In promoting its product, *Indosat* wrote 'Only *IM3* Oooredoo can make a call with IDR. 1/second, *Telkomsel*? Impossible is in one of its banner. This case is considered as the Cause Diagnosis (CD) of the existing problem. Meanwhile, *Telkomsel* has another view that its domination out of Java is not a monopoly practice, yet it is a long process which has been started since 1995. At that time, other operators were busy developing their existence in Java and some other big cities with the highest financial and commercial benefits. Differently, *Telkomsel* connection development focus was outside of Java which had smaller market opportunity. At the same time, *Telkomsel's* CAPEX (Capital Expenditure) was highly consumed. In addition, in its operation and management, the production and operational costs were actually higher than of those in Java as stated by Adita Irawati, *Telkomsel* Corporate Communication.

Nurul Yakin Setyabudi as the chairman of Indonesia Telecommunication Group (IDTUG) said that ads and rate war phenomenon that occurs between telecommunication operators in Indonesia nowadays (especially *Indosat* and *Telkomsel*) has shown the presence of indicators to negative business behavior and tend to be impudent in promoting the product they sell to Indonesian telecommunication users. Further, each operator's behavior has broken the etiquette in competing and promoting their product, which is taken as the Moral Judgement Making (MJM) from the headline of this media release.

The Treatment Recommendation (TR) for this case is to call the operators to compete in a good manner. In the case of 'Telecommunication Operator Business Fierce War', mass media has a perspective that the problem such as business competition between *Indosat Ooredoo* and *Telkomsel* may contribute to a greater challenge in Information and Communication Technology (ICT) development in Indonesia. In addition, this competition creates rate inequality between these two huge telecommunication companies in Indonesia. These problems are assumed to be mainly triggered by infrastructure gaps between Java and outer Java regions.

3.2 Indonesia Cyber Council Vacillation

The formation of Indonesian National Cyber Council was canceled because of the establishment of institution or council moratorium, which is taken as the Problem Definition (PD) in this news release. The intended establishment of National Cyber Council was just one step away. It needed the presidential regulation which would be announced by Joko Widodo, but it was canceled because of the new council formation moratorium. This case is considered as the Cause Diagnosis (CD) in the framing concept of this news release. The needs of safety related to the cyber world are aiming to protect information sources and digital data that belong to the government and public which are forbidden to be exposed to irresponsible parties.

Accordingly, the government decision in canceling the formation of National Cyber Council is the Moral Judgement Making (MJM) of the news release framing. After that, as the alternative, the role of this council was redirected to the State Coding Council which is taken as the Treatment Recommendation (TR) in analyzing the framing of the media news release.

The discussion of framing analysis results through mass media's perspective concludes that the Information and Communication Technology in Indonesia today needs to provide safety. Therefore, there must be a council that has an authority to create the safety, which is the National Cyber Council. However, because of a classic reason (moratorium of new council formation), this ideal governmental body was not established.

3.3 The Launch of Brisat and Lapan-IPB Satellite

IOP Conf. Series: Earth and Environmental Science **97** (2017) 012039 doi:10.1088/1755-1315/97/1/012039

a. Brisat Satellite has Finally Orbited

A new release page 12 is about the launching of BRI satellite controversy. Even so, the news that BRI has a satellite is considered as the Problem Definition (PD) for a news release entitled 'Brisat Satellite has finally orbited'. As the Cause Diagnosis (CD), BriSat is the first satellite that is dedicated to serve financial institution. The launching was done from a European spaceport in Kourou, with a postponement experience. Arianespace delayed the launching for about 24 hours because there was a main connection anomaly to the top part of the launcher. In spite of its controversy, the launching of Brisat satellite was a pioneer for BRI itself. Brisat is the first communication satellite in the world which is dedicated for financial institution. The fact that it will give the second band for Indonesia and South-East Asia Kuband scope becomes the Moral Judgement Making (MJM) of the media news release framing. Indonesia has contributed a big investment through BRI for the purchase of a satellite which purpose is to provide financial service for about 15 years; this case is taken as the Treatment Recommendation (TR) from this media news release framing.

In this news release, mass media see that Information and Communication Technology phenomenon can potentially become an obstacle faced by business and service stakeholders in Indonesia especially the financial sector. One of its factors is Indonesian geographical condition which is widely spread from west to east regions. Such geographic condition requires a satellite as a primary need that can be used to support many kinds of activities in the context of civil service in the country.

b. The Launch of Lapan-IPB Satellite

After the launch of BriSat, IPB also launched its satellite. The news entitled 'The Launch of Lapan-IPB Satellite for Crops and Ships Monitoring' adds treasury to the Information and Communication Technology in Indonesia. LAPAN-A3/IPB satellite launch success is considered as the Problem Definition (PD) of this news framing analysis. In the sense of Cause Diagnosis (CD), it is the launching of LAPAN-A3/IPB satellite which carried a remote sensing experimental mission to monitor crop resources. Besides, with its ability, National Aviation and Space Agencies (Lapan) – IPB will be able to identify land usage, environmental monitoring, and ship monitoring. The Moral Judgement Making (MJM) of this news release is that LAPAN-A3/IPB satellite has been launched from Sahawatish A Shriharikota Space Center in India. The Lapan-IPB sensing satellite is a 4 Bands Multispectral Imager which has 18 meters resolution complimented by 100 kilometers swath. This satellite will be used to monitor crops. Further, the Treatment Recommendation (TR) is that LAPAN-A3/IPB satellite will support data accuracy in planning crop planting period that will implicate directly to the improvement of crop endurance in the country. This matter will help the government in determining many kinds of regulations related to crops and farming.

If we see how mass media try to build perspectives from LAPAN-A3/IPB Satellite news release, we can see that mass media's judgement related to the LAPAN-A3/IPB satellite has many specific wishes which are related to crop and ship activities in the ocean. It means that the development of Information and Communication Technology is not only related to education, office, and broadcasting, but also related to crops, marine, etc. Therefore, the existence of LPAN-A3/IPB satellite is a mean to answer the need of the people in the availability of Information and Communication Technology infrastructure in Indonesia.

3.4 Head of the Indonesian Police Evaluation on Social Media

The existence of cyber world has been a polemic not only in developed countries, but also in a developing country such as Indonesia. Social media is being a serious discussion topic nowadays. This new platform has been a source of problems and crucial information sharing basis at the same time. Magazine ICT No.47-Year IV-August 2016 headlined a topic that discussed about the dynamic of social media. In this news release, the Problem Definition (PD) is that the police will control social media. The means to evaluate media based on its Cause Diagnose (CD) describes that social media is becoming a medium to provoke social problems, religion, and others. The Moral Judgement Making (MJM) from this news release is that the mean used by the Head of Indonesian Police in doing the evaluation on social media gets significant support from Indonesian Information and Communication Ministry. It is done by closing the accounts which spread provocations causing ethnic, regional, and racial unrest conflicts. Furthermore, the Treatment Recommendation (TR) is by referring to what was said by Information and Communication Minister that it is forbidden to spread hatred in cyber world as it is against Indonesian Regulations of Information and Electronic Transaction called as *Undang Undang Informasi dan Transaksi Elektronik (UU ITE)*. In this news release, mass media consider the problem in existence of social media as a part of Information and Communication Technology phenomenon. It has been considered as a new problem which can cause hostility and conflicts among Indonesian citizens.

3.5 The Arrest of Application-based Taxis

The development of Information and Communication Technology spreads into various sectors of life, not limited to transportation sector. The presence of online taxi becomes a new problem when it is not regulated by appropriate policies in the world convergence. A news release on ICT No.47-Year IV-August 2016 edition magazine which discussed a topic entitled 'The Arrest of Application-based Taxis' is taken as the Problem Definition (PD).

The Cause Diagnose (CD) of the online taxi arrest is that this online taxi is proven to be breaking the law. One of the cases argues that the taxis did not follow the vehicle testing procedures regulation. Further, to look at another aspect, that 11 units of online taxi were arrested related to the license issue is considered as the Moral Judgement Making (MJM). Then, the Treatment Recommendation (TR) from this news release is that Transportation Ministry wants this online taxi to be an alternative taxi. In the meantime, it hopes that the online taxi providers are acting proactively and responsively to fulfill all regulated requirements. The evaluation is done because of the regulations which have been approved by both government and taxi companies were not obeyed respectively during the previous Ignatius Johan leadership in the Transportation Ministry.

In this case, mass media's perspective is textually neutral. However, the reality shows that Indonesian culture (read: Jakarta) is not ready to dive in the competition between conventional taxi and the online taxi, in which the latter is becoming viral nowadays. Moreover, the challenge of Information and Communication Technology here is related to the cultural issue within the Transportation Company and government who might find the process of switching from conventional to online system uneasy.

3.6 The commotion of Pokemon Go

The using of Pokemon Go game that became viral in Indonesia is considered as the Problem Definition (PD) in this news release. Since it was launched in New Zealand and Australia on 6th July 2016, Pokemon Go has influenced many people and became global phenomenon. It can be labelled as the Cause Diagnosis (CD) in news release of this phenomenal game. In the context of Moral Judgement Making (MJM) is the fact that many people had walked along the roads, parks, and certain places to catch the Pokemon.

This game has advantages as well as its disadvantages. The impact of this game comes to the point that it was forbidden to play Pokemon Go in some places which are considered vital like the Country Palace, Military areas, and Indonesian Police areas. As a matter of fact, the game has widely spread, including the civil servants as stated by the Ministry of Administrative and Bureaucratic Reform is seen as the Treatment Recommendation (TR) of this news release. The Pokemon Go booming in Indonesia has been a unique variation in the context of Information and Communication Technology development in the country. In the reality construction, through the way Pokemon Go presented its data, it can be framed that the game may become a threat for national stability. It is caused by the fact that Pokemon Go installation can expose the users' private data to the Pokemon Go developer. It means that Information and Communication Technology trend in Indonesia can be a serious threat for its people and the country security.

3.7 The Intermittent Interconnection

The high rate of communication usage such as smartphones has enforced providers to improve their first-rate service quality for costumers each day. In terms of framing, the news entitled 'The Intermittent Interconnection' discussed that there was a competition related to the interconnection rate between providers. The Problem Definition (PD) of this news release is that the plan to implement new interconnection rate which sides with its people must be postponed from the prior schedule, per 1

September 2016. It is caused by the assessment delay of the interconnection offering list from Indonesian Telecommunication (*Telkom*) and Cellular Telecommunication (*Telkomsel*). The reasons to this postponement are taken as the Cause Diagnosis (CD). The fact that Telkom and *Telkomsel* are the operators who firmly refused the result of new interconnection rate counting is considered as the Moral Judgement Making (MJM) from this news release.

In fact, the decision to implement the new interconnection rate has decreased by 26% in average; it then became a polemic which involves many parties. Further, the Treatment Recommendation (TR) is that the mass media see the challenge of Information and Communication Technology world related to interconnection rate is still its expensive cost. However, this matter is somewhat reasonable because it requires massive investment to develop the infrastructures as being described in the news entitled 'Telecommunication Operator Business Fierce War'.

3.8 Hunting Down Google Tax

In accordance to its development, the Information and Communication Technology process results in another problem for the country, especially in the tax collection sector as an income source. The presence of a giant search engine like Google is proven to be a great challenge too. Therefore, the Problem Definition (PD) is that Google does not want to pay the tax to Indonesia which is estimated to be around IDR 5 Trillions.

The unwillingness of this giant company was identified when Google Indonesia returned a letter from the Directorate General of Taxation to check this company's tax responsibility. That was the reason why the Directorate General of Taxation planned to conduct an investigation and process Google's rejection to pay the tax by the end of September. This matter becomes the Cause Diagnosis (CD) of this news release. In reference to the news entitled 'Hunting Down Google Tax', it shows that Google tried to neglect its tax responsibility to Indonesia. It is then considered as the Moral Judgement Making (MJM) item in this news release. On Google's unwillingness to pay the tax, the Directorate General of Taxation and Ministry of Finance of the Republic of Indonesia will do their best to hunt down the tax duty of this giant internet company. Google indebtedness of tax liability is around IDR 5.2 Trillion or USD 400 Billion.

The Treatment Recommendation (TR) of this news release is that when Google is persistent in not paying the tax, the Directorate General of Taxation will fine Google as much as 400% of its current tax liability debt.

3.9 Jack Ma as the E-Commerce Supervisor

E-commerce world has become the most popular business platform nowadays. Many platforms have been introduced by many developers to develop e-commerce world in Indonesia. The E-commerce players' expansion from China, Alibaba, gets a good opportunity to develop in Indonesia because Indonesian government trusts Jack Ma to be the supervisor in the Steering Committee of National Roadmap E-Commerce implementation. In conclusion, the opinion that Jack Ma should become the E-Commerce supervisor is taken as the Problem Definition (PD) from this news release.

In terms of the news' Cause Diagnosis (CD) item, it is seen from the need of the Indonesian president to learn about the vast e-commerce to see the opportunities for Indonesia. It is considered crucial in order to help promote economic products of small and medium enterprise businesses and young Indonesian creators in China's markets. Through the discussion between Joko Widodo and Jack Ma, it is understood that both parties have the same point of view related to the development of small and medium enterprise in Indonesia. The above statement by the author is believed to be crucial as it is described from its Moral Judgement Making (MJM) that the government asked for Jack Ma's help in the E-Commerce development in Indonesia.

In the end of the discussion session with the CEO of Alibaba Group, Rudiantara as the Communication and Information Ministry stated that Indonesian E-Commerce needs development especially on enhancing Indonesian small and medium enterprise to go online. It needs full support from the government because the small and medium enterprise in Indonesia has contributed significantly the national economic growth. Therefore, there is a need for extended support especially

to export their products. Further, Rudiantara also asked Jack Ma to be a supervisor in the Steering Committee of National Roadmap E-Commerce implementation in Indonesia.
The Treatment Recommendation (TR) of this news release is that there is a drive for E-Commerce world in Indonesia to reach a value of USD 130 Million in 2020 with Jack Ma involvement in this field. Media views this news based on the reality construction that Jack Ma's contribution in Indonesian E-Commerce business is predicted to contribute positively on Indonesia E-Commerce development. As stated earlier, it is expected to be valued at USD 130 million worth of contribution to be achieved in 2020 by pointing the big boss of Alibaba to be in the supervising board.

4. Conclusion

First, the mass media view that the phenomenon of Telecommunication Operator Business Fierce War between *Indosat Ooredoo* and *Telkomsel* was triggered by the infrastructure gaps between Java and out of Java regions. Consequently, it created rate inequality among these two giant telecommunication companies in Indonesia.
Second, the mass media view that threats on ICT activities are increasing. So, the government has to provide appropriate safety and security systems related to the cyber world by forming a council. The council would be responsible on such safety and security matters and called as National Cyber Council. However, because of a classical reason (new council formation moratorium) this council was not formed.
Third, the mass media view that Indonesia has a huge area to cover. Due to such geographical condition, a satellite is needed and has become a primary needs to support many activities in the context of governmental services for its people.
Fourth, the mass media view that LAPAN-A3/IPB satellite was created in order to facilitate the needs of its people related to the availability of ICT infrastructure in Indonesia especially in the crop and marine aspect.
Fifth, the mass media view that social media as a part of ICT phenomenon and it has become a new problem which can create hostility and conflicts among people.
Sixth, the mass media view that Indonesian culture (read: Jakarta) is not ready to deal with the competition between conventional and online taxi providers which is booming nowadays. The challenge of ICT in this case is actually related to the cultural problem of the transportation companies and government, which do not want to move on from the conventional way to online system.
Seventh, the mass media views that installing Pokemon Go game application will expose users' private data to the Pokemon Go company developer. It means that ICT trend in Indonesia might become a threat for the citizens and the unity of the Republic of Indonesia.
Eighth, the mass media view that interconnection price is still arguably expensive. However, this matter might be reasonable due to the fact that infrastructure development needs significant investment funds.
Ninth, based on mean construction reality that has been created about Jack Ma's contribution in Indonesian e-commerce world, the manifestation in Indonesian e-commerce development can be projected to be around USD 130 Million in 2020. It leads to the idea of pointing out the big boss of Alibaba to be the e-commerce supervisor.

References

[1] *Beberapa taksi berbasi aplikasi (kembali) ditahan*. (2016, August). *Majalah ICT – All about ICT in Indonesia, (47),* 34.

[2] *Belanja ICT Indonesia 2016-2020 tumbuh 12.2%*. (2016, August). *Majalah ICT – All about ICT in Indonesia, (47),* 25.

[3] Berger, Peter & Thomas. (1967). *The social construction of reality: a treatise in the sociological of knowledge*. New York: A Double Day Anchor Book.

[4] Bungin, B. (2008). *Penelitian kualitatif, komunikasi, ekonomi, kebijakan publik, dan ilmu sosial lainnya.* Jakarta: Kencana Prenada Media Group.

[5] Creswell, J. W. (1998). *Qualitatif inquiry and research design*. California: Sage Publications.

[6] Entman, R. M., & Rojecki, A. (1993). Freezing out the public: elite and media framing of the U.S. anti-nuclear movement. *Political Communication, 10(2),* 151-167.

[7] Eriyanto. (2001). *Analisis wacana: pengantar analisis teks media.*Yogyakarta: LKiS.

[8] Geger Pokemon Go. (2016, August). *Majalah ICT – All about ICT in Indonesia, (47),* 4.

[9] Interkoneksi di tengah kompetisi. (2016, September). *Majalah ICT – All about ICT in Indonesia, (48),* 4.

[10] International Telecommunication Union (ITU). ICT Development Index 2015. (n.d.). Retrieved June 29, 2016, from http://www.itu.int/net4/ITUD/idi/2015/

[11] *Kementerian Komunikasi dan Informatika RI. Roadmap konvergensi infrastruktur teknologi informasi dan komunikasi Indonesia yang Disusun oleh Direktorat Jenderal Pos dan Telekomunikasi Departemen Komunikasi dan Informatika Republik Indonesi*a, Retrieved Februari 4, 2017 from *http://www.postel.go.id/content/ID/regulasi/telekomunikasi/kepdir/penganta -narasi-draft-roadmap-tik.doc*.

[12] *Kepolisian akan evaluasi penggunaan media social di Indonesia*. (2016, August). *Majalah ICT – All about ICT in Indonesia, (47),* 28.

[13] *Ketika gendering perang operator ditabuh*. (2016, June). *Majalah ICT – All about ICT in Indonesia, (46),* 4.

[14] *Kisruh taksi aplikasi vs taksi konvensional*. (2016, March). *Majalah ICT – All about ICT in Indonesia, (43),* 4.

[15]Lowenberg, J.S. (1993). *Interpretive research methodology: broadening the dialogue. ANS: Advances in Nursing Science, 16(2),* 57-69

[16] *Maju mundur badan cyber nasional*. (2016, June). *Majalah ICT – All about ICT in Indonesia, (46),* 15.

[17] *Mau dibawa kemana E-commerce kita*?. (2016, January). *Majalah ICT – All about ICT in Indonesia, (41),* 4.

[18] McLuhan, M. (1999). *Understanding media; the extension of man.* London: The MIT Press.

[19] Mengejar pajak Google. (2016, September). *Majalah ICT – All about ICT in Indonesia, (48),* 11.

[20] Moelong, L. (2007). *Metodelogi penelitian kualitatif.* Bandung: PT. Remaja Rosdakarya.

[21] Mulyana, D. (1999). *Nuansa-nuansa komunikasi: meneropong politik dan budaya komunikasi masyarakat kontemporer*. Bandung: Remaja Rosdakarya.

[22] Olivia, R. (2016, May 12). *Peluang Industri ICT di Era Globalisasi.Retrieved* June 29, 2016, from http://komunikasi.us/index.php/course/perkembangan-teknologi-komunikasi/7056-peluang-industri-ict-di-era-globalisasi

[23] *Pembangunan ICT Indonesia ternyata masih tertinggal*. (2015, December). *Majalah ICT – All about ICT in Indonesia, (40),* 12.

[24] *Pemerintah keluarkan peta jalan E-Commerce*. (2016, November). *Majalah ICT – All about ICT in Indonesia, (50),* 4.

[25] Revisi UU ITE lancar jaya. (2016, April). *Majalah ICT – All about ICT in Indonesia, (44),* 4.

[26] Rodney, C. (2005). *Modern communication technologies and the new world information order. International Journal of Communication, (2),* 147.

[27] Rogers, E, M. (1986) *Communication technology: the new media in society*. London: The Free Press.

[28] *Satelit BriSat akhirnya mengorbit*. (2016, June). *Majalah ICT – All about ICT in Indonesia, (46),* 12.

[29] Schwandt, T. A. (1994). *Constructivist, interpretivist approaches to human inquiry*. In N. K. Denzin & Y. S. Lincoln (Eds.), *Handbook of qualitative research* (pp. 118- 137). Thousand Oaks, CA: Sage.

[30] Straubhaar, J, & LaRose, R. (2004). *Media now: Undersanding media, culture and technology*. California: Thomson Learning.

[31] Warsito, B. (2008). *Teknologi pembelajaran (landasan & aplikasinya)*. Jakarta: Rineke Cipta.

The Application of Determining Students' Graduation Status of STMIK Palangkaraya Using K-Nearest Neighbors Method

Lili Rusdiana[1] and Marfuah[2]
[1] STMIK Palangkaraya, Indonesia
[2] Universal University, Batam, 29456, Indonesia
E-mail: fasliiana7@gmail.com

Abstract: K-Nearest Neighbors method is one of methods used for classification which calculate a value to find out the closest in distance. It is used to group a set of data such as students' graduation status that are got from the amount of course credits taken by them, the grade point average (AVG), and the mini-thesis grade. The study is conducted to know the results of using K-Nearest Neighbors method on the application of determining students' graduation status, so it can be analyzed from the method used, the data, and the application constructed. The aim of this study is to find out the application results by using K-Nearest Neighbors concept to determine students' graduation status using the data of STMIK Palangkaraya students. The development of the software used Extreme Programming, since it was appropriate and precise for this study which was to quickly finish the project. The application was created using Microsoft Office Excel 2007 for the training data and Matlab 7 to implement the application. The result of K-Nearest Neighbors method on the application of determining students' graduation status was 92.5%. It could determine the predicate graduation of 94 data used from the initial data before the processing as many as 136 data which the maximal training data was 50data. The K-Nearest Neighbors method is one of methods used to group a set of data based on the closest value, so that using K-Nearest Neighbors method agreed with this study. The results of K-Nearest Neighbors method on the application of determining students' graduation status was 92.5% could determine the predicate graduation which is the maximal training data. The K-Nearest Neighbors method is one of methods used to group a set of data based on the closest value, so that using K-Nearest Neighbors method agreed with this study.

Keywords: *Extreme Programming, Graduation Status, K-Nearest Neighbor, STMIK Palangkaraya*

1. Introduction

The application of data mining to overcome the existing problems at this time is so widely used as for prediction. Data mining is used for data analysis of students who have bad record so that it can be found the solution for the problem solving on those students [1]. Data mining has a variety of methods, one of which is prediction such as a research conducted to predict the results of placement for college activities [2]. The research was conducted by comparing two methods namely Fuzzy Logic and K-Nearest Neighbor

(KNN). The result was that the KNN got superior with a value of 97.33% while the Fuzzy Logic of 92.67% to predict.
The so-called research of modified KNN method is conducted to improve KNN performance by modifying it that is to use robust neighbors in training data [3]. The development is aimed to get better results.
Based on the previous research using KNN method and the research in predicting and the research using students'data and its development, hence this research also related with data mining about prediction is that the classification by using data of university students, to classify the data into passed and not passed. The grouping of data obtained is based on amount of course credits has been taken, grade point average (GPA), and mini-thesis grade. The basis of such data grouping as one of requirements at Universities such as STMIK Palangkaraya as the one of universities that apply the regulation of the calculation of course credits amount, GPA, andmini-thesis grade as the requirements of students' graduation. The aim of this research is to know the results of application constructed using KNN concept in determining university students' graduation status (passed or not passed) based on amount of course credits, GPA, and mini-thesis grade of STMIK Palangkaraya students.

2. Literature Review

Classification is widely used in a group of data to look for an inter-data relationship. One of the methods used in the classification is KNN which classifies a tuple class by finding the closest value of the training data and after the closest data known, the tuple class will follow the closest data class that previously has been known through training.
KNN method is included into data mining such as a research ever conducted to analyze medical data [4].This study was conducted because of the development of patient data stored in a hospital. The data may be hidden though it may be helpful to analyze for more accurate diagnostic results. KNN algorithm was used to obtain more accurate diagnostic results. The results of research were an application that can overcome errors in diagnosing.
KNN algorithm is used to analyze distance measurement using a set of data. Classification is a process of analyzing input and building a model for a class. KNN is used to search for the unknown class tuple and as a method to find accuracy in distance measurement [5].

2.1 K-Nearest Neighbor Method

There are many methods can be used in classification, one of which is KNN method. KNN method requires test data and training data to calculate distance. Like other methods in general that have algorithms, KNN also has algorithms that do classification by calculating the proximity between the data. A value at a distance will be used to find the value of proximity or similarity between test data and training data. If there are results that have a difference then the number data will be selected.

The KNN algorithms and its formulation as follow [6]:

a. z = (x', y'), is test data with x' and y' unknown class label
   Where:
   x' : vector/ test data attribute
   y' : class label of unknown test data
b. counting distance d (x', y'), the distance between z-test data to each training data vector, stored into D
c. Select Dz D, where K is the nearest neighbor of z
d. y' = argmax (xi,yi) Dz I(v=yi)

To handle the largest amount of data, it is usually added the use of weights to calculate the class candidates that should be taken by the test data from the nearest KNN.

The classification requires test data and training data by finding the closest data from the training data. After finding it, the closest data will be followed. The number of values will be used as training data. If the majority of data is in a class then the value obtained will go into it based on the *range* of data provided.

2.2 Extreme Programming

Extreme programming is one of agile system development with an approach to develop software through some stages as follows [7]:

a. Planning
   Planning is a stage to start defining the needs, the resulting output, the services developed in the application and its function.
b. Design
   This stage is part of designing a simple application designing in accordance with the needs of the system or application.
c. Coding
   Coding stage is a stage in prepare the code that will be used in the application development so that it can be a problem solving.
d. Testing
   This is a stage to test the services or features and functionality contained in the built application.

2.3 Screen Design Worksheet

One of the tools to design the display is by using screen design worksheet. The interface design can use it in the form of paper sheet to facilitate the documentation. It will be very useful for programmers in building a program[8]. Screen design worksheet consists of 4 parts:

a. Worksheet number
   It based on the number of display made.
b. Display
   The design of the display that will appear on the screen.
c. Navigation
   Explanations of display designed.
d. Description
   Description of the display parts.

3. Analysis and Design

3.1 Needs Analysis

Needs analysis is a discussion of some needs and or requirements related to input data, processes and output results. The needs or requirements were obtained from the data used to students' graduation status and the academic division is that the data of STMIK students in information system programs of study. Based on the obtained data, the results of needs analysis were as follows:

a. Needs of input
   The built application required input data, they were students data such as students ID number, name, course credits collected, GPA, and grade of mini-thesis.
b. Needs of process
   The process used to process input data was clustering technique using KNN algorithm.
c. Needs of output
   Output was the analysis of the results of the use of KNN algorithm so that could be known the information pattern from input data.

3.2 Data Collection Technique

In the data collection, the researcher used some techniques in the research. It helped the researcher to gather the right data and could be processed. The techniques used were:

a. Observation

   Observation was to collect data by taking samples of students' data information system programs of study in STMIK Palangkaraya and made observation on the objects under the study to find out the current system in STMIK Palangkaraya especially information system programs of study of year 2011. From the observation, the researcher found that the graduation is based on 3 things namely amount of course credits, GPA, and grade of mini-thesis.

b. Library

   Library was to collect data by reading the literature or books related to students' graduation and KNN as well as visiting websites that fit the issues discussed in this study such as visiting the website to find out the graduation requirements at the college.

3.3 Data Analysis Technique

Data analysis technique used in this study was adopting the technique contained in the Agile model is that Extreme Programming as in Figure 1.

Figure1. Data Analysis Technique

The steps of Extreme Programming used in this study as in Figure 1 as follows:

1) Planning
2) To acknowledge the scope of work as a plan to model the system, including searching for data and site information related to the needs of students' graduation data, features, and functions in accordance with system development is that students data such as name, GPA, amount of course credits, and information about grade of mini-thesis.
3) Design
4) This stage made a design of the work flow and a design that was needed in this research is that application development using Microsoft Office Excel software to design training data then it was processed using Matlab 7 software as well as the test data.
5) Coding
6) Coding stage used the codes contained in the Matlab 7 software to use training data and test data in which the test data previously were compiled using Microsoft Office Excel software.
7) Testing
8) Model testing used a simple application to know the results of model used so it got a conclusion about the services or features and functionality in the constructed application.

IOP Conf. Series: Earth and Environmental Science **97** (2017) 012040 doi:10.1088/1755-1315/97/1/012040

3.4 Design

The interface design used display worksheet for calculation view design and data analysis of the completed calculation using each method. The display worksheet in this study used Microsoft Office Word in which the worksheet itself consisted of worksheet number, display section, navigator, and description. The design facilitated preparing a simple program for the use of KNN method in general, software design considered the user needs, they are the interface display and the correctly calculation/computation process. In this study, software design was only a process and the results of calculation/computation using KNN method using students' data to know the students' graduation status. The initial design for the data used as training data was using Microsoft Office Excel 2007 software while application design and code used Matlab 7.

The interface design used display worksheet to design display of calculation and data analysis based on input and output data. The display worksheet applied Microsoft Office word consisting of number, display section, navigator, and description. The interface can make changes to the data before they entered by using KNN method such as initialization by allocating all data in a group randomly as shown in figure2 for the design and figure 3 the interface design in figure2, number 1 to 3 shows the input data and number 4 the output data.

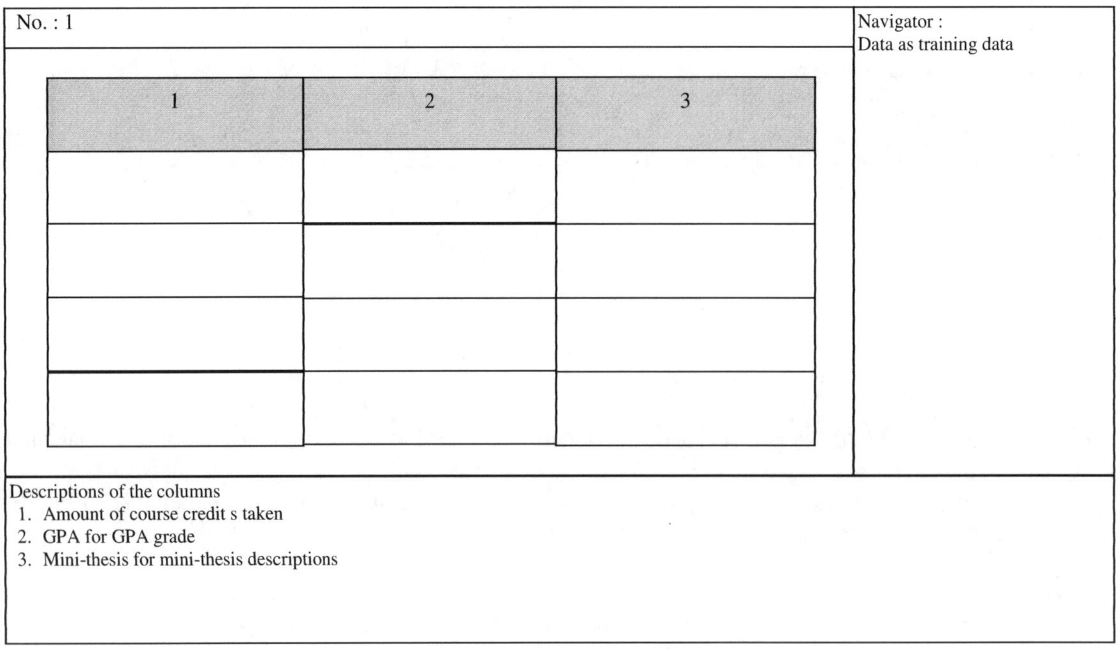

Figure2. Interface Design of Data initialization

4. Implementation and Discussion

4.1 Implementation

Interface was based on the input and output data required to implement the use of KNN method. Figure 3 shows the data from 10 training data. It would be used as the nearest search is that 0 and 1.

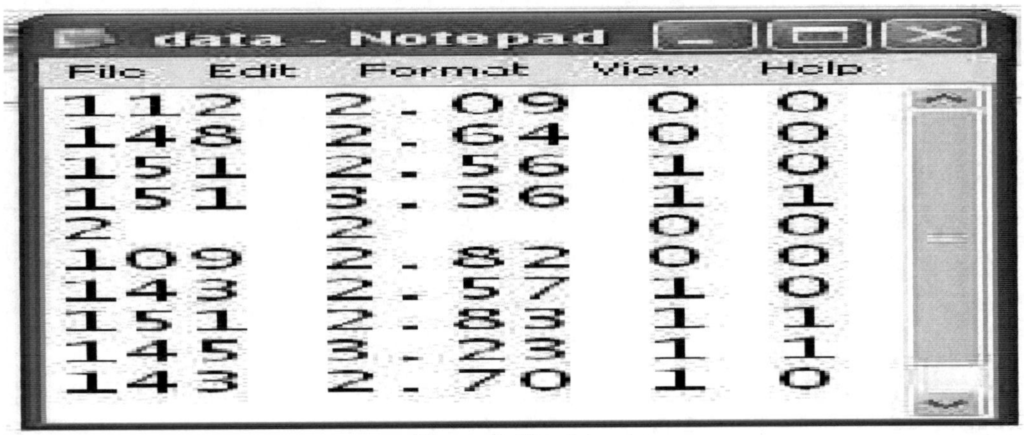

Figure 3. 10 training data

The interface was based on the design using display worksheet intended for algorithm usage and for data analysis of completed algorithm results. From the interface of Matlab itself, the researcher could make data input required before the data went to the concept of KNN method like doing initialization by randomly taking sample data of a group.

Figure 4 shows the interface of constructed application along with the samples of input and output of the application.

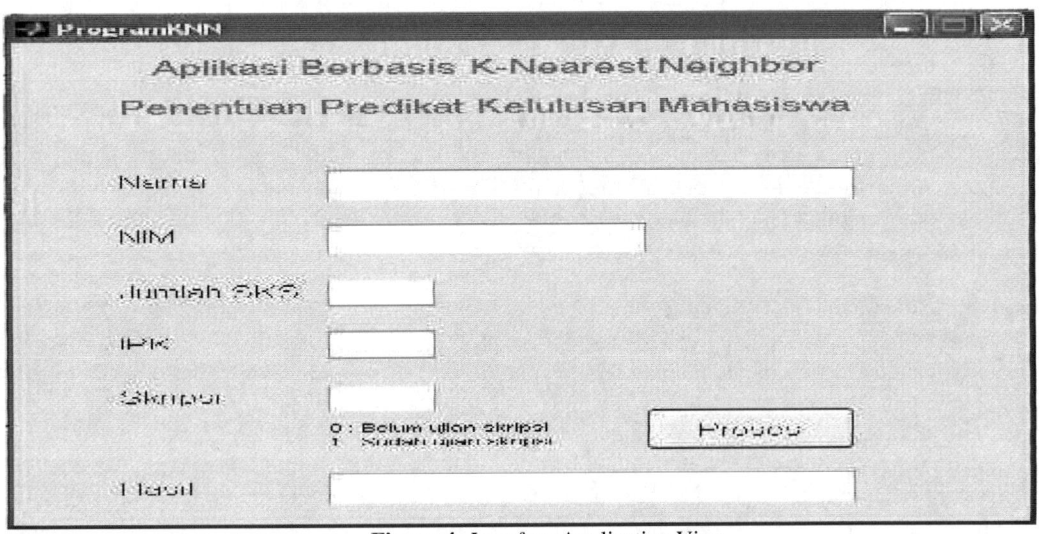

Figure 4. *Interface* Application View

Based on the process using clustering technique and KNN algorithms to determine students graduation through the input data was that amount of course credits taken, GPA achieved by the students, and mini-thesis grade then it was found the results for 10 training data and for each 50 training data given 2 phases of test that were testing for training data and all data. By the test of 10 training data, it was found that there was 10 suitable training data and none was. So, all data was recognized. While the test of all 94 data based on 10 training data was found that 88 data was suitable and 6 data was not. The test of training test based on 50 training data was 45 suitable data and 5 other was not. While the test of all data was that 94 data found that 87 data was suitable while 7 data was not. The comparison of training data testing and all data as shown in Table 1.

Table1. Comparison of data testing

Training data	10		50	
Test data	10	94	50	94
Suitable	10	88	45	87
Not suitable	0	6	5	7

Based on Table.1 it was found the percentage comparison as shown in Table 2.

Table2. Percentage Comparison of Data Testing

Training data	10		50	
Test Data	10	94	50	94
Suitable	100%	93,6%	90%	92,5%
Not suitable	0%	6,4%	10%	7,5%

At the analytical stage using KNN method, the data used as a test in this study was a combination of input and output in the form of a matrix of 3 columns. They were a column for amount of course credits, one for GPA achieved by students, and the other for mini-thesis grade description by giving 2 descriptions was that 0 (zero) and 1 (one). 0 (zero) for not passed and 1(one) for passed.

4.2 Discussion

On the clustering, the results were in the form of membership of each data on each cluster and the dominant data went into a certain cluster by finding the nearest value because the principle of KNN method is that finding the neighbors or nearby value. The use of a set of initial data as many as 136 data and after getting pre-processing the data became 94 used as data in the application of using KNN method by eliminating some data which could not be used such as double data.

Table 1 shows comparison of two tests using training data and it did not need training as in KNN method. The inconsistent value was smaller and the corresponding value was greater than the test stages performed on the more extensive data. The comparative calculation between standard testing and system testing as the calculation of accuracy using MAPE formula and the results as in table 2 showing comparison of training and testing in the form of percentage with the highest value of 100% system could recognize all test data and 0% system could recognize data from all data by using minimal training data. But through the increasing of training data the number of percentages of conformity decreased but not dramatically, the improvement remained in the test of 50 training data on 94 test data. From the results it was found that the more data trained the better suitability of data testing achieved, it could be seen from the improvement of system performance on determining students graduation status.

5. Conclusions and Suggestions

5.1 Conclusions

In this research can be concluded that the 136 initial data became 94 after experiencing pre-processing. It was served as data in the application of KNN concept. The data divides into training data and test data. The test results showed that training data could be recognized. But from the results of testing the all data, not all could be really recognizable. It indicates that the input data can be used for this research.
From the test results, it is obtained that the more data are trained and tested, the application will be better in tewting. Maximum results show 92.5% KNN-based application can recognize data from the maximum of training dara and test data used.

5.2 Suggestions

Development can be done on this research as in the following suggestions:

1) The application can be developed by adding the amount of data in which not only for one program of study, such as having others data on another program of study so the amount of data used will be even more to know the accuracy of application and method.
2) Other methods can be developed for the same case so that later it can be a new study on the comparison methods on the use of the same data to determine the quality of method used.
3) Other methods can be combined or developed in building the application in order to know comparison of results such as combining with substractive clustering method.
4) This application can be a recommendation for the parties who need such other educational institutions for research, development and use of the application itself as developed into an intelligent system or expert system.

6. References

[1] Pranav P., "*A Study of Student's Academic Performance Using Data Mining Techniques*". International Journal of Research In Computer Applications And Robotics. ISSN 2320-7345. Vol. 3, Issue 9, 2015, 59-63.

[2] Sheetal, B.M and Bakare, S., "*Prediction of Campus Placement Using Data Minng Algorithm-Fuzzy logic and K nearest neighbor*". International Journal of Advanced Research in Computer and Communication Engineering. ISSN 2278-1021. Vol. 5. Issue 6. 2016, 309-312.

[3] Parvin, H., Alizadeh, H., and B Minati, "*A modification on k-nearest neighbor classifier". Global Journal of Computer Science and Technology*. Vol.10, Issue 14, 2010, 37-41

[4] Khamis, H.S., Kriptu, W.C., and Stephen, K., "*Application of k-Nearst Neighbour Classification in Medical Data Mining*". International Journal of Information and Communication Technologi Research. Vol.4, No.4, 2014, 121-128.

[5] Mulak, P. and Nitin, T., "*Analysis of Distance Measures Using K-Nearest Neighbor Algorithm on KDD Dataset*". International Journal of Science and Research. Vol.4, Issue 7, 2015, 1201-1204.

[6] Prasetyo, E., *Data Mining Konsep dan Aplikasi* Menggunakan *Matlab*, Andi, Yogyakarta, 2012.

[7] Schach, R.S., Object-Oriented and Classical Software Engineering. Eighth Edition. McGraw-Hill. New York, 2011, 59.

[8] Santoso, I., *Interaksi Manusia dan Komputer*, Andi, Yogyakarta, 2010

The Growth of Agarwood Plants on the Different Canopy Covers Level and Fertilizer in Oil Palm Plantation

Sri Rahayu Prastyaningsih[1] and Azwin[2]
Universitas Lancang Kuning, 28265, Indonesia
Telp: +62 761 54092, Fax: +62 761 54092
E-mail: srirahayupn @unilak.ac.id

Abstract. The development of agar wood plants in oil palm plantation requires the forestry techniques in order to obtain maximum production. In an oil palm stands, the age of plant will affect the height, diameter, population and stands density. The older age of an oil palm stands will affect the canopy cover on the forest floor. Agar wood plants are semi-tolerant growth and oil palm can be used as shade. Unilak has an oil palm plantation area of 10 hectares around the campus with 10 years old and 20 years old. The soil condition at the study is Podsolik Merah Kuning (PMK) which poor nutrient and needs fertilization to increase soil fertility. This study aims to find out the effect of age of oil palm stands and fertilization for optimal growth. The split plot design with 2 main plots of the age of palm tree (10 years old and 20 years old) and five kinds of fertilizing sub plot (without fertilizer, 40 gram/plant of NPK, 80 gram/plat of NPK, 120 gram/plant of NPK and 180 gram/plant of NPK were used. The results of this research showed that the age of palm tree (canopy cover) treatment gave non-significant influence on the growing of agar wood until it reaches 4 months of growth. The canopyy cover by 10 years old of oil palm tree produce the best response on height (15 cm) and diameter (0,4 cm) growth of agar woods..Fertilizing treatment di not give any significant influence on the heigh and diameter growth of agarwood plants until reach 3 months. The interaction by 10 years old of palm with fertilizing gave non significant results.
Keywords: Agar wood plants, palm oil, canopy cover, fertilizer

1. Introduction

A Unilak oil palm stand has the potential to be developed with agroforestry system that will plant with crops of gaharu-plants. The oil palm planted spread in several locations. Oil Palm stands near the UPT Agriculture, beside of Faculty of Administration, around the arboretum of forestry faculty and along the main road Lancang Kuning University. The total area is ± 10 ha. Based on the condition, oil palm planted in Lancang Kuning University has different stands age. 20-year-olds oil palm stands around the UPT Agriculture, while 11 years old beside of Faculty of Administration and around the arboretum of forestry faculty and along the main road Lancang Kuning University. The older palm oil plants have few densities and affect the size of the canopy cover, so that the entry of sunlight on the forest floor will be reduced. With the character of the semi-tolerant agar wood planting, it is necessary the influence of oil palm trees or the age of the oil palm above it.

Agar wood-planting was done by comparing the growth of the different age of oil palm stands around the Lancang Kuning University by ± 20 years old (around UPT agriculture) and ± 11 years old (beside Faculty of Administration). Agar wood-planting that semi-tolerant species will get a shade or shade from stands above it. Fertilization is an effort and optimizes the growth of agar wood plants. The type

of soil at the study was PMK where less fertile or low pH, so the cultivation of agarwood-planting requires fertilizer to improve soil fertility.
This study aims to know the effect of shade level by age of oil palm stands and fertilization levels by agroforestry combination of oil palm and agarwood-planting, so that the growth can be optimal. Fertilizer is added into agarwood-plants to optimize growth. This research is expected to be useful as information about forestry technique on agroforestry of gaharu-plants among oil palm stands.

2. Method

This research was conducted in December 2016 until April 2017. Materials were gaharu-planting (*Aquilaria malacensis* Lamk) from 12 months old natural seedlings, labels, hoes and machetes, NPK fertilizer, poles, lux meter, pH meter, sewing meter and caliper.
Treatments
This research was designed with Split Plot design with main treatment is shade level by age of oil palm plantation (10 years and 20 years old) and fertilization as plot small treatment. Fertilization treatment of 5 levels of 0 grams, 40 grams, 80 grams, 120 grams and 160 grams. NPK fertilization is done on each plant per period every 3 months, with treatment of five replications. Gaharu-planting from 4 meters oil palm stands.
Measurements:
1) Agarwood-planting high of growth (cm).
2) Agarwood -planting stem diameter (cm).
3) Environment : Ppercentage of crown cover and canopy openness (%),ttemperature, light humidity, light intensity, soil pH and altitude

Statistical analysis.
The growth data of agar wood-planting is tabulated and averaged on treatment and replication. Analysis of variance (anova) was done on every measured parameter to determine the significance of difference between means of treatments. Means for each parameter were separated by Duncan's Real Differences Test.

3. Result

The results of analysis of variance (anova) in Table 1 and Table 2 shows that the treatment of canopy cover by the different age of oil palm stands, fertilization and interaction did not give a significant effect on the increase in diameter and high of agar wood-planting until of 4 months.

Table 1. The results of analysis of variance (anova) increase the diameter of agarwood-planting until of 4 months.

Analisis sidik ragam	Db	JK	KT	Uji F test		Sig
				FHit	FTab5%	(95%)
Blok	4	296,52	74,13	0,7554	2,67	Ns
Umur Sawit	1	92,48	92,48	2,3454	7,71	Ns
Error (a)	4	157,72	39,43			
Pemupukan	4	369,92	92,48	0,9424	2,67	Ns
Umur Vs Pemupukan	4	135,52	33,88	0,3452	2,67	Ns
Error (b)	32	3140,24	98,1325			
Total	49	4192,40				

Information:
Ns = Not significantly different at 95% confidence level.

Table 2. The results of analysis of variance (anova) increase the high of agarwood-planting until of 4 months.

Analisis sidik ragam	Db	JK	KT	Uji F test		Sig
				F Hit	F Tab5%	(95%)
Blok	4	0,0548	0,0137	0,3981	2,67	Ns
Umur	1	0,0800	0,0800	2,3247	4,15	Ns
Error (a)	4	0,0740	0,0185			
Pemupukan	4	0,0928	0,0323	0,6742	2,67	Ns
Umur x Pemupukan	4	0,0400	0,0100	0,2906	2,67	Ns
Error (b)	32	1,1072	0,0346			
Total	49	1,3688				

Information :
Ns = Not significantly different at 95% confidence level.

The result of increase diameter of agarwood-planting until of 4 months can be seen in Figure 1 and Figure 2 below:

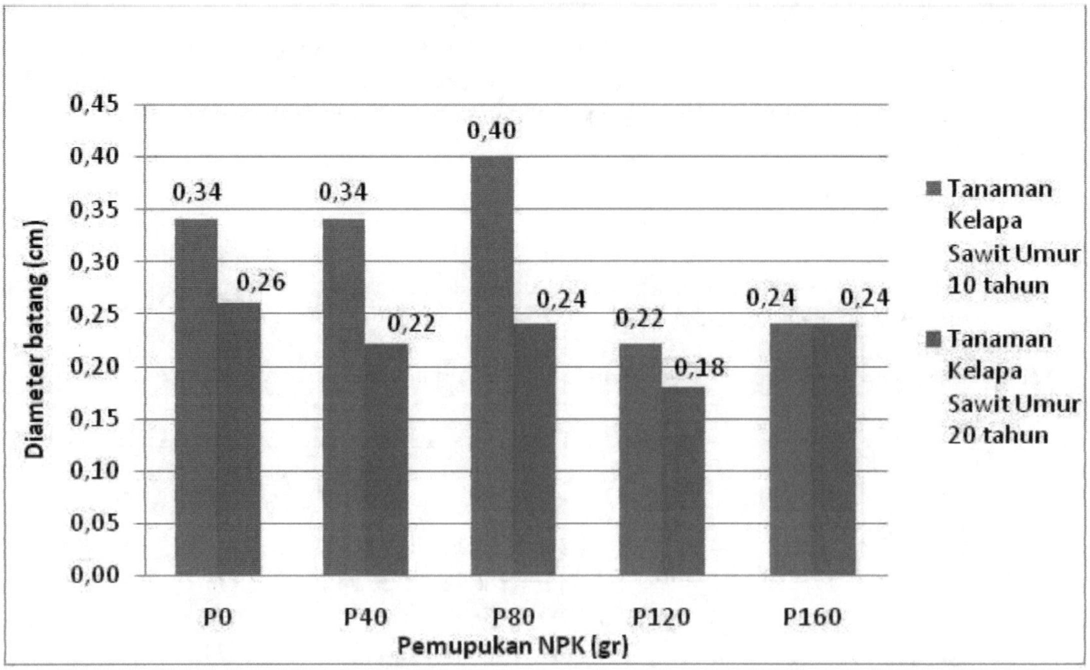

Figure 1. Diameter of agarwood-planting until of 4 month under oil palm stands of 10 and 20 years old

The increase in the diameter of agarwood-planting under 10-year-old palm stands is greater than the increase in the diameter of agarwood-planting under 20-year-old palm stands. The fertilizer on agarwood-planting under 10-year-old oil palm stands increase in diameter between 0.22 cm and 0.40 cm, while in 20-year-old stands of oil palm between 0.18 cm and 0.26 cm.

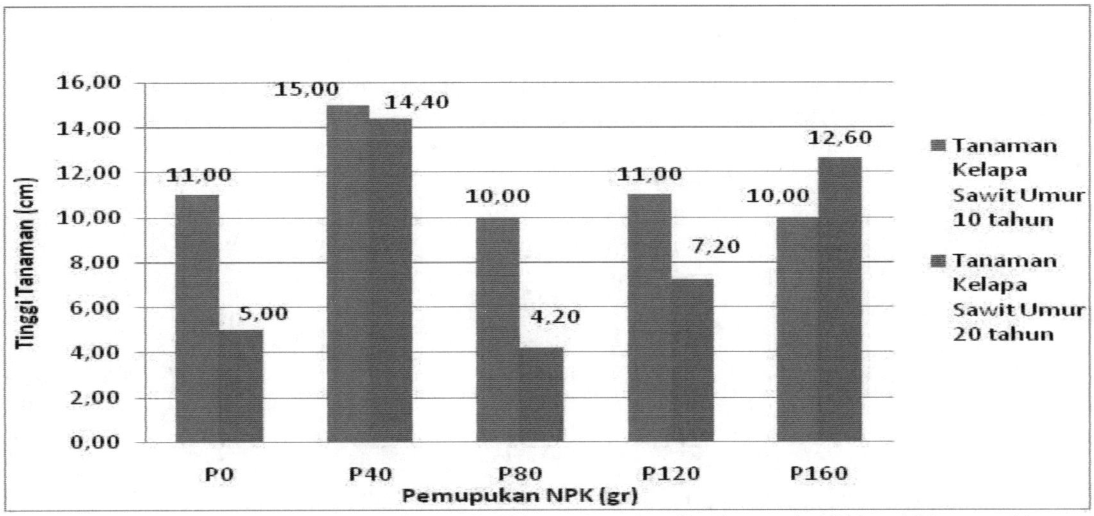

Figure 2. High of agarwood-planting until of 4 month under oil palm stands of 10 and 20 years old

The high growth of agarwood- planting under 10-year-old oil palm stands is greater than 20-year-old oil palm stands. The fertilizer on agarwood- planting under 10-year-old oil palm stands increase in diameter between 10 cm and 15 cm, while in 20-year-old oil palm stands between 4.2 cm and 14.4 cm. Although the effect by statistical is not real, but agarwood-planting under 10-year-old oil plant stands have higher growth than 20-year-old stands. Thegrowth at 10 years are 0.308 cm and 11.4 cm, while the 20-years old is 0.228 cm and 8.68 cm. When visually, agarwood-planting was under 10-year-old oil palm stands have a higher and larger appearance than 20-years-old.

4. Discussion

The plot (1) was located near Faculty of Administrative (10-years-old oil palm stands) at U 00 ° 34[1] 36.49[11] with 30 m dpl (2) the UPT Agricultural (20 years old oil palm stands) at U 00 ° 34[1] 25.97[11] with 38 m dpl. Ttopography was flat, yellow podzolic soil. In table 3, there are differentt vegetation and microclimate between 10 and 20 years old oil palm stands.

Table 3. Vegetation and microclimate by 10 and 20 years old oil palm stands

	Vegetation and microclimate	Oil palm stands	
		10 years old	20 years old
1	Vegetation -high (m) - diameter (cm) - canopy cover (m2)	 7 70 28,26	 4 57 78,5
2	Canopy cover -lengtt (m) -canopy cover (%)	 4 0,7 (4,375%)	 4 9,4 (58,75%)
3	Light intensity (Luxmeter)	1.065,5	354,68
4	Temperature (°C)	39,7	36,58
5	Humidity (%)	47,7	54,3
6	Soil pH	6,2	5,4

Source: Primary Data 2017

Growth of plants influenced by genetic factors and several environmental factors i.e. climate (temperature, humidity, sunlight and others), soil conditions or nutrients contained therein. Sunlight is a source of energy plays a role in the process of plant photosynthesis. Agarwood-planting known to have a slow growth, at the time of semi-tolerant, so in the cultivation of special treatment is required to growth. Sri Haryati (2010) writes that the leaves are the center of photosynthesis where sunlight captured and absorbed by the leaves through chlorophyll. At high light intensity it can decrease the rate of photosynthesis due to the rapid photo-oxidation of chlorophyll so as to damage chlorophyll, but the intensity of light that is too low will also limit photosynthesis and tend to be more widely used than stored. In the research of Suhartati and Wahyudi (2010), the 9-month-old agarwood-planting can growth 76 cm height, while the age of 12 months can growth 104 cm. Growth phase of agarwood-planting at the age of 3-9 months, agarwood -planting require shade or lower light intensity less than 40%. Oil palm leaf cover did not affect the increase until of 4 months (tables 2 and 3) was suspected because the seedlings were planted in a 12-month-old with a diameter of ± 0.65 cm and a high of ± 69 cm. Suhartati (2013) writes that agarwood-planting 12 months will change of nature from semi-intolerant to intolerant that greater light intensity.

Based on vegetation and microclimate, the temperature and humidity is 36.58 ° C and 54.5% in 20-years-old oil palm stands, lower than the age of 10 years -old oil palm stands ie 39.7 ° C and 47 , 7%. The diferent of temperature and humidity was caused by canopy coverings so % in 20-years-old oil palm stands that sunlight does not reach the ground. The percentage of canopy cover was inversely proportional with the light intensity. If the percentage of canopy cover is low, so the intensity of the light is high. The percentage of crown cover and light intensity at 10-year-old palm stands was 4.375% and 1,065.5 lux, while the age of 20 years was 58.75% and 354.68 lux. The greater the intensity value of incoming light, the greater the light is absorbed by the plant. Agarwood-planting in the middle of palm trees ± 4 meters and get full sunlight.Toumey & Korstia in Simorangkir (2000) suggest that the growth of plants diameter related the rate of photosynthesis proportional to the amount of sunlight intensity and respiration. Marjenah (2001) suggests that stem diameter growth is faster in open areas than in shelter so that open plants tend to be short and stocky. Inhibition of stem diameter growth due to photosynthetic product and solar spectrum less stimulates hormone activity in the process of meristem cell formation toward stem diameter especially at low light intensity (Daniel in Milliang, 2011). The measurement of soil pH between 5.4 - 5.6, so it is categorized acid. The optimal soil pH for plants is about 5.6 to 8. In acid soil, the P element will be bound by Al and Fe, so it is not available to the plants (Kartasaputra, 2005). If the soil pH is lower than 4.0, there is an increase of Al3 + in the soil solution and physically damaging the root system causing the growth to be inhibited (Leiwakabessy, 2003). Hardjowigeno (2003) suggests that acid soils can cause P elements to be absorbed because they are bonded by Al, Fe and Mn to become unavailable to plants.

According to Hardjowigeno (1987), NPK nutrients are needed plants because it can improve growth and simplify the physiological process. Low soil fertility requires additional nutrients to change their physical and chemical properties. Fertilizer will provide better compared with plants that are not fertilized. NPK fertilizer is an inorganic material, when added to the soil or plant will add nutrients and can improve the physical, chemical and biological properties of soil or soil fertility. Nitrogen (N) element is useful the process of forming chlorophyll, photosynthesis, protein, fat and other organic compounds. Phosphorus (P) elements are useful for forming roots, protein base ingredients, strengthening plant stems and assisting assimilation and respiration. The element of potassium (K) is useful for the formation of proteins, carbohydrates and helps the various physiological processes of plants. Suhartati (2010) writes the dose of NPK 150 grams / plant can increase the diameter of 34.5% and 77.65% high. Furthermore Suhartati (2013) states that with a spacing of 4 m from oil palm trees and the provision of 159 grams of NPK fertilizer per plant will significantly increase the diameter and height of gaharu-producing plants. Mile (2004) showed that inorganic compound NPK 30 gram with combination of liquid bokashi fertilizer did not give significant difference to the growth of sengon until the age of 4 months. After the second fertilization, the dose is raised to 100 grams and combined with liquid bokashi fertilizer then the results become significantly different at the age of 8 months. The NPK fertilizer response at higher seedlings is more optimal than the increase in seed diameter due to vertical growth faster than the horizontal growth (Lewenussa 2009 in Prastyaningsih 2014).

The acid soil in oil palm area needs to be neutralized with calcium (high pH or alkaline). Suhartati (2010) stated that the calcium with a dose of 1 kg / plant can increase 125% diameter and 42.5% high, neutralize soil acidity from pH 4 to pH 6. Hardjowiguna (1987) reported that neutral soil pH can facilitate the absorption of elements Nutrient by plants, reduce toxic elements and accelerate plant growth. Suhartati (2013) states that fertilization of gaharu plants is done at the age of 1, 2 and 3 years with the dosage of NPK fertilizer in the first year 50 gr, second year 100 gr and third year 150 gr per tree. In addition, compost fertilizer can be used with a dose of 10 kg per planting hole and dolomite for soil that has acid soil pH. Dolomite is given together with basic fertilizer before planting, with doses of 150 to 250 g per planting.

Many factors affect the slow growth of diameter and high of seedlings of agarwood-planting. In chemical properties, soil physical properties also affect plant growth. The soil structure provides better growth compared to solid soil structures. The soil with a clay texture is different from the good sandy clay sand soil (Aris, 2011). The result of soil observation in the research location shows that the physical soilunder the stand of palm trees 20 years more clay, whereas in the stand of 10 year old palm is more crumbly.

5. Conclusion

1) Canopy cover with the different of age of oil palm stands was not significant at the increase in diameter and hhigh of agarwood-planting at 4 months of age.
2) NPK fertilizer with different doses has no significant effect on the increase of diameter and hhigh of growths the 4-month agarwood-planting.
3) The interaction of the age oil palm stands with NPK fertilizer had no significant effect on the growths of diameter and height of the 4-month agarwood-planting.
4) Diameter gain at the age 10 years of the oil palm stands with 80 gr fertilization of 0.4 cm, while the high increase in fertilization 0 gr is 15 cm.
5) The best diameter gain at the age of palm 20 years with 0 gr fertilization is 0.24 cm, while the high increase in fertilization 0 gr is 14.4 cm.

6. References

[1] Badan Litbang Kehutanan, 2014. *Kajian Pohon Gaharu Dalam Mendukung Upaya Konservasi Dan Pelestarian Alam di Provinsi Riau.*Pekanbaru
[2] Hardjowigeno, 1987. *Ilmu Tanah*. Mediatama Sarana Perkasa. Jakarta.
[3] Haryati, Sri, 2010. *Pengaruh Naungan Yang Berbeda TerhadapJumlah Stomata dan Ukuran Porus Stomata Daun Zephyrantes Rosea Lind*
[4] Iriansyah M *et al.* 2006. *Gaharu Komoditi Masa Depan Yang Menjanjikan*. Balai Penelitian dan Pengembangan Kehutanan Kalimantan. Samarinda.
[5] Jiang, *S.* et al, 2011. *Journal of Chinese Pharmaceutical Sciences, 609-614*
[6] Kartasaputra, Sutejo Mulyani, 2005. *Teknologi Konservasi Tanah dan Air*. PT Rineka Cipta. Jakarta
[7] Ponirin S. 2003. *Budidaya Gaharu*. Balai Penelitian dan Pengembangan Kehutanan. Departemen Kehutanan. Jakarta.

[8] Prastyaningsih, 2014. *Respons Pemberian Pupuk NPK terhadap pertumbuhan Bibit Pohon Penghasil Gaharu Jenis Aquilaria mallacensis Lamk.* Prosiding Simposium Penelitian Bahan Obat Alami (SBOA)XVI & Muktamar XII Perhipba 2014. Surakarta.

[9] Prastyaningsih, S.R 2015. *Potensi Pohon Penghasil Gaharu di Kabupaten Kampar Provinsi Riau*. Jurnal Wahana Foresta Vol **10** No 2 hal 88-100. Fakultas Kehutanan. Universitas Lancang Kuning.

[10] Sahromi, Wahyuni, Sri dan Pribadi, Didit Okta, 2010. *Analisis Pengaruh Pemupukan Terhadap tingkat Kesintasan dan Pertumbuhan Bibit Gaharu (Aquilaria mallacensis Baill)*. Bulletin Kebun Raya Vol **13** No 1 Januari 2010.Bogor.

[11] Setyawati, Titik. 2011. P*otensi dan Kondisi Regenerasi Alam Gaharu (Aquilaria malacensis Lamk di Provinsi Lampung dan Bengkulu. Isi Buku Gaharu.* https://ml.scribd.com/doc/.../Isi-Buku-Gaharu-5-Januari-2011-Full-Size. *Diakses 12 Desember 2014.*

[12] Simorangkir, 2000. Analisis Riap Dryobalanops pada Lebar Jalur yang Berbeda di Hutan Koleksi Universitas Mulawarman. Lempake. Frontir Nomor 32. Kalimantan Timur.

[13] Sofyan Agus, Sumadi Agus, Kurniawan Agus dan Nurlia Ari (2010). *Pengembangan dan Peningkatan Produktifitas Pohon Penghasil Gaharu Sebagai Bahan Obat di Sumatera*. Laporan Hasil Penelitian. Kementrian Kehutanan. Balai penelitian Kehutanan. Palembang.

[14] Sudomo, Aris dan Midawati, Nina, 2011.*Tekno Hutan Tanaman*. Vol 4 No 3 Desember 2011, 111-118. Bogor

[15] Suhartati, 2010. P*engaruh Pupuk NPK Terhadap Pertumbuhan Gaharu di areal Perkebunan Sawit.* Prosiding Seminar bersama BPK Arek Nauli, BPK Palembang dan BPHS Kuok. Pekanbaru, 4-5 November 2010

[16] Suhartati dan Wahyudi, Agus. 2011. *Jurnal Penelitian dan Konservasi Alam*. Volume **8** No 4: 363-371. Bogor

[17] Suhartati, 2013. *Budidaya Tanaman Gaharu (Aquilaria mallacensis Lamk) di lahan kebun kelapa sawit dengan aplikasi teknik silvikultur*. Info Teknis Eboni. Vol 1 No 1 Mei 2013: 37-47.Bogor

[18] Sumarna, Yana. 2009. *Budidaya Gaharu*. Seri Agribisnis. Jakarta. Penebar Swadaya.

[19] Sumarna, Yana, 2007. *Pengaruh Diameter dan Luas Tajuk Pohon Induk Terhadap Potensi Permudaan Alam Tingkat Semai Tumbuhan Penghasil Gaharu. Jurnal Penelitian Hutan dan Konservasi Alam.* Vol **V** No. 1 hal: 21-27, 2008. Bogor

Acknowledgment

Acknowledgments to LPPM of Lancang Kuning University which has provided the financing of the main research scheme no. contract: 253 / Unilak-LPPM / B.07 / 2016 dated December 5, 2016.

Geographic Information System (Gis) for Culinary in Pekanbaru using Herversine Formula

Yogi Yunefri[1,] Mariza Devega[2,] Dwi Kristanto[3]
Universitas Lancang Kuning, Pekanbaru, 28265, Indonesia
[1]yogiyunefri@unilak.ac.id,
[2]Marizadevega87@gmail.com
[3]DwiKristanto@gmail.com

Abstract: Pekanbaru is one of a big city in Indonesia with 897.767 inhabitants' population on 2010. There are various cultures of the population. That diversity presents the difference of culinary in Pekanbaru, so Pekanbaru be one of the cities which appropriate for culinary that give charm circumstance and worth to taste. One of the obstacles that often occur is the tourists still difficult to find the right and nice place to eat, close to tourism place and find information about culinary is difficult as well. Therefore a web-based GIS application built to give information about culinary in Pekanbaru. This application built through some steps, i.e. system analysis, design system, implementation, and testing. This application built using PHP as a programming language and harversine Formula as a method to fine the closest distance. After it built the application, the data of culinary tested used black box. The result shows that testing using this application is similar with the manual test. Thus the application has been built correctly.

1. Introduction

Pekanbaru is one of a big city in Indonesia with 897.767 inhabitants' population on 2010. There are various cultures of the population. That diversity indirectly presents the diversity of culinary in Pekanbaru, so Pekanbaru be one of the cities which appropriate for culinary that give charm circumstance and worth to taste. Exploring culinary nowadays is very popular when we visited the place. If exploring culinary, people who are just looking for a satisfying need for their hunger, are now beginning to consider other factors in determining where they will eat such as distance, menu variations and other supporting information. (Sugeng Purwantoro et al, 2012).

The problems that often occur from the many culinary tourism options in Pekanbaru is still lack of the information about the location and the menu which is offered. The software which containing map of a region that can be accessed using web-based system is expected to overcome the problems faced by tourists in getting information about culinary in Pekanbaru. On a web-based system a digital map that can be used is Google Map. Google maps are accessible with browsers on both computer and mobile devices. Google Map has a Geolocation feature that can be used to identify locations marked by latitude and

longitude. Harversine Formula is a method which is used for this application. Harversine Formula is a method of knowing the distance between two points by counting that the earth is not a plane but is a field that has a degree of curvature.
Harversine formula calculates the distance between 2 points by the length of a straight line between 2 points on longitude and latitude. (Ryan Herwan DWI Putra Et Al, 2015), this application is expected to provide information about culinary as well as culinary location in Pekanbaru City.

2. Method

In this research, the researcher used the method which developed by W. Royce that is waterfall method; the stages in this method can be seen on figure 1

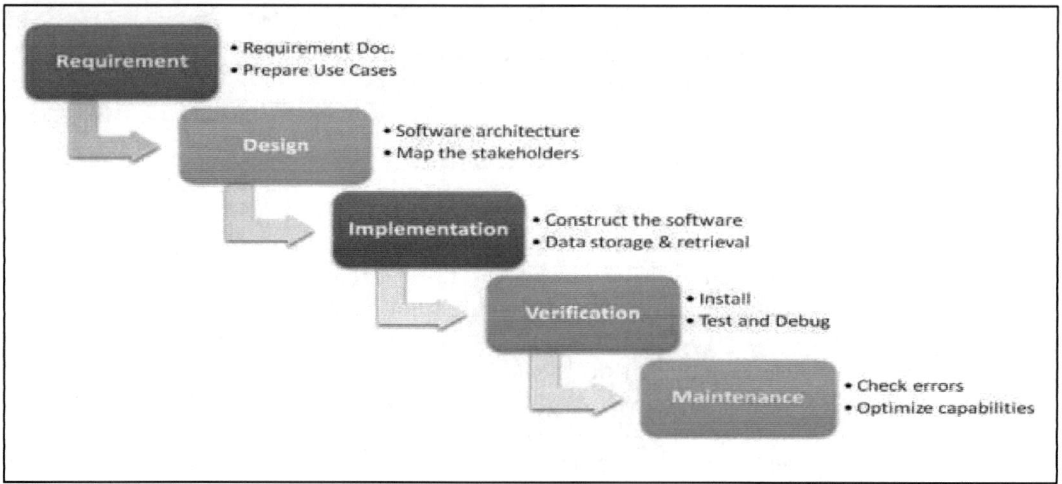

Figure 1. Research method

2.1 Need of Analysis

This step is an analysis of system requirements. Data collection in this phase can do a research, interview or literature study. Data collection in this stage can be done from a study, interview or literature study. Systems analyst digs as much information from the user so it created a computer system that can perform tasks desired by the user. This stage will generate user requirement document that is used as data related to what users want in making system. This document will be a reference system analyst to convert, it into the programming language.

2.2 Design System

This step, giving the opinion and system planning toward problem solving by using software of design system, such as flow data diagram, entity relationship diagram, structure and data material.

2.3 *Writing the Program Code*

Writing code or coding is a translation design in a language that can be recognized by the computer. Performed by the programmer who will translate the transaction requested by the user. This stage is the real step in working on a system. In the sense that the use of computers will be maximized in this stage. After the coding is complete it will be testing the system that has been made earlier. The purpose of testing is to find errors on the system and then fix it.

2.4 Program Testing
The final stage in which the new system is tested its ability and effectiveness so that the system deficiencies and weaknesses are found and then the review and improvement of the application becomes better and better.

2.5 Implementation and Maintenance
The software that has been submitted to customers usually change. The change could be due to an error because the software must adapt to a new environment (peripheral or new operating system), or because the customer needs a functional development.

3. Study Literature
This research is carried out not apart from the results of previous studies that have been done as a comparison and study materials. As the results of research that made the comparison cannot be separated from research topics related to information both from the web and mobile.
Bambang Yuwono, et al (2015) conducted research on geographic information system based on android for tourism in Magelang used Grapple method. Grapple method has 5 stages, namely: Requirement Gathering, Analysis, Design, Development, and Deployment. ut in this study, the authors only do stages of development or development. This system is built using eclipse software as an editor for coding android, Dreamweaver as coding editor of Code Igniter framework, MySQL as database server, and PHP as its server programming language.

3.1 Geographic Information System
Eko Budiyanto (2015), defined GIS (Information System) or Geographic Information System is a system of mutual series with each other. Bako Sutarnal (2016) describes GIS as an organized collection of computer hardware, software, geographic data and personnel designed to acquire, store, repair, manipulate, analyze and display all forms of geo-referenced information. Thus, the analytical basis of GIS is spatial data in digital form obtained through satellite data or other data digitized. GIS analysis requires experts as interpreters, computer hardware and software support.

3.2 Harversine Formula Method
The haversine formula method can be used to calculate the distance between two points, based on latitude and longitude. The method of the Haversine Formula is now undergoing development, by using a simple spherical law of cosine formula, which by calculating the computer can provide a very accurate sprecision level between two points.

Haversine Formula :
x = (lon2-lon1) * cos ((lat1+lat2)/2);
y= (lat2-lat1);
d= sqrt(x*x+y*y)*R
Keterangan:
x = Longitude (Lintang)
y = Lattitude (Bujur)
d = distance
R= Radius = 6371 km
1 degree = 0.0174532925 radian

4. Result and Discussion

After the design, then the application on the system does. The results of the implementation can be seen in Figure 2 below.

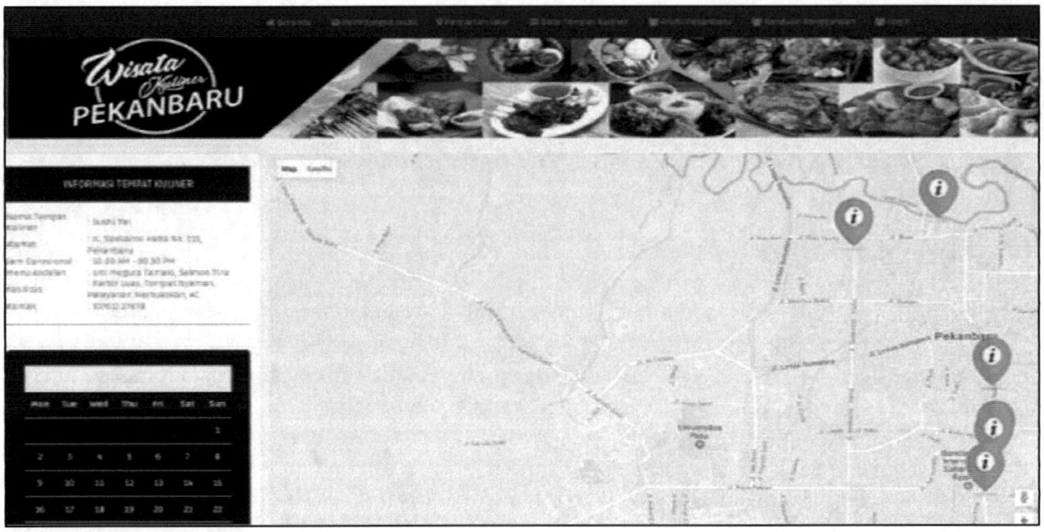

Figure 2. The implementation

This system is designed to determine the closest distance between the user positions to the location where the culinary is. Then this system will calculate the distance of each point where the culinary location of the user's position is. Manual calculations performed by calculating the distance of the user to the culinary place by using the method of Haversine Formula, as for the formula of Haversine Formula are:

d = acos (Sin 1 . Sin 2 + Cos 1 . Cos 2 . cos). R

Steps to determine the distance between two points with Haversine Formulla are as follows:

a. Determining starting point with coordinates (-7.561678783, 110.7965773)
b. Specifies the destination point with coordinates (-7.564725, 110.80341)
c. Convert the coordinates of start and destination points into radians.

$$\text{Radlat 1} = \frac{-7.561678783}{180} * 3.14 = -0.131909285$$

$$\text{Radlong 1} = \frac{1.107.965.773}{180} * 3.14 = 1.932.784.738$$

$$\text{Radlat 2} = \frac{-7.564725}{180} * 3.14 = -0.131962425$$

$$\text{Radlong 2} = \frac{11.080.341}{180} * 3.14 = 193.290.393$$

d. Enter the starting point and ending point into the Haversine Formula.
   c = acos (sin 1 · sin 2 + cos 1 · cos 2 · cos)
   c = acos (sin(Radlat1) * sin(Radlat1,) + cos(Radlat1()) * cos(Radlat2) * cos (Radlong21)
   c = acos (sin(−0.131909285) * sin(−0.131962425) + cos(-0.131909285) * cos(−0.131962425) * cos(1.19192 * 10^5)
   c = acos (1)
   c = 0.00013

e. Convert the result into Km by multiplying by R = 6371.
   d = - R x c
   d = 0.00013 x 6371 Km
   d = 0.82539971 Km

Thus, the distance between the starting point and the end point based on Haversine Formulla is 0.82539971Km.

5. Testing

In the final stage the authors test the results of the system to the respondents who numbered 10 people to use the system then provide results or experience using this system by answering the questionnaire.

6. Conclusion

Based on the results of the implementation and testing system "Culinary Search in Kota Pekanbaru", can be drawn conclusion that is Haversine Formula for the calculation process determines the location of nearby Culinary Place can be applied and the calculation of the nearest culinary tour location done by manual and produced by the system is the same.

7. References

[1] Wolf, Paul R & Ghilani, Charles D. 2002. *Elementary Surveying : An Introduction to Geomatics.* Prentice Hall. New Jersey.

[2] Agus Suryantoro. 2013. *Integrasi Aplikasi Sistem Informasi Geografis.* Penerbit Ombak, Yogyakarta.

[3] Bambang Yuwono dkk. 2015. *Sistem Informasi Geografis Berbasis Android Untuk Pariwisata Di Daerah Magelang.* UPN "Veteran" Yogyakarta.

[4] Danny MAnongga dkk. 2009. Sistem Informasi Geografis Untuk Perjalanan Wisata Di Kota Semarang. Jurnal *Informatika.* Vol. **10** No. 1

[5] Dwi Prasetyo dkk. *Penerapan HAversine Formula pada Aplikasi Pencarian Lokasi Dan Informasi Gereja Kristen Di Semarang Berbasis Mobile.* Universitas Dian Nuswantoro.

[6] Eko Budiyanto. 2015. *Sistem Informasi Geografis Menggunakan Arc View GIS.* Penerbit Andi, Yogyakarta.

[7] Hamidi. *Aplikasi Sistem Informasi Geografis Berbasis Web Penyebaran Dana Bantuan Operasional Sekolah.* Jurnal *Masyarakat Informatika.* Vol. **2** No. 3

[8] Ika Ratna dkk. 2015. Sistem Informasi Geografis Lokasi Pabrik Di Kecamatan Ngoro Berbasis Android. Jurnal *Teknik Informatika.* Universitas Muhammadyah Sidoarjo.

[9] Nuraniah Muslim dkk. 2012. *Sistem Informasi Geografis Berbasis Web Pemetaan Potensi Panas Bumi Di Indonesia Menggunakan Google Maps.* Jurnal Dasi. Vol. 13 No. 2

IOP Conf. Series: Earth and Environmental Science **97** (2017) 012043 doi:10.1088/1755-1315/97/1/012043

Detection of White Root Disease (*Rigidoporus Microporus*) in Various Soil Types in the Rubber Plantations Based on The Serological Reaction

Cici Indriani Dalimunthe, Radite Tistama, and Sri Wahyuni
[1]Sungei Putih Research Center, Indonesian Rubber Research Institute
Galang, Deli Serdang – North Sumatera, PO BOX 1415 Medan 20001
[2] Quarantine Belawan, Jl. Sulawesi II Belawan
email: cc_dalimunthe@yahoo.com

Abstract
The Conventional detection of White Root Disease (Rigidoporus microporus, WRD) still uses the visual method based on an abnormal color of leaf or mycelium growth on the tap root neck. The method was less effective and less efficient. The serological technique uses yolk chicken antibodies induced by immunization with mycelium extract. The purpose of this research was to examine the consistency of selected antibodies in detecting root fungi at various soil types in the rubber plantations. This research used a Completely Randomized Design non-factorial with twelve (12) treatments and two (2) replications. The results showed that the antibodies could detect WRD in various soils types. The serological detection was higher precisely than visual observation. The development of WRD mycelium varies depending on the soil types and it was different in the each estate area. In addition, this research is expected to get a serology kit to detect early symptoms of WRD in the rubber plants.

Keywords: *Hevea brasiliensis*, White Root Disease, Early Detection, Serologycal Technique,

1. Introduction

White root disease caused by the fungus *Rigidoporus microporus* is one of the most important diseases in rubber plants in Indonesia. The financial loss due to this disease is very high, especially in smallholder rubber plantations (Setyawan, et al., 2013). Therefore, the preemptive effort through early detection will be more effective and economically viable than curative approach. Early detection of WRD disease symptoms is still difficult to be done. Observation of leaf symptoms is the first method to identify plants primarily for young plants. The plant canopy color change indicator indicates that the pathogen attack has already reached to the heavy stage.

The Conventional method for WRD detection has been developed since 1930. Napper (1932) and Pichel (1956) recommended detection by soil removal around the collar to expose a length of the tap and lateral roots. Declert (1960) proposed the use of log traps to detection of WRD mycellium development. Martin (1964) reported good and economical results with detection by mulching, which induces rhizomorphs to grow from the superficial contaminated rhizosfere. The result of Guyot and Fluri (2002) research explain than a good detection of diseased trees is needed especially in young crops (< 2 years after planting). Detection by infrared photography to characterize spectral changes between the reflectance of healthy and infected rubber tree crowns is unsuccessful (Nandris et al., 1985).

The serology test by using antibodies in the serum and it can detect pathogen microorganism (Fang & Ramasamy, 2015). The specific discovery of antibodies to target diseases is essential to facilitate diagnosis. Dalimunthe et al., (2016) has obtained antibodies that can recognize the presence of WRD in the soil and WRD infections through WRD infected leaf extract. Further testing of these antibodies is warranted for the technology. The purpose of this research was to examine the consistency of selected antibodies in detecting root fungi at various soil types in the rubber plantations.

2. Material and Methods

The research was conducted in March - June 2017 at the Laboratory Plant Protection, Sungei Putih Research Center. The soil samples were taken from four rubber estate area of PTPN III: Labuhan Haji, South Merbau, Rantau Prapat and Tanah Raja fields respectively.

The research design was a Completely Randomized Design non-factorial with twelve treatments and two replications. The soil was taken 1 m from the neck of the roots at a depth of 0-20 cm. The soils were sterilized at 121^0C for 30 minutes and incubated at the room temperature for 2 days before it was transferred into each jar bottle weighing ± 250 gr. Pure culture of R. microporus diameter 1 cm2 was inoculated into each soil and incubated for ± 3 months. The observed growth of the JAP mycelium was performed visually and qualitatively by using light (+), moderate (++) and heavy (+++) categories.One gram of soil-inoculated WRD mycelium was extracted used General Extract Buffer (GEB) with the addition of 80 mg of polyvinyl pyrrolidone (PVP). The extracts were centrifuged at 11,000 rpm for 15 min at 4OC. The supernatant was used as an antigen source and was given Ag1-Ag12 code. The antibodies used were antibodies immunized with WRD extract (antibody-induced WRD mycelium extract/M.Ab) and have successfully detected WRD mycelium in soil and leaf (Dalimunthe et al., 2016) The antigen is diluted 100 times with PBST for antigen and carbonate coating buffer for antibody dilution. After that, the each antigen and M.Ab were reacted into each cuvet of 750 µl. The antigen-antibody reaction was read by a spectrophotometer at a wavelength of 405 nm. The data were analyzed statistically using ANOVA.

3. Results and Discussion

WRD development in the soil visually was varied both between plantations and within the plantation itself. The highest WRD developments were in the field of Labuhan Haji and the lowest in Tanah Raja field (Table 1). The development of WRD depends on pH, organic matter content, moisture and aerate soils. WRD grow well on high humidity above 90%, high content of organic matter and good aeration. If this condition is appropriate, WRD may spread as far as 30 cm within 2 weeks (Sinulingga & Eddy, 1989). The result of artificial inoculation WRD shows that mycelium was grown varies on different soil types (Figure 1).

Table 1. The development level of white root disease attack qualitatively with artificial inoculation

Location of Sample Soils	Development level of WRD
Soil in Labuhan Haji Field	+++
Soil in Labuhan Haji Field	+++
Soil in Labuhan Haji Field	+++
Soil in South Merbau Field	+++
Soil in South Merbau Field	+++
Soil in South Merbau Field	++
Soil in Rantau Prapat Field	+++
Soil in Rantau Prapat Field	+
Soil in Rantau Prapat Field	+++
Soil in Tanah Raja Field	+
Soil in Tanah Raja Field	++
Soil in Tanah Raja Field	++

Note: + (light), ++ (moderate) dan +++ (heavy)

Figure 1. Artificial inoculation of white root disease in twelve soil types as a source of antigen (Ag). WRD's were cultured in the room condition for three months.

The reaction of selected antibodies with Ag1 - Ag12 was measured at a wavelength of 405 nm can be seen in Table 1. All of the antigens reactions were higher five times than negative controls. Increasing of absorbances indicated that of antigen-antibody bonds. The reaction proved that the antibodies recognize WRD mycelium in the soil. Nevertheless, all antigens-antibody reaction was lower than antibody-mycelium extract reactions (Ab-MAg) as positive controls. Based on these reactions can be predicted the development of mycelium in the soil between gardens were varies. Even the development of WRD in the same field there was also varied such as Labuhan Haji, South Merbau, Rantau Prapat and Tanah Raja field. Development of WRD mycelium the highest is known in the soil of Labuhan Haji field with an average absorbance of 0.1430 and the lowest in Tanah Raja field with an absorbance rate of 0.1243. The variation was thought to be related to the chemical and physical properties of the soil.

This level of serology reaction has corresponded to the number of mycelium that develops in the soil. The higher the rate of development of mycelium in the soil the higher the absorbance at 405 nm. Thus, MAb can be accurately recognized mycelium in the soil.

The use of mycelium extracts to produce polyclonal and monoclonal antibodies has been reported to detect the types of mycorrhiza (Gobel et al. 1995) and Candida famata (Pisa et al., 2007). The results of Dalimunthe et al. (2016) showed that antibodies of mycelium (Ab) were able to recognize mycelium antigens as positive controls with the absorbance of 0.1307. The next results of studies using selected antibodies were also able to detect mycelium antigens with the absorbance of 0.223 as a positive control and significantly different from negative control with the absorbance of 0.026 (Table 2).

Table 2. The reaction of antibodies (Ab) with antigen (Ag) of various soil types inoculated WRD based on absorbance value 405 nm.

Location of Sample Soils	Antigen	Absorbance 405 nm.
Negative Controls	AbBufer	0,026d
Positive Controls	AbAgM (mycelium extract)	0,223a
Labuhan Haji Field	AbAg1	0,1520b
Labuhan Haji Field	AbAg2	0,1350bc
Labuhan Haji Field	AbAg3	0,1420bc
South Merbau Field	AbAg4	0,1440bc
South Merbau Field	AbAg5	0,1340bc
South Merbau Field	AbAg6	0,1310bc
Rantau Prapat Field	AbAg7	0,1420bc
Rantau Prapat Field	AbAg8	0,1230 c
Rantau Prapat Field	AbAg9	0,1360bc

Tanah Raja Field	AbAg10	0,1195c
Tanah Raja Field	AbAg11	0,1240c
Tanah Raja Field	AbAg12	0,1295bc

Figures followed by different letter in the same column are significantly different based on Duncan multiple range test at 5% significant level

4. Conclusions

Serology techniques by utilizing polyclonal antibodies could detect the development of WRD mycelium in various soil types appropriately. Base on the result, the serological test could be developed as an early warning system for WRD infection. The short term output of this research is a serological kit that can use for detecting of WRD early symptoms on infected rubber plant.

5. Acknowledgements

The authors are grateful to Ministry of Research, Technology and Higher Education of the Republic of Indonesia through the Incentive Program of National Innovation Research System (InSinas) 2016 which has allocated funding of research at Sungei Putih Research Center, Indonesian Rubber Research Center with Letter of Agreement for Research Grant No. 293 / SP2H / LT / DRPM / III / 2016 on March 10, 2016.

6. References

[1] Dalimunthe, C.I., R. Tistama., H.S. Darwis., & S. Wahyuni. (2016). *Development of Serology Technique for Early Detection of White Root Disease (Rigidoporus microporus) in Rubber Plants*. Final Report. Sungei Putih Research Center.

[2] Declert, C., (1960). *Une technique nouvelle de detection des agents!des pourridies: La b! uchette-piege.Son application a l' etude de! Leptoporus lignosus*. Rev. Myc. 26, 119–127.

[3] Fang Y. & R.P. Ramasamy. (2015). Current and Prospective Methods for Plant Disease Detection. *Biosensor*, 4 : 537-561.DOI:10.3390/bios5030537

[4] Gobel, H., Hahn, A., & Hock, B. (1995). *Production of polyclonal and monoclonal antibodies against hyphae from arbuscular mycorrhizal fungi*. Crit. Rev. Biotechnol., 15(3-4), 293 – 304.Doi:10.3109/07388559509147415.

[5] Guyot, J & A. Flori. (2002). *Comparative study for detecting Rigidoporus lignosus on rubber trees*. Crop Protection (21). Elsevier.461-466.

[6] Martin, R. (1964). *Sur la propension de Fomes lignosus, agent de la pourriture blanche des racines d'hevea, a la fonction rhizomorphique" hors du milieu souterrain*. Rev. Gen. Caout. Plast. 41 (2), 279–284.

[7] Martin, R. R., James, D., & Lévesque, C. A. (2000). *Impacts of molecular diagnostic technologies on plant disease management*. Annual Review of Phytopathology, 38, 207-239. Doi: 10.1146/annurev. phyto.38.1.207

[8] Nandris, D., Tran Van, C., Geiger, J.P., Omont, H., Nicole, M., (1985). *Remote sensing in plant diseases using infrared colour aerial photography: application trials in the Ivory Coast to root diseases of Hevea brasiliensis*. Eur. J. For. Pathol. 15 (1), 11–21.

[9] Napper, R.P.N., (1932). *A scheme of treatment for control of Fomes lignosus in young rubber areas*. J. Rubber Res. Inst. Malayas. 4, 34–38.

[10] Pichel, R.J., 1956. *Les pourridies de l'h! ev! ea dans la cuvette congolaise.! Publ. INEAC*, Serie technique, No. 49, 480 pp.

[11] Pisa D., Ramos, M., Molina, S., Garcia, P., & Carrasco, L. (2007). *Evolution of antibody response and fungal antigens in the serum of a patient infected with Candida famata*. J. of Medical Microbiology, 56, 571–578. Doi: 10.1099/jmm.0.47042-0.

[12] Sinulingga, W & Eddy, S. (1989). *Control of White Root Disease on Rubber Plants. Indonesian Rubber Research Institute*, Sungei Putih.

Efficiency of Public Service in Pekangbaru City With E-Government

Zamzami[1], Fajrizal[2], Mhd Arief Hasan[3]
Universitas Lancang Kuning, Pekanbaru, 28265, Indonesia
E-mail: zamzami@unilak.ac.id

Abstract: The use of technology in the field of computers today is increasing, along with the development of science and globalization of information that demands the creation of a state of computerization. E-Government is now the right solution for local and central government, it is because it can assist the government in managing government data and able to provide information to the community quickly, precisely and efficiently. The current service system at the Office of Population and Civil Registration of Pekanbaru City still uses a manual system that takes a long time and cost a lot. In this E-Government application will contain application feature profile information of the Office of Population and Civil Registration of Pekanbaru City, public service requirements, office address of Pekanbaru City Sub-district, community complaint, KK (Family Card), and birth certificate online. The general purpose of making E-Government application system is to provide a new alternative for the Office of Population and Civil Registration of Pekanbaru City in improving its public service system. And the specific purpose is to facilitate the delivery of information, the formation of the system of birth certificate and death online and realize the excellent service for the Office of Population and Civil Registration Pekanbaru and the public.

Keywords: E-Government; Office of demographic affairs; Service; Government

1. Introduction
The use of technology in the field of computers today is increasing, along with the development of science and globalization of information that requires the creation of a state of computerization. The role of information technology in supporting the operational system in government agencies today is felt very important. Because basically the majority of government services to the public are matters relating to the collection, processing, and provision of various data information, knowledge, and policies and dissemination to the community. Utilization of information technology and technology is known as E-Governmet (Electronic Government.). The advances in information and communication technology and the invention of modern communicational and information devices in today's world have created new forms of doing things and giving services, thus the accuracy speed and transparency have been increased, waste of time and costs have been reduced and doing things and giving services have been improved quantitatively and qualitatively (Grove, 1999)

E-Government is now the right solution for local and central government, it is because it can assist the government in managing government data and able to provide information to the community quickly, precisely and efficiently. So as to enhance the relationship between government and society. E-government also allows the customers to cooperate in different activities such as socialactivities (Hazlett & Hill, 2003). Globalization makes governments establish e-government in order to sell goods andservices and export their culture to other countries (Lane & Lee, 2001). The aim of this study is to investigate therole of e-government establishment in the effectiveness of government organizations.

The current service system at the Office of Population and Civil Registration of Pekanbaru City still uses a manual system that takes a long time and cost a lot. By filling the blank manually at the local village office, the residents have to pay a lot of money. Coupled with the limited number of dispatchers who cause delays in the inputting process data citizens. So that causes the service handling becomes slow. With the E-Government application is expected to simplify the process of managing services and improve service systems for the better.

In this E-Government application will contain application features profile information of the Office of Population and Civil Registration of Pekanbaru City, public service requirements, office address of Sub-District Head of Pekanbaru City, community complaint, KK (Family Card), KTP (Identity Card) And birth certificates online.

Based on the above description the author tries to discuss and pour it in a study entitled "E-Government Applications On Office of Population and Civil Registration in Pekanbaru.

2. Basic Theory

2.1 E-Government

E-government allows governments to use new technology, which has access to government and service information, modifies their quality, and provides opportunities to cooperate in democratic processes and symbols (Hazlett & Hill, 2003). E-government is the use of easy-to-use information technology to offer Government services to customers directly and throughout the day. It also makes changes to anything that interacts with government Like citizens and commercial centers (Okotuma & cafery, 2000). In this case, an effective strategy significantly improves the government sector. These strategies include facilitating the process of providing services to citizens, removing additional levels of government management, preparing citizen accessibility, merchants, and employees Government information and services, facilitating institutional executive processes, reducing costs through Integration, removing additional systems, promoting the effectiveness of government operations to meet the needs Citizens, and ultimately achieve accountability, transparency and so on. Therefore, e-government is a collection of electronic Communications between government, corporations, and citizens (West, 2000).

2.2 Implementation of E-Government in Indonesia

In the instruction of the President of the Republic of Indonesia Number 3 Year 2003 on National Policy and Strategy of e-Government Development, mandates every Governor and Regent / Mayor to take the necessary steps in accordance with their respective duties, functions and authorities for the implementation of e-Government development national.
In general, mentioned about the things that become consideration of the implementation of e-Government in Indonesia, such as:

- The advancement of rapid communication and information technology and its widespread utilization potential, opens opportunities for accessing, managing and utilizing large volumes of information quickly and accurately.

- The utilization of communication and information technology in government process (E-Government) will increase efficiency, effectiveness, transparency, and accountability of government administration.
- To organize good governance and improve the effective and efficient public services it is necessary to have E-Government policy and development strategy.

3. Results

3.1 Company Overview

Office of Population and Civil Registration Pekanbaru City as one of the regional tools that help the Mayor in organizing the wheels of government, especially in the field of Population and Civil Registration have a very strategic role in organizing the orderly administration of population.

The series of population events which include resettlement of residents, Family Card ownership and Identity Card as well as important events which include birth, death, marriage, divorce, adoption / recognition and other important events concerning the population must be integrated in the national population database as well as the validity And the truth of the demographic document which is regulated in the framework of the realization of the discipline of population administration both from regional to national level, as mandated by Law Number 23 of 2006 which has been amended by Law Number 24 Year 2013 on population administration.

3.2. System of Proposals

The procedure of the proposed design is as follows:

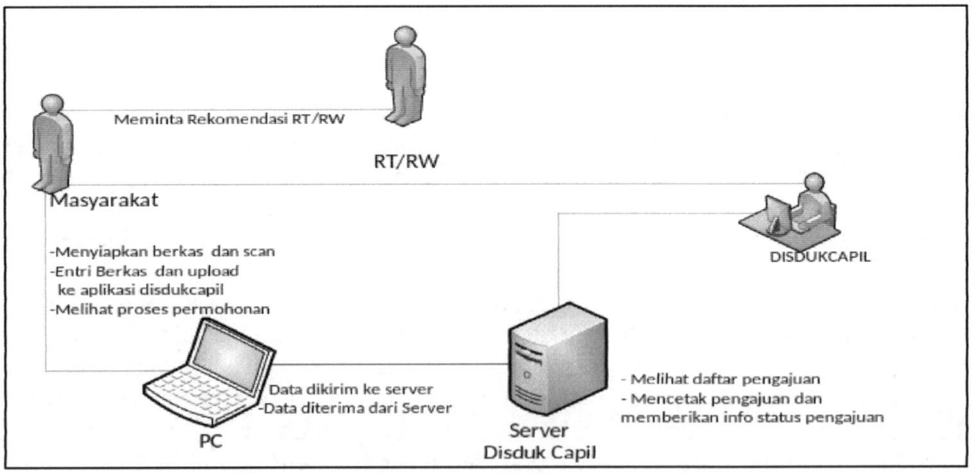

Figure 1: Design System

Explanation of operational system of procedure of handling submission of Birth Certificate and New Death on Disdukcapil Pekanbaru City that is:

- The residents asked for recommendations from local RT / RWs by completing the existing files.
- The public uploads all existing files through their pc units to the Disdukcapil Pekanbaru City web service system.
- The Disdukcapil officer can view the draft file submission through the system service. And provide status information submission.
- Officers can print the file submitted and then validate the file.
- The public can view the status information filing through the given tracking code. When it is finished the community comes directly to DISDUKCAPIL Pekanbaru City to retrieve the file.

3.3 System Design
At this stage, the authors provide information about the system design that the authors propose to tackle the problems of the system mentioned earlier.

- *Use Case Model Diagram*

The actors involved in this E-Government application can be classified into 2 actors namely operators and residents. For more details, the actor roles that exist in the system can be seen in table1 identification of actor.

Actor	***Description***
Operator	Operators are responsible for managing submissions from residents (see filing and status updates).
People	Submitting filings for filing Birth and Death certificates. View status, and update data.

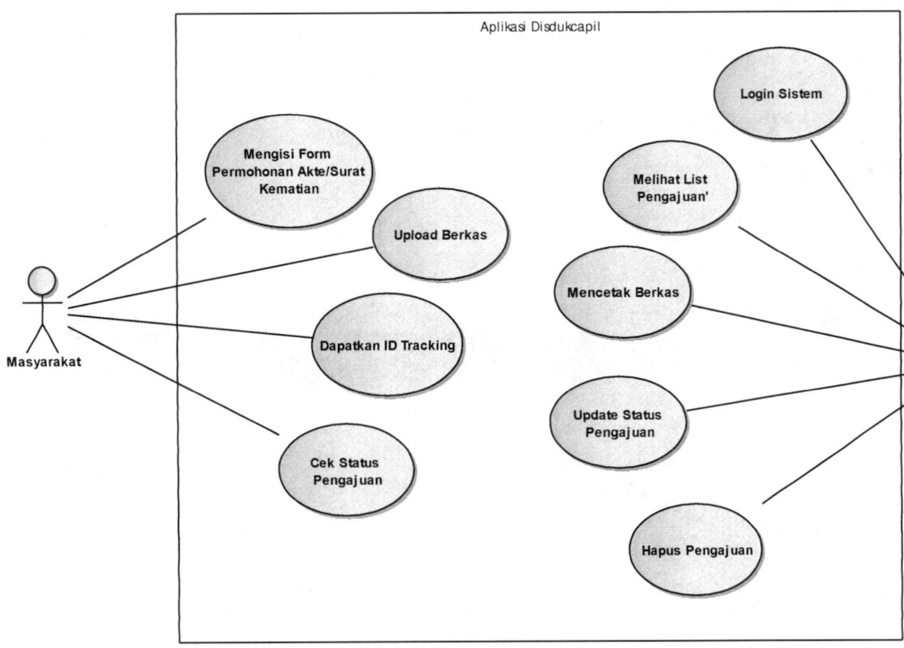

Figure 2. Use Case Diagram

- *Class Diagram*

Visualization of the proposed system object structure, described in Class diagrams. The proposed system object structure can be described in figure 3 below.

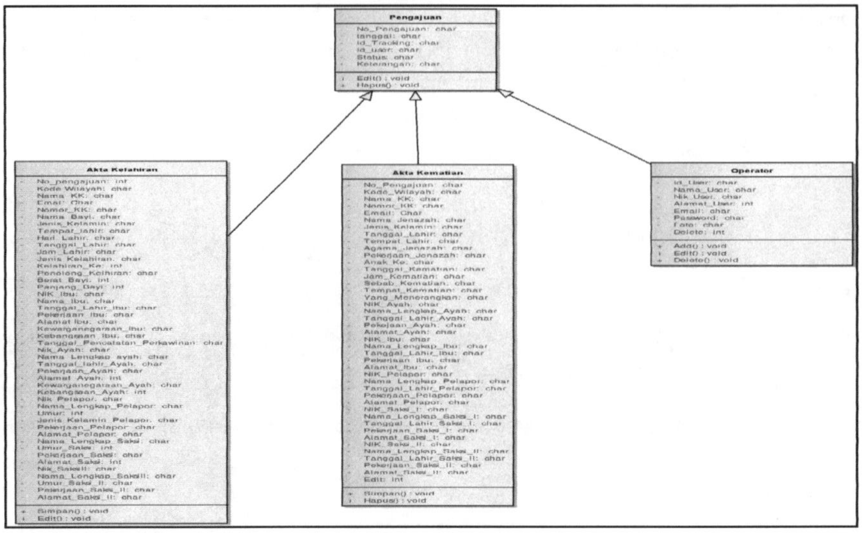

Figure 3. Class Diagram

- *Page Form Design Process Death Data*

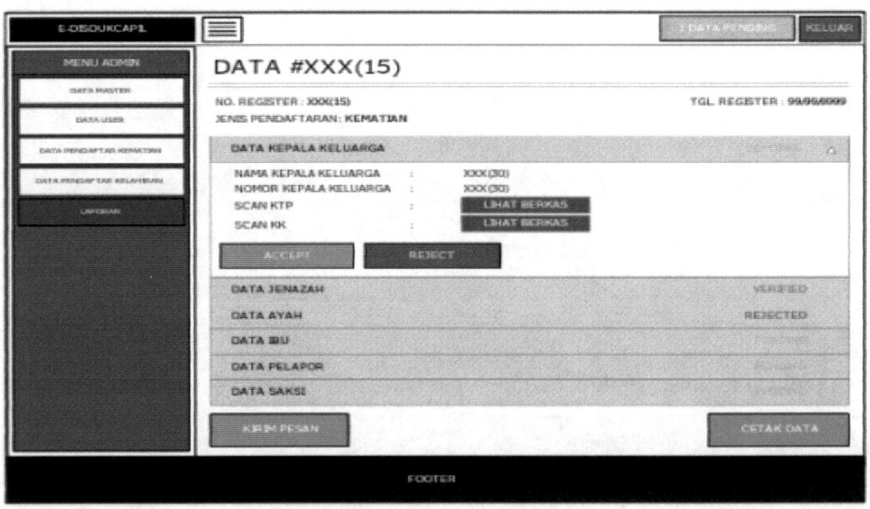

Figure 4 Death Form Design

In this form the operator officer validates from the submission by the residents. Later the operator will provide information whether the information provided is complete or not and will also be given the status is completed if the Deed and the new KK has been printed.

3.4 E-Government Website Effectiveness Test

The effectiveness of e-Government website in Pekanbaru City Civil Registry Office is known from the time of service and pruning of the bureaucracy. It also looks effective from the distance traveled by the community to go to the Office of Population and Civil Registration of Pekanbaru City. Busy people do not need to be bothered to visit the Office of Population Civil Registration City Pekanbaru. The effectiveness test is also seen from the operator side. Operators do not need to be bothered with stacks of application files because all files have been uploaded via the website.

4. Data analysis

Evaluation is done through several activities, namely: (1) first experimental test and the first prototype of e-Government website by expert, (2) prototype practicality test by operator and community (3) prototype effectiveness test through time duration of service. The results of the analysis of the evaluation activities are:

4.1 Test the Validity of E-Government Websites

E-Government websites that have been designed on the computer are validated by seven validators (Appendix 1) using validity test sheets (Appendix 2). Based on the results of the validator's validation of the validation sheet, all validator answers are distributed to the Guttman table.

After all validator answers are compiled into Guttman's table, the next step calculates the economies of scale by using a scalogram analysis. For this we need to count the number of errors (e). If the calculated cells are empty of the "Yes" answer deviating from the questions 1 through 19 in table 2, the number of errors is 2. After the error response is obtained, the next step is to calculate the reproducibility coefficient (Kr) which shows the protability level and the prototype Use the following formula:

$$Kr = 1 - \frac{2 \quad 2}{19 x 7 19 x 7}$$
$$Kr = 1 - \frac{2 \quad 2}{133 133}$$
$$Kr = 1 - 0,015$$
$$= 0,985$$

The coefficient of reproducibility or Kr = 0.985 indicates that the E-Government website on the Population and Civil Registry Office of Pekanbaru developed has been very valid as above the validity requirement of 0.90. While the coefficient of scalability or Ks, with c is the possibility of getting the correct answer. Because the answer is yes and no, then chances are 0,5

n = Number of questions x number of respondents = 133
Tn = Number of answer options "Yes" = 119

$$Ks = 1 - \frac{2 \quad 2}{0,5\ (133-119) 0,5\ (133-119)}$$
$$Ks = 1 - \frac{2 \quad 2}{0,5 x 14 0,5 x 14}$$
$$Ks = 1 - \frac{22}{77}$$
$$= 1 - 0,28$$
$$= 0,72$$

Ks or Coefficient of Scalability is 0.72. This reinforces the reproducibility coefficient and states it is very valid since it is large and equal to 0.60. Based on the results of validity test analysis said that e-government website on Pekanbaru City Government has been declared valid by the validator.
Having obtained the degree of validity of this e-government website, the next step is to make a revision in accordance with advices given validator through the validation sheet. The suggestions provided by the validator in the process of e-government website development process are as follows:

1) First Validator
The first validator, providing input for the page on the learning website in accordance with website resolution standards that is 1024 x 768 pixels for e-government website can be received on two different monitor display technology resolution in general that is 800 x 600 pixels and 1600 x 1200 pixels.

The selection of navigation and toolbar of the website follow the concept of the letter L inverted, so that the availability of a proportional page for e-government website. In addition, the first validator also suggests coloring that supports website content such as text and graphics on e-government websites, as well as making layout consistencies so that users can easily find the information they are looking for.

2) Second Validator
The second validator gives inputs to the end of the series of activities on the e-government website material provided guidance in order to facilitate the community to proceed to the next procedural or vice versa. To support the layout of e-government website the second validator also provides input to be made interesting and not too formal in grammar.

3) Third Validator
The third validator gives input so that in the use of language in the material progress instructions are made in the sentence to draw the public's attention.

4) Fourth Validator
The fourth validator provides input in the development of e-government website so that each sub menu displayed is given introduction so that the purpose of the service can be known by the community

5) Fifth Validator
The fifth validator advises that the e-government website also comes with a discussion forum as a forum for questioning about procedural issues.

6) The Sixth Validator
Validator six commented that the use of e-government website at the Office of Population and Civil Registry should be used for the benefit of the general public.

7) The Seventh Validator
Validator seven does not provide any suggestions or comments about the e-government website.
Based on the suggestions given by the validator, the first prototype was revised so that the revision result would be worthy to be tested as an e-government website in the civil service apparatus and civil registration of Pekanbaru City as the operator and the community in Pekanbaru City.

4.2 Practice Test of E-Government Website Disdukcapil Pekanbaru

1) Test the practicality of the operator / apparatus

In practical test for operator / apparatus begins by doing observation afterwards conducted training use of e-government website to operator / apparatus. To see the practicality is done through a questionnaire of practicality by the operator / apparatus after trying to use the website. The results of the assessment of the practicality are distributed into the Guttman table.

$$Kr = 1 - \frac{1 \quad 1}{10 \times 310 \times 3}$$

$$Kr = 1 - \frac{1\ 1}{3030}$$

$$Kr = 1 - 0{,}033$$

$$= 0{,}967$$

Reproducibility coefficient or Kr = 0.967 to see the product termination, also need to be reviewed The coefficient of e-Government website scalability in Disdukcapil Pekanbaru developed.

IOP Conf. Series: Earth and Environmental Science **97** (2017) 012044 doi:10.1088/1755-1315/97/1/012044

$$Ks = 1 - \frac{1 \qquad 1}{0,5\ (30-23)0,5\ (30-23)}$$
$$Ks = 1 - \frac{1 \quad 1}{0,5\ x\ 70,5\ x\ 7}$$
$$Ks = 1 - \frac{1\ 1}{3,53,5}$$
$$= 1 - 0,285$$
$$= 0,715$$

From the value of Ks (coefficient of reproducibility) is good to use, because the value of Ks below 0.715. From the results of the test of practicality test on the operator can be said that e-Government website practical on the test operator / apparatus.

2) Test the practicality to the community

In the small group practicality test obtained through a questionnaire of practicality by the community. The results of the assessment of the practicality of the community are distributed into the Guttman table.

$$Kr = 1 - \frac{1 \quad 1}{10\ x\ 610\ x\ 6}$$
$$Kr = 1 - \frac{1\ 1}{6060}$$
$$Kr = 1 - 0,017$$
$$= 0,983$$

The coefficient of reproducibility or Kr = 0.983 indicates that e-government website developed is very practical because of the above requirement of practicality is 0.90. While the coefficient of scalability or Ks. With c is the possibility of getting the correct answer. Because the answer is yes and no, then chances are 0,5

n = Number of questions x number of respondents = 60
Tn = Number of answer choices "Yes" = 54

$$Ks = 1 - \frac{1 \qquad 1}{0,5\ (30-23)0,5\ (30-23)}$$
$$Ks = 1 - \frac{1 \quad 1}{0,5\ x\ 70,5\ x\ 7}$$
$$Ks = 1 - \frac{1\ 1}{3,53,5}$$
$$= 1 - 0,285$$
$$= 0,715$$

Ks or Coefficient of Scalability is 0.67. This reinforces the reproducibility coefficient and states practically that is 0.60. From the results of the analysis of moderate group practicality test can be said that e-government website is practical on the test community.

4.3 Test of Effectiveness

The effectiveness of E-Government website new data obtained after this system if it has been fully implemented in Pekanbaru City. The test is how long the service process presented by using the application compared to the manual service that has been going on.

5. Conclusions and Suggestions

First year's research has yielded according to the type of outcome. The main output that will be generated is E-Government Application to help Efficiency of public service in Civil Registry Office of Pekanbaru City. The outcomes will be achieved by implementing specific research objectives with the following indicators of success: It has been identified the types of services that need to be electronically the modeling for e Disdukcapil for the handling of birth certificates and deaths. The design has been made for further to be developed into application program that can be applied fully in Disdukcapil Office Pekanbaru City

6. Suggestion

The need for further testing of the developed model to be right on target. The program must be implemented immediately to provide great benefits for the people of Pekanbaru City.

7. References

[1] Grove, A. S. (1999). *Only the paranoid survive: How to exploit the crisis points that challenge every company*. USA: Crown Business.
[2] Hazlett, S-A., & Hill, F. (2003). E- government: *In the instruction of the President of the Republic of Indonesia Number 3 Year 2003*
[3] *The realities of using IT to transform the public sector. International Journal of Managin Service Quality*, 13(6), 445–452.
[4] Layne, K., & Lee, J. (2001). *Developing fully functional E-government: A four stage model. Government Information Quarterly*, 18(2), 122–136.
[5] Martz, W. (2008). *Review of the book: The handbook of leadership development evaluation*. American Journal of Evaluation, 29, 584–589.
[6] Okot-Uma, R., & Caffrey, L. (2000). *Trusted Services and Public Key Infrastructure*. London.

Counseling Model Application: A Student Career Development Guidance for Decision Maker and Consultation

Irwan[1] and Gustientiedina[2] and Sunarti[3] and Yenny Desnelita[4]
[1, 2, 3, 4] STIKOM Pelita Indonesia, Pekanbaru, 28265, Indonesia.
Telp: +6276124418, Fax: +6276135508
E-mail: Irwan@lecturer.pelitaindonesia.ac.id

Abstract: The purpose of this study is to design a counseling model application for a decision-maker and consultation system. This application as an alternative guidance and individual career development for students, that include career knowledge, planning and alternative options from an expert tool based on knowledge and rule to provide the solutions on student's career decisions. This research produces a counseling model application to obtain the important information about student career development and facilitating individual student's development through the service form, to connect their plan with their career according to their talent, interest, ability, knowledge, personality and other supporting factors. This application model can be used as tool to get information faster and flexible for the student's guidance and counseling. So, it can help students in doing selection and making decision that appropriate with their choice of works.

Keywords: Application model, counseling, career development guidance, decision maker

1. Introduction

In the field of education, career guidance and student ca reer development are the services guidance and counseling programs which aimed to help or to facilitate the students and to achieve their self-development. At school, students are required to master a variety of abilities or competencies, both related to the subject, as well as those related to personal, social, and career development. In order to achieve the optimal development, the optimal service is required from basic element of education in school. The elements of the education are management, leadership, learning, and elements of student coaching.

Guidance and Counseling is an activity that comes from human life. The fact shows that man in his life faces some problems. Human being are not equal to each other, both in nature and ability. There are people who can solve the problems by himself, but some others cannot solve the problem by himself. Related to the student's careers and how to make a decision for their life, some students decided their career based on their friends and parents sugestions, not by themselves. This illustrates that student decision making is more influenced by personal component is the narrow knowledge of career orientation than the support of parents.

Expert System is a system designed to be able to do reasoning like an expert in a certain area of expertise. In the expert system there is a knowledge base that contains the knowledge in problem solving. Implementation of knowledge base can be used in various fields, testing the accuracy of the system by looking at the suitability of interest with the recommendations of the system [1]. Then one example of application from use of the backward chaining method is used in designing a prologue for an expert system [2]. Expert System one of the mainstay technology in Knowledge Management, using knowledge management system in education can share information effectively and relevant [3]. The relevance of integration between information and communication technology in technical, vocational education and training on knowledge management [4]. The purpose of this study is to produce a software model for counseling guidance focused on career counseling and career development of students both at school and college-based Knowledge Management System.

2. Literature Review

This researchers analyze based on the multiple perspectives and different point in accordance with the interests and the concentration of each researcher. This study has been reviewed several previous research. Although this review is not exhaustive, considering the limited resources and time, but could represent a review that conducted on several sources. The result this research is based on the phenomenon confusion of students at SMK to Career which will be taken. The education that is being pursued by many is not in line with the actual career desired. The purpose this research is to know the profile career maturity of students at SMK, Conducting study on career guidance program at school, knowing career counseling by BK, knowing the activity of career guidance program done. The career guidance program proves to be effective for improving student maturity so that it is feasible to apply in counseling and guidance services [5]. In the career guidance program published in Agustinus that is the result of the assessment of counseling guidance experts and the designer of the program of counseling guidance program, the product of this development is considered feasible, shown by the high research result from the counseling counselor and the result of the expert assessment of the counseling guidance program. The empirical results of the expert test scores indicate that, after revision, the product of this development can meet acceptability [6].
The development of interactive multimedia career guidance to enhance the skills of making career decisions make career decisions on the student's specialization program, how interactive multimedia career guidance improves students 'career-making skills decisions that research objectives produce interactive multimedia career guidance to enhance students' career decision making skills [7]. Research by [8] on career selection by title *Linguistic and educational socialization of Sudanese-background high-school students in Australia.* In this study analyzed the interest of talent and providing solutions to the problem of career selection. The method used mixed method. The results of this study, knowing the problems of doubt in the selection of careers against young Australians. While knowledge-based decision making [9], this study aims to help students to choose the best suitable career choice program at the Higher Learning Institute in Kenya. The current flow of approaches in Kenya is where students get career advice from teachers, motivators and career counselors. The basis of the approach in this study is to design a smart artificial system model using the JESS plugin in an eclipse IDE that receives KCSE results. Variables required values and subjects.

3. Method

The method used is System Development Life Cycle (SDLC).

1) Project Identification and Selection. In this phase, direct observations are made to experts or counselors. The results of the observations are then summarized and analyzed to see the weaknesses, advantages, opportunities, and threats that exist. Thus illustrated the possibility of the aspects that should be developed towards software guidance and career development.
2) Project Initiation and Planning
   The second phase takes initial action (initialization) and develops a plan for the selected development project in the first phase above. The prefix action is to focus more on observations on the knowledge of student career guidance and development.
3) Analysis
   From some existing technological tools, student guidance and career development programs will be applied to web-based applications.

4) Design
   The guidance and development program will be implemented in web-based applications, the interface design of this software and career development using a text editor added with CSS and Javascript, then to design the database using PhpMyAdmin
5) Code Generation
   The process of making coding or coding is a translation of design into a language that can be recognized by the computer, programming language that is used is the programming language Php
6) Testing
   Testing techniques to test the software guidance and career development is by using black box testing techniques. Because with this technique testing is done to determine whether the program is made in accordance with its functional requirements. Tests are performed on all program modules.

4. Result and Discussion

This chart is used to describe users of apps and user behavior against apps. The user is represented by the actor while the behavior represented by the use case is shown in figure 1.

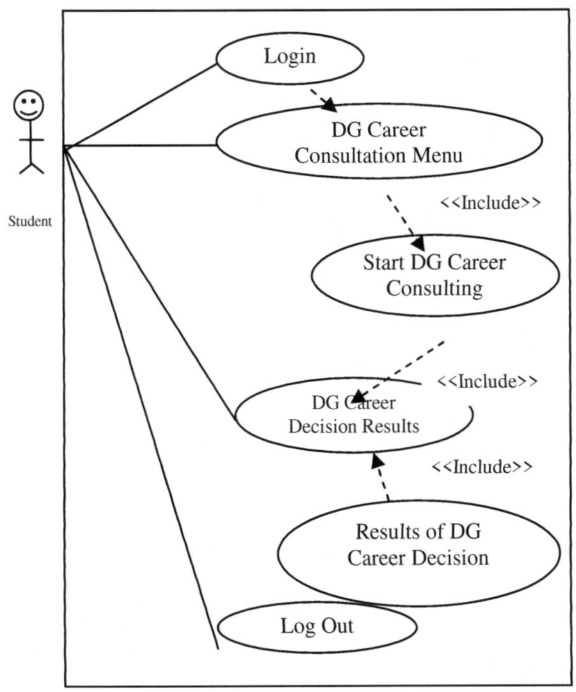

Figure 1. Use Case Diagram of DG Career Counseling

Where the use description diagram guidance career development students can be seen in the table below.

IOP Conf. Series: Earth and Environmental Science **97** (2017) 012045 doi:10.1088/1755-1315/97/1/012045

Table 1. Description Use Case Diagram login Career Development Guidance

Use Case Name	Login Student Career Development Guidance
Requirment	A1-A6
Goal	Students successfully enter the Career Development Guidance page (DG Career)
Pre-condition	Students may start Career Development Guidance (DG Career)
Post-condition	Shown the results of the decision DG Career
Failed end condition	Students do not do DG Careers
Primary Actor	Student
Main Flow/Basic Path	1. Students can see the results of DG Career decisions 2. Students can print DG Career results 3. Students exit the DG Career application

4.1. Knowledge Base

Knowledge base is formed on facts in the form of information about how to generate a new fact from a known fact. This knowledge is a representation of the knowledge and the way of a thought approach of a counselor. This software is designed 3 main parts of the knowledge base (the place of actual information storage), Inference Engine for the reasoning process to find solutions and conclusions that data sent by the user and the facts stored on the knowledge base, and user interface.

4.2 Application Model Counseling Guidance Career Development

The application model of counseling career development counseling students for decision-making and consultation to improve information management skills on career knowledge is illustrated in Figure 2.

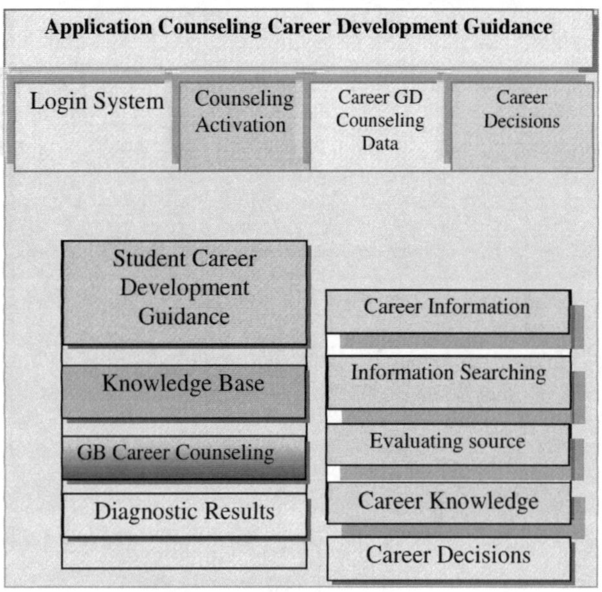

Figure 2. The Model Application Counseling

4.3. Conceptual Framework

The conceptual framework of making application counseling model: student career development guidance for decision making and consultation integrates application counseling design Career Development Guidance illustrated in figure 5

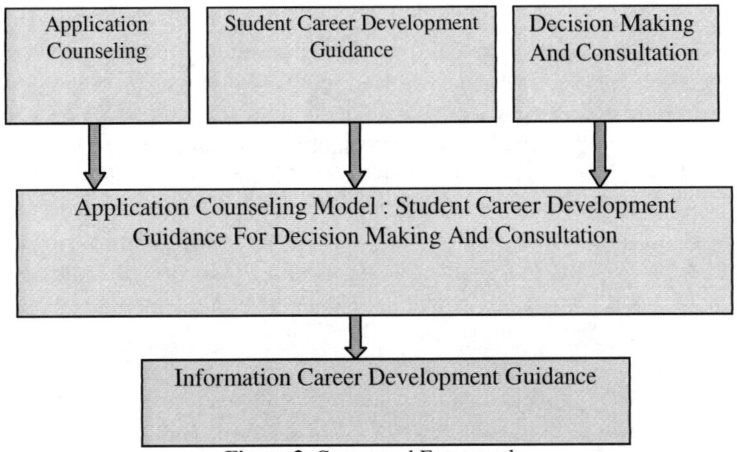

Figure 3. Conceptual Framework

4.4. Application Architecture

Figure 6 shows the application architecture of Career Development Counseling. The application consists of four modules: (i) Self Potential module, (ii) Career, (iii) Decision Making and Consultation, and (iv) Career Knowledge. Furthermore, application module Counseling Guidance Development of career carrier (figure: 6) contains applications to help students make decisions whether to continue to the next level of education or work. This application can be used as counseling service provision of Career Development Guidance indirectly because counselor do not directly face to face with counselee (student). Counseling services Career Development through Application (Information Technology) is expected to reach students. This represents a small contribution to the advancement of education in Indonesia.

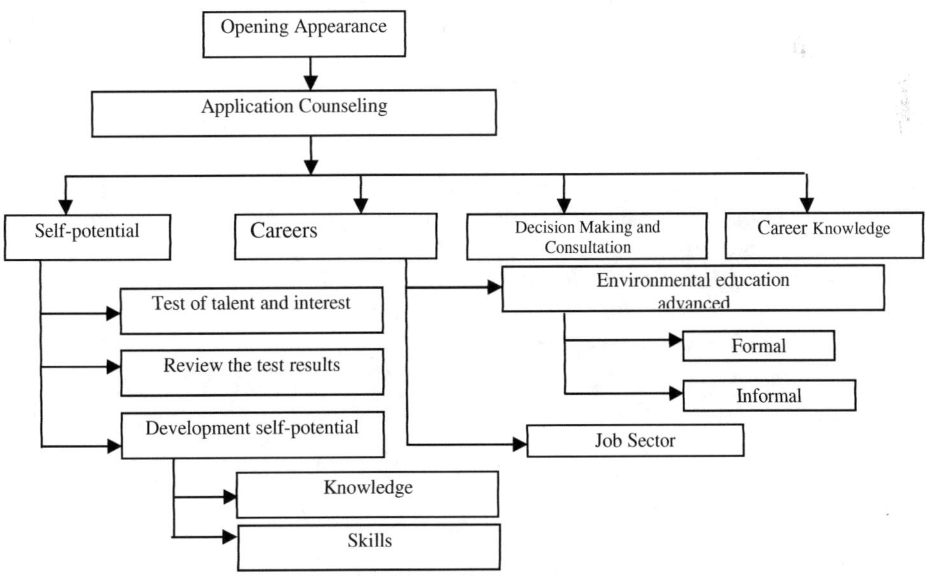

Figure 4. Application Archittecture Counseling Guidance Develoopment Career

5. Conclusion

This paper shows**,** though the design of the model result counseling application: student career development guidance for decision making and consultation in an effort to assist students and counseling teachers in handling the careers of students they will take. This counseling application can be an alternative to student career development guidance, among them: 1) This application model is created to help the students to get immediate results from career counseling guidance counseling application, as well as to assist teachers counseling in directing students for career development; 2) This application is designed web-based, making it easier for students in its use and information can be seen on the spot; 3) This counseling application has the ability to store the knowledge base and archive data with a database that is owned and can be done data backup at any time.

6. Acknowledgements

This work is supported by Ministry of Research, Technology and Higher Education Republic of Indonesia and Sekolah Tinggi Ilmu Komputer (STIKOM) Pelita Indonesia Pekanbaru Riau.

7. References

[1] Levi JH, et al. 2015. *Rancang Bangun Aplikasi Penjurusan Minat Bakat Menggunakan Metode Forward Chaining (Studi Kasus : Universitas Multimedia Nusantara)*

[2] Li Guoqi, et. al. 2014. *An IPC-based Prolog Design Pattern For Integrating Backward Chaining Iference Into Applications Or Embedded System.* Chinese Journal Of Aeronautics.

[3] Anumnu, S. I. 2014. *Knowledge Management And Development Of Etrepreneurial Skills Among Students In Vocational Technical Institutions In Largos, Nigeria.* The Electronic Journal of Knowledge Management Volume 12 Issue 2 (pp144-154) available online at www.ejkm.com.

[4] M.S. Saud.et al. 2011. *Effective Integration of Information and Education and Training (TVET) Toward Knowledge Management in The World of Work.* African Journal of Business Management Vol 5(16), pp. 6668-6673, 18 august, 2011. Available Online at http://www.academicjournals.org/AJBMISSN:1993-8233

[5] Juwitaningrum. I. 2013. *Program Bimbingan Karir untuk Meningkatkan Kematangan Karir Siswa SMK, Jurnal Bimbingan dan Konseling.* PSIKOPEDAGOGIA, 2013. Vol. II. No 2. Prodi Bimbingan dan Konseling UAD. ISSN: 2301-6160

[6] Agustinus. R.K. 2009. *Pengembangan Program Bimbingan Konseling Di Universitas Slamet Riyadi Surakarta*

[7] D.M. Leksana and M.E. Wibowo And I. Tadjri. 2013. *Pengembangan Modul Bimbingan Karir Berbasis Multimedia Interaktif Untuk Meningkatkan Kematangan Karir Siswa. Jurnal Bimbingan Konseling*

[8] A. Hatoss, et,al. 2012. *Career choices : Linguistic and educational socialization of Sudanese-background hingh-school students in Australia.* Elsevier Inc. All rights reserved

IOP Conf. Series: Earth and Environmental Science **97** (2017) 012046 doi:10.1088/1755-1315/97/1/012046

Web Based Cattle Disease Expert System Diagnosis with forward Chaining Method

Ahmad Zamsuri, Wenni Syafitri, Muhamad Sadar
Universitas Lancang Kuning, Pekanbaru, 28265, Indonesia
Telp. 0811 753 2015
e-mail : ahmadzamsuri@unilak.ac.id, wenni20@gmail.com, sadarzen@unilak.ac.id

Abstract
Cattle is one of the livestock who have high economic potential, whether for livestock, cattle seed, or even for food stock. Everything that comes from Cattle is a treasure for example the Milk, the Meat, and Cattle-hide. The factor that cause Cattles to die is the spread of disease that could crock up the Cattle's health. So that the Expert system is needed to utilize and analye the Cattle's disease so it could detect the disease without going to the veterinarian. Forward chaining method is one of the correct method in this expert system wherein began with Symptoms to determine the illness. From this matter, we built a web based expert system application on Cattles disease to ease the disease detection and showing the brief information about the Cattles itself.
Keyword: Expert System, Web, Forward Chaining

1. INTRODUCTION
Indonesia has high potential livestock for example dairy cattle and beef cattle, Livestock breeding products are growing and concentrated in the development area of the production center. With large amounts of production, the need for animal protein in Indonesia is increasing with increasing public awareness of the importance of nutritional intake. Therefore, the health of livestock raised by farmers is essential to meet the nutritional needs and in addition to income for the livestock owners themselves. Of the various types of livestock that many reared by rural breeders are Cattles.

This paper discusses the techniques of reasoning, namely problem-solving techniques by presenting problems into the knowledge base, and making reasoning to find solutions. Reasoning to find solutions to livestock is a method of knowledge-based approach, to study and solve problems based on data collected and stored.

Application Cattle disease detection is a useful software for designing an expert system application program, which is able to provide an accurate diagnosis of the likelihood of cattle suffering from disease and how to treat it. By Using Mobile Android breeders can directly send the data- Cattle data to the server so that consumers can see the appropriate criteria of Cattles desired. From the description above, the authors are interested to build an application "Android Based Cattle Disease detection application design using Certainty Factor".

2. Theoritical Basis
The theory of this expert system consists of the definition of expert systems, expert system architecture, Mechanism of inference, explanation facility, comparison of expert system with Conventional system, and inference techniques.

2.1. The Definition of Expert system
Expert systems are computer-based applications that are used to solve problems as the expert thinks. Experts in question here are people who have special skills that can solve problems that

can not be resolved by the layman. For example, the doctor is an expert who is able to diagnose the illness suffered by the patient and can provide the solution of the disease.
Expert system is an artificial intellegence program that combines knowledge base (Knowledge Base) with inference system. This is a high-level specialization software section that seeks to duplicate the function of an expert in one area of expertise. This program acts as an intelligent consultant or advisor in a particular skill environment, as a result of a collection of knowledge that has been collected from several experts. Thus even a layman can tap the expert system to solve various problems that he faced.

2.2 Metode Certainty Factor

Certainty Factor introduced by Shortlife Buchanan in creating of MYCIN. *Certainty Factor* (*CF*) is a clinical parameter that given by MYCIN to show the certainty value. According to Giarrantano dan Riley [2]. *Certainty Factor* defined as:

$$CF(H, E) = MB(H, E) - MD(H, E)$$

Where:
CF (H,E) : *Certainty Factor* from hypotesis H affected by *evidence* E. The magnitude of CF range from -1 to 1. Value -1 shows absolute distrust, whereas 1 shows absolute trust.
MB(H,E) : *Measure of increased belief* to hypotesis H affected by evidence E.
MD(H,E) : *measure of increased disbelief)* to hypotesis H affected by evidence E
By exploring the results of interviews with experts, the value of CF (Rule) obtained from the interpretation of the expert to a specific CF value can be seen in table 2.1 below:

Tabel 1. Value CF Interpretation

Certainty Term	**MD/MB**
Do not know/None	0.2
Maybe	0.4
Most likely	0.6
Most certainly	0.8
Certainly	1.0

(Source : Turban and Frenzel, 1992) [3]

2.3 Cattle

Cattle are livestock as a source of meat, milk, labor and other needs. Cattle produce about 50% of the world's meat needs, 95% of milk needs, and 85% of skin needs. Cattles are from bovide families, such as bison, bull, buffalo (Bubalus), African buffalo (Syncherus) and Anoa.

2.3.1Type of Cattles

The types of cattle that are present in Indonesia today are the original Indonesian cattle and imported cattle. Of the types of cows, each has distinctive properties, both in terms of the outer form (body size, color of feathers) and from the genetic (growth rate). The types of cattle, among others, will be discussed below:

1) Balinese Cattle
2) Simmental Cattle
3) Limousine Cattle
4) Brahman Cattle

2.3.2 Cattle Disease

The types of Cattle disease that will be discussed is :

1) Brucellosis
2) Infection Bovine Rinotracheitis
3) Johnes's Disease
4) Antrax
5) Mad cow disease
6) Bovine Viral Diarrhea

3. Analysis and Design

3.1 System Analysis

In the theoretical chapter chapter has described expert systems in diagnosing cow disease. In this final project the author will use Forward Chaining and Certainty Factor method, forward chaining is a reasoning that starts from the facts to get a conclusion (Conclusion) from that fact.

The method of certainty factor aims to predict the value of cow disease uncertainty, through reasoning for symptoms experienced by animals, and also with suggestions or information required in connection with the prediction of the diagnosis. So to deal with the uncertainty values in cow disease based on symptoms in this system in conducting the diagnostic process using Certainty Factor (Certainty Factor).

From the analysis of the problems that have been done, the authors can build applications that can perform the diagnosis of cattle disease by transferring the knowledge that experts have into the computer system.

In developing this expert system required knowledge and information obtained from several sources, ie from experts, as well as books about existing cattle disease. So it can solve problems that can not be solved by most people, or can solve problems in a more efficient way. Therefore the scope of the discussion of cattle disease will not deviate from the knowledge of the experts. The system to be built is an expert system-based application to diagnose cow disease in general by using Forward Chaining method. This application aims to help the community or cattle ranchers to be able to make a diagnosis of cow disease and how to treat cattle disease

3.2 Production Rules

Production rules are one of the many forms of knowledge representation Used in the development of expert systems. Representation of knowledge with the rules of production, Basically a rule (rule) in the form of IF-THEN.

The following is a reprentation of expert system application knowledge to diagnose Cattle illness with production rules. The production rules are outlined below:

a. Rule 1
IF High Fever
AND Weak Body
AND Weight Loss
AND Have an abortion
THEN Brucellosis

b. Rule 2
IF Have an abortion
AND Nerve Disorder
AND Reproduction Disorder
AND Diarrhea
AND Death
THEN Infection Bovine Rinotracheitis

c. Rule 3
IF Weight Loss
AND Diarrhea
AND Milk Production Decreased
AND Trembling body
AND Death
THEN Johnes's Disease

d. Rule 4
IF Hard to breathe
AND Death
AND Darker Eyes
AND Depretion
AND Rapid Breathing
AND Increase Pulse Rate
AND Seizure spasm
AND Stagger
AND Out of Saliva
THEN Antrax

e. Rule 5
IF Milk production decreased
AND Fetal infection
AND Respiratory system disorders
AND Decreased Appetite
THEN Mad cow desease

f. Rule 6
IF Diarrhea
AND Death
AND Experiencing appetite decreased
AND Experiencing Blood Out Of The Nose
THEN Bovine Viral Diarrhea

3.3 Model To Calculate Certainty Factor from Rule

There are two stages of the model that are often used to calculate the level of confidence (CF), of a rule is as follows:

3.3.1 By Digging out the results of interviews with experts.

By extracting from the results of interviews with experts, the value of CF (Rule) is derived from the interpretation of the term 'from experts' to the value of MD / MB Specific.

Table 2. Value MD/MB

Certainty Term	Interpretation Value MD/MB
Do not Know/There is no any	0.2
Maybe	0.4
Most likely	0.6
Almost definite	0.8
Certainly	1.0

(Source : Turban and Frenzel, 1992) [3]

3.3.2 Manual calculation of CF Value

Difficulty in writing this final task is the process of determining the value of certainty or certainty factor symptoms of cow disease. Because in the medical field there is no standard provision of cervical disease certainty value. In this study, the determination of certainty factor figures for each of the symptoms of cow disease, the order is based on the main symptoms of the disease to n symptoms. $CF[h,e] = MB[h,e] - MD[h,e]$

With :

CF[h,e] = Certainly Factor

MB[h,e] = Size of Trust to Hypothesis H MD[h,e] = Size of Unbelief The first case example on the main symptoms of Brucellosis disease can be seen in Table 3 Tabel 3. Keputusan Symptoms Disease Brucellosis

Disease	Disease CF Value	Symptoms	Symptoms CF Values
Brucellosis	0.04	High Fever	0.5
		Body Weakness	0.2
		Body Weight Loss	0.3
		Have an Abortion	0.6

Certainty Value On High Fever Symptoms

MB(h,E1) = (CFsymptoms-CFdisease)/(1-CFdisease)
= (0.5-0.04)/(1-0.04)
= 0.46/0.96
= 0.479

MD(h,E1) = (0.04-0.04)/(0-0.04)
= 0

CF(h,E1) = MB(h,E1)-MD(h,E1)

= 0.479 – 0
= 0.479

Value of Certainty In Symptoms of Weak Bodies

MB(h,E2) = (CFsymptoms-CFdisease)/(1-CFdisease)
= (0.2-0.04)/(1-0.04)
= 0.16/0.96
= 0.166
MD(h,E2) = (0.04-0.04)/(0-0.04)
= 0
CF(h,E2) = MB(h,E2)-MD(h,E2)
= 0.479 – 0
= 0.166

Certainty Value On Symptoms Weight Loss

MB(h,E3) = (CFsymptoms-CFdisease)/(1-CFdisease)
= (0.3-0.04)/(1-0.04)
= 0.26/0.96
= 0.270
MD(h,E3) = (0.04-0.04)/(0-0.04)
= 0
CF(h,E3) = MB(h,E3)-MD(h,E3)
= 0.270 – 0
= 0.270

Certainty Value In Symptoms Abortion

MB(h,E4) = (CFsymptoms-CFdisease)/(1-CFdisease)
= (0.6-0.04)/(1-0.04)
= 0.56/0.96
= 0.583
MD(h,E4) = (0.04-0.04)/(0-0.04)
= 0
CF(h,E4) = MB(h,E4)-MD(h,E4)
= 0.583 – 0
= 0.583

CFcombination (CF1, CF2, CF3, CF4) = CF(h,E1)
+ CF(h,E2) + CF(h,E3) + (1-CF(h,E1))
CFkombinasi = 0.479 + 0.166+0.270+0.583+(1-0.479)
= 0.521 * 1.498
= 0.710

Then the calculation value of the formula 4 above shows that the value of certainty Brucellosis cow disease with the level of certainty 0.710. After the process will appear the value of CF, which based on the calculation of MB and MD values of the selected symptoms. And the result of the diagnosis process isCFdisease is 0.710 and

The condition of the degree of CF is **almost certain**

4. Implementation

System implementation phase is a design translation phase based on the results of analysis into a particular programming language and application of software built in the real environment. Adapaun implementation discussion consists of software builder, builder hardware, implementation constraints and interface implementation.
From the description of matters relating to the expert system, then the diagnosis of cow disease by using certainty factor based on this android, so it needs to be implemented with the following explanation:

A. Main Interface

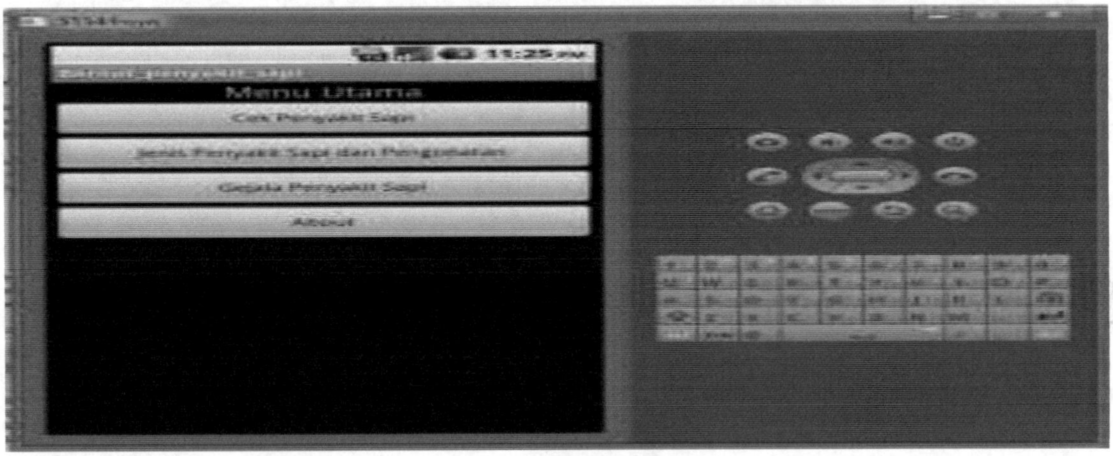

Gambar 1

1. Main Interface
This screen explains the forms of cow disease checks, cow disease form and treatment forms, cattle's symptoms formulations and form about.

B. Consulation Interface
This screen is a screen of consultation on the cow disease check form, on the screen of the celebration where the user is asked to answer the questions indicated about the symptoms experienced by the cattle. The Question screen can be seen in Figure 2 below.

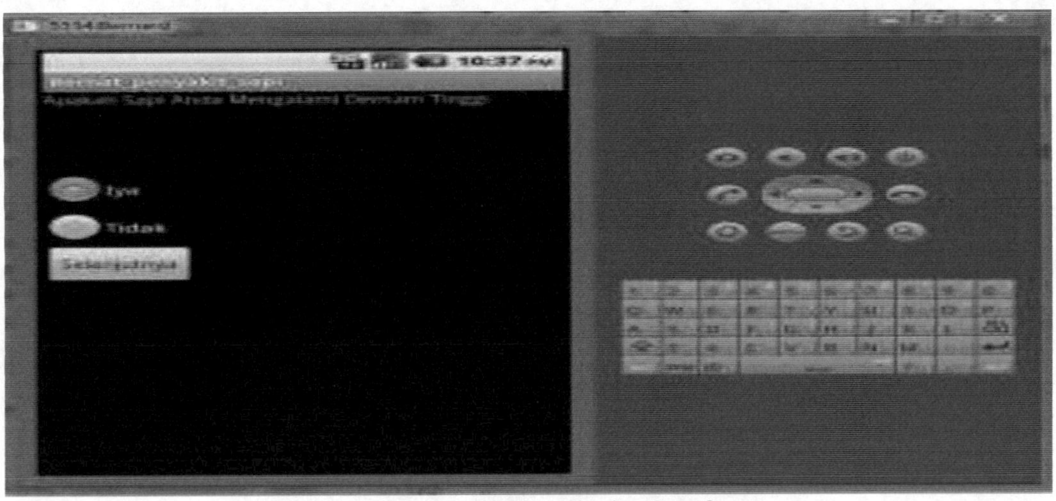

Gambar 2. Consultation Interface

C. Consultation Result Interface

On the Screen of the Consultation Result is a screen containing the results of the results of the deaf, based on the user's choice in answering the questions indicated about the symptoms experienced by the farm animals.

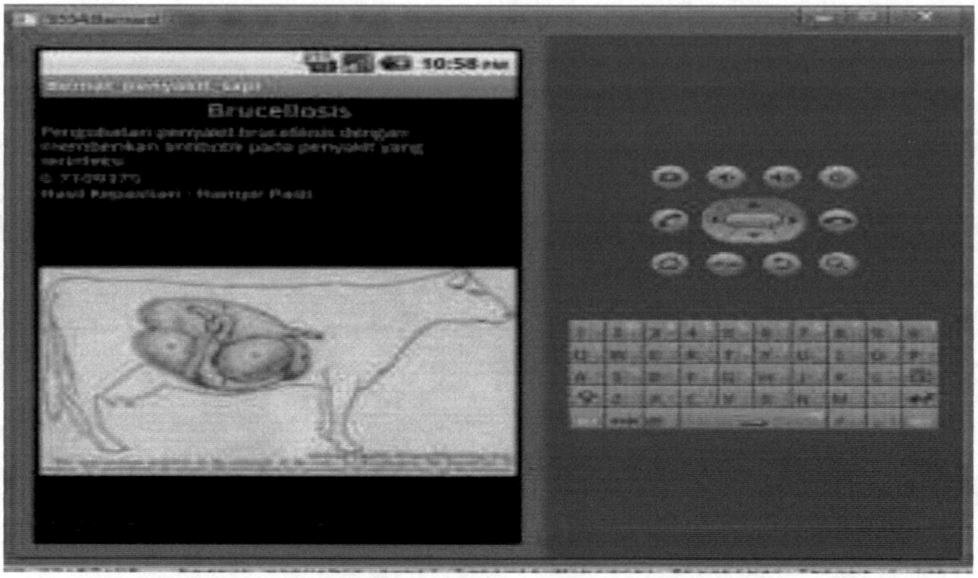

Gambar 3. [1] [2] [3] [4] [5] [6] Consumer Results Menu Screen Display

5. Conclusions and recommendations

Conclusion of design result from user Expert systems to diagnose cow disease are:

1) The prototype of this expert system has been completed covering various aspects of cow disease. Rules that have been made have been in accordance with the expert system, expert system prototype is designed to be easily operated by the user (user friendly)
2) Prototype of this expert system has been using forward chaining and certainty factor, which is used as a knowledge base in asking questions to farmers.

Suggestions that can be given for Development of this expert system, in order to be more useful and effective in diagnosing cow disease in android are:

1. The addition of new rules or rules about cow disease on the knowledge base, to improve accuracy in diagnosing.
2. Can be added various kinds of new cow disease, so the type of disease and symptoms disease can update the latest data with automatic update online.

6. Reference

[1] S. Suparman, Mengenal Artificial Intelligence, Yogyakarta: Andi Offset, 1991.

[2] J.Giarratano and G. Riley, Expert Systems : Principles and Programming, Boston: PWS Publishing Company, 2005.

[3] F. Turban, Expert Systems and Applied Artificial Intelligence, Prentice Hall Professional Technical Reference, 1992.

Learning Methodology for Early Child Education Education (Paud) in the Recognition of Legal Capitals based on Android

Didik Siswanto
Universitas Lancang Kuning, Pekanbaru, 28265, Indonesia
Tel/Fax : +6271 52324
E-mail :, didik@unilak.ac.id,

Abstract : School as a place to study require a medium of learning. Instructional media containinginformation about the lessons that will be used by teachers to convey a lesson. School early childhood education Al-Kindy Pekanbaru interms of learning the letter hijaiyah still use conventional learning media. But with the conventionalmedia is not very attractive to use, so the need for an exciting learning medium that can make childrenbecome interested in learningThe purpose of this study was to create a Media Learning Introduction Letter Hijaiyahmultimedia form and benefit from the introduction of letters Hijaiyah Learning Media is a renewal of themedium of learning in School early childhood education Al-Kindy Pekanbaru.In this study the authors tried to make the learning application that contains the basicknowledge of letters hijaiyah dsertai with animation, audio and explanation how to read the letters inorder to complete the learning media letters hijaiyah more interactive.

Keywords: *Interactive Learning, Hijaiyah, Mobile, Android,(LTSA)*

1. Introduction

Information is a very important need for organizations, educational institutions and government agencies and everywhere. Excellence in processing and getting information is the ease of access, data up date and data security. With the development of the era in the modern era and this technology, many students and Muslims or muslimat especially those who have used technology tools that are often called "smartphones / gadgets" at the present time. Therefore, I have one of the designs to make the introduction of Hijaiyah letters introduction application of fundamental basic Islam. In order for the development of this age, we can always be in the rules and rules of Islam that have been taught through the Qur'an and Hadith. In the world of education today many parents who provide an understanding of religion by way of telling and teaching with the conventional media of reading books and hijaiyah letters.

It is very ineffective and efficient in learning and teaching activities, which will be difficult for children / students to grasp the meaning that is being conveyed by teachers / parents. With the implementation of religious learning, especially the introduction of hijaiyah letters in Indonesian society since early, may be more easily acceptable especially applied at an early age (Santi I.T. and Sukadi., 2014). Android provides users with extensive access to use the necessary libraries and tools that can be used to \ build better applications. Android has a bunch of tools can be used to help developers in improving productivity when building applications that are created(*Source:* *http://www.android.com/about/* (Purnapraweti, 2015).

IOP Conf. Series: Earth and Environmental Science **97** (2017) 012047 doi:10.1088/1755-1315/97/1/012047

2. Literature Review

Icon application is the icon used application learning letters hijaiyah on Android-based Smartphone as a button to use the application. This button must be touched the user to run the application "learn hijaiyah" and connect to teaching materials paud teacher in school Al-Kindy Pekanbaru this. The image icon is like in figure 1 below.

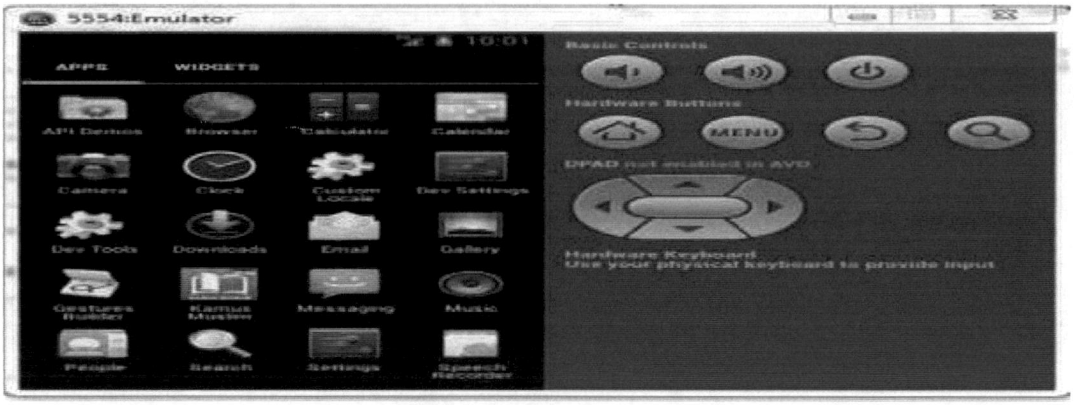

Figure 1. Image Icon

System is a unity of component procedures that are related to each the other, work together in accordance with the rules applied so as to form a common goal, where in a system if there is one part that is not working or damaged then a goal can occur error results or output, the system is a collection of elements that mutually (Ardiansyah, 2013), While the system is an order or alignment consisting of a number of functional components (with units of functions or special tasks) that are interconnected and together aims to fulfill a process or work certain. In the main menu view shown in Figure 5.2 above the user will go to the front page menu which automatically appears itself as a link from the front view to the layout view 1 garden hijaiyah and layout 2 garden hijaiyah

Figure 2. Hijaiyah apps on desktop

Testing Testing is done in two main stages, namely testing emulator and Smartphone. Emulator testing is done by testing the attributes and methods that exist in the classes built. Testing the handset is done by way of Acceptance Testing is to test whether the system developed in accordance with what is contained in the functional specifications of the system on the Smartphone. The following presented the results of the system implementation:

Emulator Testing and On Gadged Application Icon The results of the Application Icon we can see fit as in the picture below.

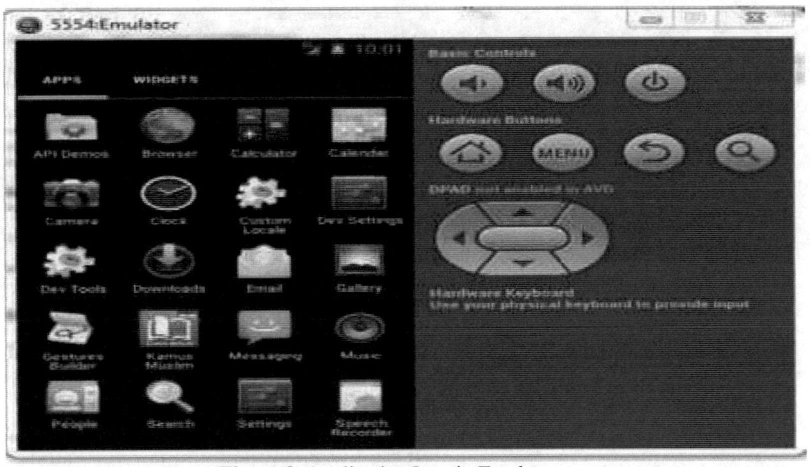

Figure 3. Application Icon in Emulator

Taman Hijaiyah In this layout each student paud can choose the form of hijaiyah letter one by one by shifting the screen and can also listen to the spelling of the sound that came out of this application, and students can also learn how to pronunciation and spelling of good and true according to religious teachings Islam.

Figure 4. Layout of *Taman Hijaiya*

Kebun Hijaiyah in above picture Hijaiyah Gardens, each pupil paud can see the shape of the letters of hijaiyah one by one and can also listen to the spelling of the sound that came out of this application, and students also can learn how to pronunciation and spelling is good and true according to the teachings of Islam.

3. Methodology

In this test using Android Smartphone with Android operating system version 5.0 (Lolipop). So that app can be used in Smartphone then application. "Learning Hijaiyah" must be installed on the Smartphone. here's how to install this application hijaiyah letter. Copykan apk application file that has been created "Learning Hijaiyah Letters", then look at the Smartphone (see it on the main screen "My Files") in the picture below.

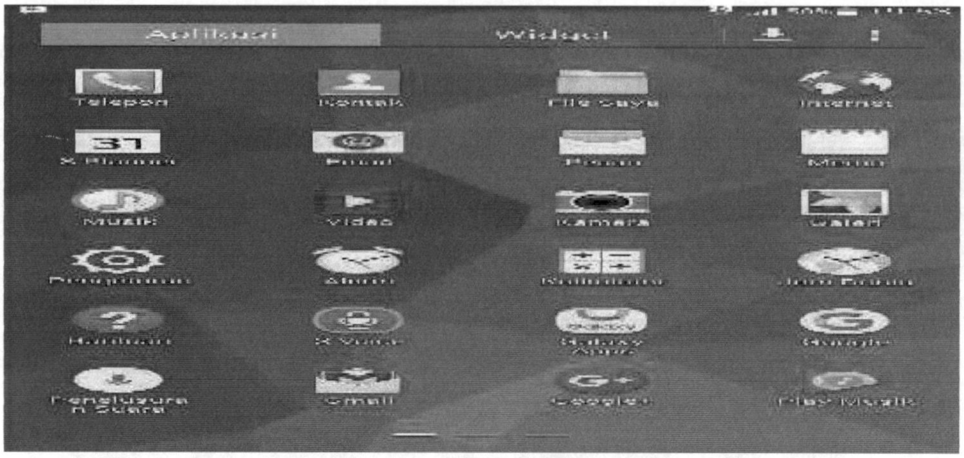

Figure 5. Layout *of Learning Hijaiyyah*

4. Results and Suggestion

After the application is successfully installed then the application learn hijaiyah letters can be used and its use is very simple and easy to understand by the user in this case as the user is paud students and teachers paud who are in school paud Al-kindy Pekanbaru Riau.

4.1 Fungsionality Test

In this test, distributed questionnaires to (User) were 60 respondents, that is from the founder foundation element along with the structural and assembly of teachers and guardian of paud students and parents of paud students themselves. Results from the distribution of the test questionnaire functionality for learning application design letters hijaiyah based Android Smartphone is as follows: the questionnaire were designed in native language of the kindergarten students and found that. The use of the application of learning methods based on the letters hijaiyah android is easier for teachers to provide understanding of their students to be able to learn at home respectively.

4.2 Suggestion

See the reality faced by teachers and students paud as well as consideration of the quality of learning, and the need for development in subsequent research then in writing suggestions as follows: Teaching and learning process that still use props to switch to using Android-based applications because students today are no longer taboo with the rapid development of technology today, especially in the Android-based digital animation display. To the foundation and schools to be able to play an active role to motivate the educators in developing their skills and creativity to develop learning by using the Android-based Applications and learning methods as learning tools and facilities, and apply multimedia learning mechanisms and interactive learning methods in providing stimulus process learning to motivate learners to follow the learning process. Fuzzy Multi Criteria Decision Making by utilizing Software Tools Web Browser this system can we apply only as a tool for decision makers in determining the best result.

6. References

[1] Purwati I,. 2013. *Perancangan aplikasi pembelajaran huruf hijaiyah berplatform android untuk madrasah baca tulis al quran al-fattah desa widodaren kabupaten ngawi* vol 2 no 1 – maret 2013 issn: 2302-1136 - seruniid.unsa.ac.id ika purwanti_ I_kapoor@ymail.com

[2] Santi I.T. dan Sukadi., 2014. *Pembuatan Game Pembelajaran Pengenalan Huruf Hijaiyah Di Taman Kanak-Kanak (TK) Az-Zalfa Sidoharjo Pacitan* Isma Trisna Santi, Sukadi ismashanty@gmail.com Journal Speed – Sentra Penelitian Engineering dan Edukasi – Volume 6 No 2 - 2014 - ijns.org

[3] Zahrotun L. dan Ricky Z., 2015. Telematika, vol. 12, no. 02, juli, 2015, pp. 75 – 81Issn 1829-667x *Media pembelajaran pengenalan huruf hijaiyahUntuk anak usia dini 2-3 tahun* Lisna zahrotun(1), zaka ricky soleh Teknik informatika,universitas ahmad dahlan Jl.Prof.Dr.Soepomosh,janturan,umbulharjo,yogyakartaE-mail : lisna.zahrotun@tif.uad.ac.id(1)(2), zakaricky1@gmail.com

[4] Supriono H., Saputra A.N., Sudarmilah E., and Darsono R., 2014. Jurnal informatika vol. 8, no. 2, juli 2014 __12 *Rancang bangun aplikasi pembelajaran hadis Untuk perangkat mobile berbasis android*_Heru supriyono1; ardhiyatama nur saputra2, endah sudarmilah; Dan ruswa darsono 3Program studi teknik elektro fakultas teknik universitas muhammadiyah surakarta Program studi teknik informatika fakultas komunikasi dan informatika universitas Muhammadiyah surakarta 3Fakultas agama islam universitas muhammadiyah surakarta Jl. A. Yani tromol pos i pabelan kartasura, surakarta, jawa tengah, indonesia 57102 Kontak email: heru.supriyono@ums.ac.id.

[5] Saputra A.N., Supriyono H. And Darsono R., 2014 . Teknologi Informasi 2014 (SNATI 2014) Yogyakarta, 17-18 Juni 2014 ISSN: 1907-5022. *Rancang bangun aplikasi pembelajaran hadis untuk Perangkat mobile berbasis android Program studi teknik informatika Fakultas komunikasi dan informatika*. Ardhiyatama Nur Saputra, Dr. Heru Supriyono, M.Sc. dan Ruswa Darsono, S.T.

[6] Pratiwi H. dan Wardijo A.B., 2014. Prosiding seminar ilmiah nasional komputer dan sistem intelijen (kommite 2014) vol.8 Oktober 2014 Universitas gunadarma–depok–14–15oktober2014. issn : 2302-3740 *Aplikasi belajar huruf arab dasar dan iqra lengkap Dengan suara untuk perangkat mobile berbasis android* 1Hana pertiwi, st1Dr bheta agus wardijono, ssi., mt 2Teknik informatika, fakultas teknik industri, universitas gunadarma 12 Sistem informasi, stmik jakarta sti & k Apetraaini@gmail.com, 2 Bheta@jaks-stik.ac.id

[7] Holla S. dan Katti M. M., 2012. International journal of computer trends and technology-volume3 issue3- 2012 *Android based mobile application Development and its security* #Suhas holla #1,Mahima M Katti #2department of information science & engg, r v college of engineering Bangalore, India

[8] Jannah N. M., dan Hakim N. F., 2015 . *Jurnal Teknologi Informasi dan Komunikasi, ISSN:2087-0868, Volume 6 Nomor 2 September 2015. Rancang bangun aplikasi huruf hijaiyah berbasis android (studi kasus :taman baca al quran raziev reinezhwa Banjarmasin kalimantan selatan)Mifta nurul jannah, fitro nur hakim Program studi teknik informatika stmik provisi semarang* Miftaenjhe@gmail.com, dan Masfitro@gmail.com

Reconstruction of Power Supply System 20 kV Distribution to Compare Power Rate and Fall Voltage PT. PLN (Persero) Area Dumai

Abrar Tanjung[1], Monice[2]
University Lancang Kuning, Pekanbaru, 28265, Indonesia
Email :abrar@unilak.ac.id, monice@unilak.ac.id

Abstract: Electricity in Bagan Siapi city fire is channeled through a feeder distribution system of 20 kV. The main supply of Bagan Siapi-api city comes from PLTD unit Bagan Siapi fire which is Â± 1.5 kms from the load center and Duri Substation is Â± 102 kms from Bagan Siapi-api city through Hubung Ujung Tanjung. The long distances between the Duri Mainstation and Bagan Siapiapi city resulted in a 14.85 kV end-voltage and a 988.7 kW loss. Voltage losses resulted in ineffective service to the consumer and large network power losses being uneconomical for power delivery operations. The result of end voltage calculation is 10.42 kV and the power loss is 988.7 kW. After the New Substation operates, reconfiguration-1 produces the lowest end-voltage calculation of 16.21 kV and a power loss of 136.59 kW, while reconfiguration-2 produces a low end stress calculation of 17.37 kV and a power loss of 56.93 kW.

Keywords: feeder 20 kV, voltage drop, power loss, substation, reconfiguration.

1. Introduction

PT PLN (Persero) Dumai Area with Rayon Bagan Siapi-api unit as the implementer is a unit that provides electricity service directly to consumers in Rokan Hilir district. The source of electricity comes from PLTD Bagan Siapiapi with capacity of 12,340 MW and Duri substation (90 MVA) through feeder distribution system 20 kV to Ujung Tanjung relay station then to Bagan Siapi fire town with total length 102 kms, and electrical system embrace radial system pattern. The magnitude of the load supplied can increase the voltage drop and the power losses in the system feeder 20 kV.

This is due to the parameters that determine the voltage drop and power losses other than the impedance is the current flowing on the system coming from the load. Power loss is a function of the square of the current on the system, so the greater the current will increase power losses and voltage drops. Therefore, to reduce the voltage drop and power losses by shortening the distance of the load center and the source of its power supply. The way to shorten the channeling distance between the power supply and the load center is to reconfigure the load after the build-up of the substation. The purpose of this study is to analyze whether the development of Bagan Siapi-fire substations by shortening the distribution can be one of the solutions in improving the characteristics of 20 kV feeder between the Ujung Tanjung Transmission Bridge to Bagan Siapi-api City, which includes voltage drop, load current, and loss - loss of power.

2. Literature Review

Muhammad Nasir Malik stated that the large drop of stress and total power loss in the primary distribution network of feeder Adyaksa Substance Panakukang Makassar with descriptive research method [1]. Danang Ramadhianto stated that the magnitude of the shrinkage of energy that occurs in power distribution network and to analyze the shrinkage value of energy, through the method of measurement and calculation and then compare the results of both [2]. Fery Jusmedy, stated that the stresses and losses on the 115 kV PT. Chevron Pacific Indonesia using ETAP 4.0 software [3].
A.V.Sudhakara Reddy states the Dragon Fly Meta-heuristic Algorithm (DFA) is used to restructure and identify the optimal Switches to minimize real power losses In the distribution network. Strategies have been tested The bus system 16 bus, 33-bus and 69-bus IEEE to demonstrate Achievement and adequacy of technical proposals. The result indicates that a significant reduction of loss and increase in real power voltage profile [4].

Anita Gupka, et al declare to reduce losses which is less or less and then determine the reduction of losses in terms of units and annual savings in the form of rupees Because of the proposed method. To examine the feasibility of the proposed work, the annual savings and return periods of the proposed method are also determined [5].

Based on the results of the analysis of the calculations using the ETAP 7.5 program the lowest voltage on the Caterer Feeder channel is found in TR.1064 transformer of 16.27 kV whereas The lowest voltage on the Idiac Feeder on the SP.26 transformer is 19.87 kV with a total power loss of 731.04 kW. After reconfiguration, then reconfiguration 2 then the lowest voltage of Caterer Feeder is on transformer TR.1102 of 18,70 kV, at Feeder Langgam channel that is on SP.26 transformer equal to 19,05 kV and on channel of white sand feeder on transformer ST.262 Of 18.78 kV with a total power loss of 410.65 kW. Total power loss savings were obtained at 320.39 kW [6].

Julen Kertoni et al stated that Reconfiguration was carried out by load transfer by means of status change and changes to Load Break Switch layout simulation results based on ETAP 12.6 on the lowest four-voltage reconfiguration on Stork feeder Sakti is 18.664 kV with a power loss of 0.598 MW and 0.787 MVAR [7].

2.1. Power System

Electric power system consists of three main components, namely power generation, transmission and distribution. At the center of electricity is done by generating electricity by utilizing synchronous generators. The electrical power generated at the center will be transmitted through the transmission system. Before transmitted, the voltage is increased by a step-up transformer at the center of the power. The voltage is raised to a high voltage level (between 70 kV to 150 kV) or extra high voltage (above 150 kV). The transmission system can be a system Air or ground cables.

2.2. Distribution Network

The distribution network consists of two types, medium voltage distribution network and low voltage distribution network. For medium voltage distribution network there are three types of configuration. Among them are radial configuration, lup configuration and spindle configuration.

2.3. Radial Configuration Distribution Network

Radial configuration is an interconnection between substations of distribution, where several distributed series of distribution substations supplied by a busbar Substation. This configuration consists of several repeater out of the substation and the source of the voltage only one direction saja.Dalam penyulang there are substations that are equipped distribution By voltage-lowering transformer to low voltage. This configuration is the simplest and easiest configuration in operation.

2.4. Distribution Network Loop Configuration
The configuration of the loop is the interconnection between the distribution substations that form a loop. In this configuration there can be more than one busbar of the Substation, and each of the repeater forms a closed circuit with the Substation. The advantage of this magnetic configuration is the power supply Of the substation is more secure.

2.5. Spindle Configuration Distribution Network
The spindle configuration is a series connection between a distribution substation which both ends are connected by the busbar of the Substation and Connector, which is the hallmark of this network is the presence of an express feeder. This express repeater serves as a repeater reserve that will supply the load power when one of the feeders is interrupted. In this spindle network there are several repeater supplied by Substation and ends in a substation.

2.6. Voltage Drop
Voltage falling in general is the voltage used at the load. The voltage drop is caused by the current flowing through the wire resistance. The voltage drop V on the conductor is greater if the current I in the conductor is greater and if the conductor resistance R? The greater the value. As a result the value of the voltage on the recipient side will be different from the voltage value of the sender. The equation of voltage drop can be seen in equation 1.

$$\Delta V = I \times Z \qquad [1]$$

Informasi :

ΔV = Voltage drop (volt)

I = current (Amper)

Z = Impedance (ohm)

In theory this is meant by the voltage drop (ΔV) is the difference between voltage sent (V_s) to receive a voltage (V_R), the voltage drop can be defined in equation 2.

$$\Delta V = V_S \times V_R \qquad [2]$$

information :

ΔV = Voltage drop (volt)

V_S = Send voltage

V_R = Voltage received

2.7. Power Loss
Power losses are the amount of power lost in a network, which is equal to the power supplied from the source less the amount of power received. Selection of cable types to be used in the distribution network is an important factor to be considered in the planning of a power system. Types of cables with small resistance values be able to minimize power losses. The amount of power losses in the distribution network can be written as follows:

$$\Delta P = I^2.R \text{ (watt)} \qquad [3]$$

Infomation :

ΔP= Power loss on the network (watt)

I = Flow current on the network (Ampere)

R = Pure prisoners

For power losses on three-phase networks expressed by the equation:

$$\Delta P = \sqrt{3}.\ I^2.R \text{ (watt)} \quad [4]$$

By ignoring capacitive currents in the system, the currents along the length of the wire can be considered equal and the magnitude is equal to the current at the receiving end.

$$I = \frac{P}{\sqrt{3}VI\cos\varphi} \text{(ampere)} \quad [5]$$

The amount of power in the three-phase system is

$$P = \sqrt{3}\ V\ I\ Cos\ \varphi \text{ (watt)} \quad [6]$$

Information :

P = Load power at the receiving end of the system(Watt)

V = Fasa Voltage (Volt)

Cosφ = Load power factor

Voltage is also very influential on power losses, the greater the voltage on a system, the smaller the current on the system. While the current is one of the factors that affect the magnitude of power losses on a system.

3. Methodology

Reconfigure the distribution network by resetting the network configuration by opening and closing the switches on the distribution network to reduce the loss of power in the distribution network and / or to improve the reliability of the distribution system so that the efficiency of the power supply is increased and the consumer can be served well. Network Reconfiguration is an attempt to change the shape of a distribution network configuration by operating remote controlled switches on the distribution network without causing risky consequences on the operation and shape of the entire distribution network system. Under normal operating conditions, network reconfiguration is performed for two reasons.

- Reduce loss of power in the system (loss reduction).
- Get a balanced load to prevent overloading on the network (load balancing).

4. Results and Discussion

4.1. Impedance Calculation

The distribution system of 20 kV Duri Substation to City Bagan Siapi-api uses AAAC (All Allumunium Alloy Conductor) wire with 240 mm2 cross-section for main system, and 150 mm2 with 70 mm2 for branching system. The three wire conductor is arranged 19 fibers. The system configuration is horizontal with distance between phases 75 cm. The average temperature of the distribution system environment used temperature 300 C. The dc resistance at temperature 20 0C and the temperature coefficient (α) At temperature 20 0C as follows :

Table 1. Resistance dc and Temperature Coefficients

Wire Delivery AAAC	Dc resistance (temperature $20\ ^0C$)	Temperature Coefficient at $20\ ^0C$ (α_{20})
70 mm^2	0,438	0,00360 / ^0C
150 mm^2	0,210	
240 mm^2	0,137	

Source: Standard of State Electricity Company (SPLN 41-8:1981)

Then the impedance can be calculated as

a. For cross-section area 240 mm^2

Dc Resistance :

$$R_{dc}\,(30\,^0C) = R_{dc}\,(20^0\,C) \times [1 + \alpha_{20}\,(30^0\,C - 20^0\,C)]$$
$$= 0{,}137 \times [1 + 0{,}00360\,(10)]$$
$$= 0{,}137 \times [1 + 0{,}0360]$$
$$= 0{,}141932\ Ohm/km$$

The amount of skin effect factor (mr) is:

$$mr = 0{,}0118 \times \sqrt{f} = 0{,}0118 \times \sqrt{50}$$
$$= 0{,}08343$$

Resistance ac :

$$R_{ac} = \frac{1}{2} R_{dc} \times \left[\sqrt{1 + \frac{(0{,}08343)^4}{48}} + 1\right]$$
$$= 0{,}5 \times 0{,}141932 \times 2$$
$$= 0{,}141932\ Ohm/km$$

$$GMR = 0{,}4345 \times \sqrt{A}$$
$$= 0{,}4345 \times \sqrt{2{,}4}$$

$$= 0{,}67312\ \text{cm}$$

$$GMD = \sqrt[3]{D_{rs} \times D_{st} \times D_{tr}}$$
$$= \sqrt[3]{(75) \times (75) \times (150)}$$

$$= 94{,}5\ \text{cm}$$

$$X = (2.\pi.f) \times (2 \times 10^{-4}) \times \ln\left(\frac{GMD}{GMR}\right)$$

$$= (2 \times 3{,}14 \times 50) \times (2 \times 10^{-4}) \times \ln\left(\frac{94{,}5}{0{,}67312}\right)$$

$$= 0{,}3105\,Ohm/km$$

$$Z = 0{,}1419 + j\,0{,}3105 \qquad = \sqrt{(0{,}1419)^2 + (0{,}3105)^2}$$

$$= 0{,}34139\,Ohm/km$$

b. For cross-sectional area 150 mm^2

DC Resistance:

$$R_{dc}\,(30\,^0C) = R_{dc}\,(20\,^0C) \times [1 + \alpha_{20}\,(30^0\,C - 20^0\,C)]$$
$$= 0{,}21 \times [1 + 0{,}00403\,(10)]$$
$$= 0{,}21 \times [1 + 0{,}0403]$$
$$= 0{,}21846\,Ohm/km$$

The amount of skin effect factor (mr) is :

$$mr = 0{,}0118 \times \sqrt{f} = 0{,}0118 \times \sqrt{50}$$
$$= 0{,}08343$$

Ac Resistance :

$R_{ac} = \frac{1}{2} R_{dc} \times \left[\sqrt{1 + \frac{(0{,}08343)^4}{48}} + 1\right]$

$= 0{,}2187 Ohm / km$

$GMR \quad = 0{,}4345 \times \sqrt{A}$

$= 0{,}4345 \times \sqrt{1.5}$

$= 0{,}53215$ cm

$GMD \quad = \sqrt[3]{D_{rs} \times D_{st} \times D_{tr}}$

$= \sqrt[3]{(75) \times (75) \times (150)}$

$= 94{,}5$ cm

$X \quad = (2.\pi.f) \times (2 \times 10^{-4}) \times \ln\left(\frac{GMD}{GMR}\right)$

$= (2 \times 3{,}14 \times 50) \times (2 \times 10^{-4}) \times \ln\left(\frac{94{,}5}{0{,}53215}\right)$

$= 0{,}32527 Ohm / km$

$Z \quad = 0{,}21847 + j\, 0{,}32527$

$= \sqrt{(0{,}21847)^2 + (0{,}32527)^2}$

$= 0{,}39183 Ohm / km$

c. For cross sectional area 70 mm^2

DC Resistance :

$R_{dc}\,(30\,^0C) \quad = R_{dc}\,(20\,^0C) \times [1 + \alpha_{20}(30^0\,C - 20^0\,C)]$

$= 0{,}438 \times [1 + 0{,}00403\,(10)]$

$= 0{,}438 \times [1 + 0{,}0403]$

$= 0{,}45565 Ohm / km$

The amount of skin effect factor (*mr*) is :

$mr \quad = 0{,}0118 \times \sqrt{f} = 0{,}0118 \times \sqrt{50}$

$= 0{,}08343$

Ac Resistance

$R_{ac} = \frac{1}{2} R_{dc} \times \left[\sqrt{1 + \frac{(0{,}08343)^4}{48}} + 1\right]$

$= \frac{1}{2}\,(0{,}45565) \times \left[\sqrt{1 + \frac{(0{,}08343)^4}{48}} + 1\right]$

$= 0{,}45565 Ohm / km$

$GMR \quad = 0{,}4345 \times \sqrt{A}$

$= 0{,}4345 \times \sqrt{0{,}7}$

$= 0{,}36352$ cm

$GMD \quad = \sqrt[3]{D_{rs} \times D_{st} \times D_{tr}}$

$= \sqrt[3]{(75) \times (75) \times (150)}$

$= 94{,}5$ cm

$X = (2.\pi.f) \times (2 \times 10^{-4}) \times \ln\left(\frac{GMD}{GMR}\right)$

$= (2 \times 3{,}14 \times 50) \times (2 \times 10^{-4}) \times \ln\left(\frac{94{,}5}{0{,}36352}\right)$

$= 0{,}34920\,Ohm / km$

$Z = 0{,}45565 + j\,0{,}34920$

$= \sqrt{(0{,}45565)^2 + (0{,}34920)^2}$

$= 0{,}57437\,Ohm / km$

4.2. Determining the Value of Flow Current in Voltage 20 kV

To determine the value of current on the primary side of the transformer distribution, authors first examine the transformer load value on the secondary side. To determine the value of the current balanced authors calculated the average value of the sum of the load measurement phase Amper R = 106, S = 46 and T = 85 Amper Amper as follows:
For transformer T.001 secondary side flows

Balance stream:

$I_{SC} = \frac{106 + 46 + 85}{3}$

$I_{SC} =$ 79 Amper

After obtaining the current value at equilibrium condition, then to calculate the value of current on the primary side using the way the secondary side current value in balanced condition divided by the ratio of the voltage regulation in the tap changer or calculated as follows:

For transformers T.001 primary side currents;

$I_{PR} = \frac{79}{(18000/380)} = 1.67$ Amper

4.3. Determine Voltage Drop and Power Loss Existing Conditions

In this condition the voltage drop is calculated, so it will get the value of transmitted and received voltage and inter-channel power losses calculated as follows.

Voltage Drop $= \sqrt{3}$ x Arus x Total impedansi

= 1.732 x 273.07 A x 13.65 Ohm/km

= 5.607 V

So :

Send Voltage = 20.400 V
Voltage Received = 20.400 V – 5.607 V

= 14.792 Volt

Total Loss of existing power condition are:

$= 1.198.417 - [\,(77.25 + 33.9)^2 \times 5.676 \times 3]$

= 1.198.139 – 210.369
= 988.048 Watt

4.4. Condition of New Substation Operating (Reconfiguration 1)
In this condition, reconfiguration of medium voltage network has been done so that the configuration of the network where the power is not supplied again from the substation of thorns but has been supplied from the New Master Substation. So obtained the voltage drop and power loss as follows:
End Voltage at

OGF A at Bangko Pusako = 16.212 V

OGF B at Bagan Siapi-api = 19,130 V

Total total power losses OGF A and OGF B

=118.219 Watt + 18.372 Watt

= 136.591Watt

4.5. Condition of New Substation Operating (Reconfiguration 2)
In this condition, the reconfiguration of the medium voltage network has been done so that the network configuration is changed which the load of Banko Pusako is supplied again from the Duri Substation and the load to the Siam-Fire Channel is supplied from the New Master Substation. Calculation The total reconfiguration power losses of 2 are the sum of OGF A, OGF B, Power Supply from the Substation of Duri reduced the power losses of Feeder Polres and the Feeder of the White Land, so the calculation is as follows:
= (41.110 + 18.373 +207.818) – [$(77.25+33.9)^2$ x5.676 x 3]
= 267.301– 210.369
= 56.932 Watt

4.6. Comparison Of Losses And Voltage Each Configuration
From the calculation of losses and intermediate voltage of the medium-voltage network for the existing conditions, reconfiguration 1 and reconfiguration 2 are shown in table 2 as follows:

5. Conclusions
From the result of the discussion, it can be concluded that the voltage profile and power loss condition of power distribution from *Ujung Tanjung* to *Kota Bagan Siapi-api* with the result of end stress calculation is 11,542 kV and the power loss is 988.7 kW. After the new substation is in operation the reconfiguration-1 produces the lowest end voltage calculation of 16.21 kV and a power loss of 136.59 kW, while reconfiguration-2 produces the lowest end voltage calculation of 17.37 kV and a power loss of 56.93 kW

Table 2 Results Calculation of Voltage Drop And Power Loss Each Configuration

Configu-ration	Power loss	Send Voltage	Edge Voltage	Total Trafo (V_p< 18 kV)
	Watt	Volt	Volt	Unit
Eksisting	988.048	20.400	11.642	112
Reconfiguration 1	136.591	20.400	16.212	54
Reconfiguration 2	56.932	20.400	17.372	26

5. References

[1] Muh. Nasir Malik. 2009. *Analisa Losses Jaringan Distribusi Primer Pada Penyulang Adhyaksa Makassar*, Media Elektrik Volume 4 Nomor 1, Fakultas Teknik Universitas Negeri Makasar.

[2] Danang Ramadhianto. 2008. *Study of Energy Loss On Electricity Distribution System Through Measurement and Calculation Analysis*, Engineering Faculty University of Indonesia.

[3] Fery Jusmedy. 2007. *Power System Flow Study 115 kV PT. Chevron Pacific Indonesia*, Engineering Faculty, University of Sumatera Utara

[4] A.V.Sudhakara Reddy, M. Damodar Reddy, *Optimization Of Distribution Network Reconfiguration Using Dragonfly Algorithm*, Journal Of Electrical Engineerin, India.

[5] Ankita Gupta, Harmeet Singh Gill, Isha Bansal. 2012. *Effectiveness Of High Voltage In Distribution System: High Voltage Distribution System*, IOSR Journal Of Electrical And Electronics Engineering (IOSRJEEE) ISSN: 2278-1676 Volume 1, Issue 5 (July-Aug. 2012), PP 34-38

[6] Tanjung, Abrar. 2014. *Reconfiguration Of Distribution System 20 kV Teluk Lembu of Substation Books And Plow Langgam Power To Reduce Power Rule And Drop Voltage*, Jurnal Sains, Teknologi Dan Industri, Vol. 11, No. 2, Juni 2014, Pp. 160 - 166 Issn 1693-2390

[7] Julen Kartoni S, Edy Ervianto. 2016. *Analysis Of Fertilizing Reconstruction To Reduce Loss - Power Response On 20 kV Distribution Channel*, Jom FTEKNIK Volume 3 N0.2, Pekanbaru

IOP Conf. Series: Earth and Environmental Science **97** (2017) 012049 doi:10.1088/1755-1315/97/1/012049

The Characteristics Of Malay House Spatial Layout Of Pekanbaru In Accordance With Islamic Values

Boby Samra
Universitas Lancang Kuning, Pekanbaru, 28265, Indonesia
Tel/Fax: +62761 52324
E-mail: boby@unilak.ac.id

Abstract
House is not only is a place to get rest and do activities bu also treated as a pride for the Malay community. The values contained in the spatial layout of the house have specific meaning to the owners. This makes the Malay house becomes the symbol of pride to uphold the "tuah" and dignity of the owner. This research is conducted using qualitative approach through management and data management available through several methods such as observation, interview, documentation and group discussion. This is expected to provide understanding of the perception of Islam dealing with the characteristics of the spatial layout of the Malay house of Pekanbaru.
Keywords: spatial, symbol, islam, house

1. Introduction

Some traditional Malay expressions refer to the house as "the living light of the earth, the place to preserve the custom, to gather relatives, the place of the last trade stop, the debt of the parents to their children. " The Malays also crave a good and perfect residence, that is, the physical building which fulfills the customary and the needs of its inhabitants, while from the spiritual side, the house can bring happiness, comfort, peace and tranquility (Mudra, [6]).

Malay culture contains its symbols and philosophies into every part of the building, ranging from stairs, poles, spaces and roof tops. The art of Malay house starts s from the main buildin, then other spaces are added. The function and its form can not be exceeded from the main building of the house which reflects the philosophies of "the house has its owner,the village has its leader, the land has its king" or a saying says "the chicken has its mother, lemongrass has its foliage, the net has its bud, the bound has its knot" (Effendy,[3]).

It is natural for every creature to build a place to live that is used as a place to get rest and protect themselves in different shapes and sizes according to the ability and needs of each creature itself. For animal, its house is called a nest, for humans it is called house. The Qur'an introduces two terms to call house Firstly, it is mentionedin Surah an-Nahl [16]: 68

وَأَوْحَىٰ رَبُّكَ إِلَى ٱلنَّحْلِ أَنِ ٱتَّخِذِى
مِنَ ٱلْجِبَالِ بُيُوتًا وَمِنَ ٱلشَّجَرِ وَمِمَّا يَعْرِشُونَ

It means:
"And your Lord revealed to the bees:" Make the nests (bait) in the hills, on trees, and in places which made man. "

The word nest (bait) in this surah means a place to spend the night. The house is called "bait" because it is used to get rest from the bustle. It is also similar to animals'life, like birds returning to their nests in the evening to spend the night and rest. In addition, the house serves to protect the owner from various external interference, such as heat, cold, and other creatures attack as mentioned in Surah al-Baqarah [2]: 125

وَإِذْ جَعَلْنَا ٱلْبَيْتَ مَثَابَةً لِّلنَّاسِ وَأَمْنًا

It means:
"And (remember) when We made the House (Baitullah) a gathering place for humans and safe place ......"
Another name that Allah introduced to call house is "maskan" as can be seen in Surah An-Naml [27]: 18

حَتَّىٰٓ إِذَآ أَتَوْا۟ عَلَىٰ وَادِ ٱلنَّمْلِ قَالَتْ نَمْلَةٌ يَٰٓأَيُّهَا ٱلنَّمْلُ
ٱدْخُلُوا۟ مَسَٰكِنَكُمْ لَا يَحْطِمَنَّكُمْ سُلَيْمَٰنُ وَجُنُودُهُۥ وَهُمْ
لَا يَشْعُرُونَ ﴿١٨﴾

Meaning:
"Till, when they reached the valley of ants and an ant said: O ants, go into your nest in order you are not stepped on by Solomon and his soldiers while they perceive not."
In Surah at-Taubah [9]: 72 Allah also said

وَعَدَ ٱللَّهُ ٱلْمُؤْمِنِينَ وَٱلْمُؤْمِنَٰتِ جَنَّٰتٍ تَجْرِى مِن تَحْتِهَا
ٱلْأَنْهَٰرُ خَٰلِدِينَ فِيهَا وَمَسَٰكِنَ طَيِّبَةً فِى جَنَّٰتِ عَدْنٍ ۚ
وَرِضْوَٰنٌ مِّنَ ٱللَّهِ أَكْبَرُ ۚ ذَٰلِكَ هُوَ ٱلْفَوْزُ ٱلْعَظِيمُ ﴿٧٢﴾

It means:
"Allah has promised to men and women who have faith, (will get) paradise beneath which rivers flow, they eternally therein, and (got) a nice places in heaven` Eternity. And the pleasure of Allah Is bigger; It is a great fortune. "

The word "maskan" comes from the word "sakana" which means calm, peaceful, and happy. Therefore, the house in the view of the Qur'an does not only serves as a place to spend the night, a place to get rest and shelter but also a place of seeking for peace and happiness. In the house (maskan) is the human family creates "sakinah" family, the family order that brings happiness and peace of mind.
If the house is only treated as a house, it will be felt like living in hell. That is what God described in Surat al-Ankabut [29]: 41

وَإِنَّ أَوْهَنَ ٱلْبُيُوتِ لَبَيْتُ ٱلْعَنكَبُوتِ ۖ لَوْ كَانُوا۟ يَعْلَمُونَ

It means:
"... And verily the weakest house (the fragile) is the house of the spider if they know."

Spider's house is not only fragile in structure, because it is unable to protect its occupants from all sorts of external interference such as heat, cold and so on. However, the spider's house is also fragile from the side of its inhabitants. Therefore,we should make our home a "maskan", a place to find peace and happiness of life. A house is not just a shelter, a place to stay or shelter (temple), as done by animals. No matter how good the house is, but if it doesn't bring peace and happiness, it is also not good. It is better living in a simple house, but it can gives peace and happiness. Thus, the house is really a maskan.

2. Research Methods: Constructing Field Data

The research is done qualitatively so that it can be seen directly and it get the data and the problem as a whole. Architectural, cultural and social change researchers will tend to bring researchers to the field to get direct data that can be grouped into: physical data, community interpretation, and written documents.
Recording public opinion by interviewing was also done so that it could give information and knowledge through past experiences. In the survey, the following are the steps

- Recording physical data from the adjustment with the actual conditions in the field.
- Interviewing community leaders, customary leaders to gain local history and identity of the local cultures.
- Taking three-dimensional documentation using handycam, photographs and field sketches.

3. Discussion and Results

The spatial layout of Pekanbaru Malay house from previous research is divided into two forms: traditional building, i.e building which uses " telo" and no "telo". The building is also seen from the height of the floor of the ground if the building is on the edge of the river then the floor will be very high, if the building is located in areas far from the river or on the area with contour of the land which is relatively high then the floor of the building is not too high. So the space under the floor of the building can be functioned for public activities.

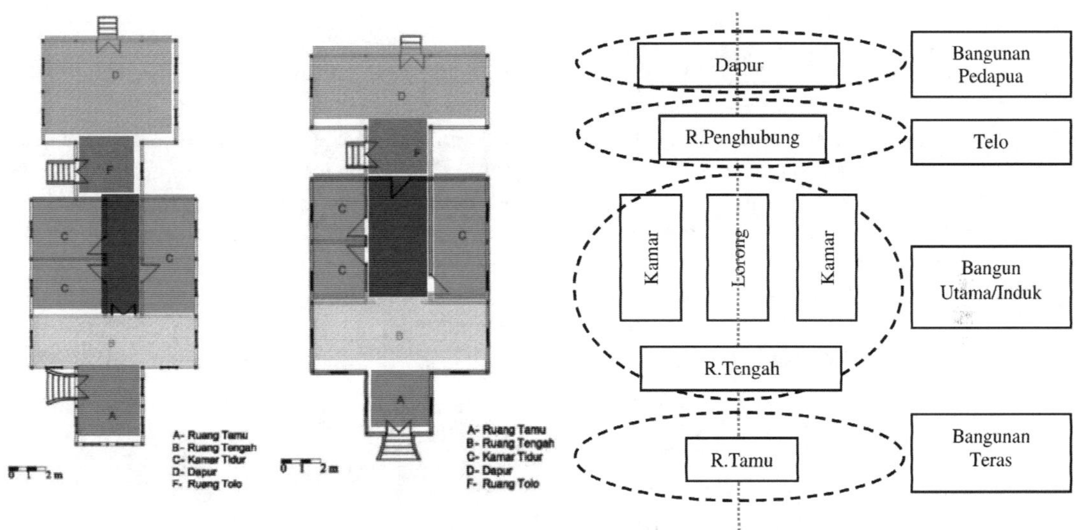

Figure 1: Spatial Shape With "Telo", Malay Home of Pekanbaru

The Spatial from its shape, the Malay house is divided into several parts, the terrace, the main building, the "telo" and "pedapuan." With that spatial characteristic pattern, it is divided into public area , private area and service area. In Islamic law such separation is very important to know how the function of the space can accommodate all activities of the owner who is restricted on every activity in the unity of Malay house. Firstly, the terrace is functioned to receive male guests, neighbors, honorable and elderly people, this function is very much in line with Islamic values because the guests who are not "muhrim" to the host are not allowed to enter the private area

IOP Conf. Series: Earth and Environmental Science **97** (2017) 012049 doi:10.1088/1755-1315/97/1/012049

Figure 2: Delimiters among spaces

The main building is functioned for private activities. In this area, there are a living and some rooms. The living room functions for family activities in "muhrim" bond. In this part, delimiter is placed in the form of a curtain or door to go to the rooms. Delimiter in Malay house has three rooms. one is used by mother and father and the other two rooms are used each for boys and girls. In Malay terms they are called inner chamber, middle chamber and back chamber. This is an implementation in Islamic teachings that boys and girls should be separated from their beds. In the implementation in earlier time, boys are recommended to sleep in the front room or in "surau."

"pedapuan" is functioned for the service activities that are used for cooking and preparatory activities in carrying out daily activities. This room is connected by a connecting building or "telo" to separate rooms and is also used to put working tools and in this space there is an exit of the house.

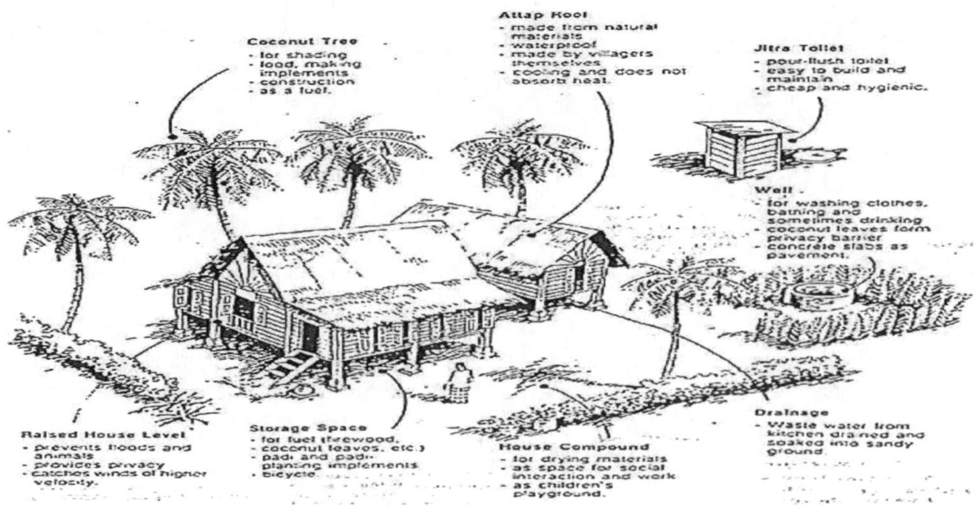

Figure 3: Traditional Malay House Environment

IOP Conf. Series: Earth and Environmental Science **97** (2017) 012049 doi:10.1088/1755-1315/97/1/012049

In the service area there is also a bathroom that can be connected or separated from this building. This separation is in accordance with the teachings of Islam, because Islam teaches that every activity done in the bathroom should be started with prayers such as:

اللَّهُمَّ إِنِّي أَعُوذُ بِكَ مِنَ الْخُبُثِ
وَالْخَبَائِثِ

Meaning: *O God, I take refuge in You from the male and female shamans (Bukhari and Muslim)*
Islam is really religion of Allah SWT beautiful and perfect religion of Allah SWT. In every step we take, Islam teaches its people to always pray to Him. In Malay house every person who entered the house is also recommended to wash the feet with the availability of a place of water in the rock / vat beside the front steps of the house.
The hygiene of a house in Islamic teachings is recommended by buiding a house away from dirty places, such as garbage dumps, puddles because cleanliness and sanctity are part of faith. It is obligatory for a Muslim to pay attention to the cleanliness and sanctity of his home, his environment, and himself, because the environment also shows the personality of the inhabitant, "Zhahir" is the reflection of his inner being.
From Abu Malik Al-Ash'ariy radhiallahu'anhu that the Prophet sallallaahu alaihi wasallam said:

الطَّهُورُ شَطْرُ الإِيمَانِ

"Cleanliness is part of the faith." (Muslim)

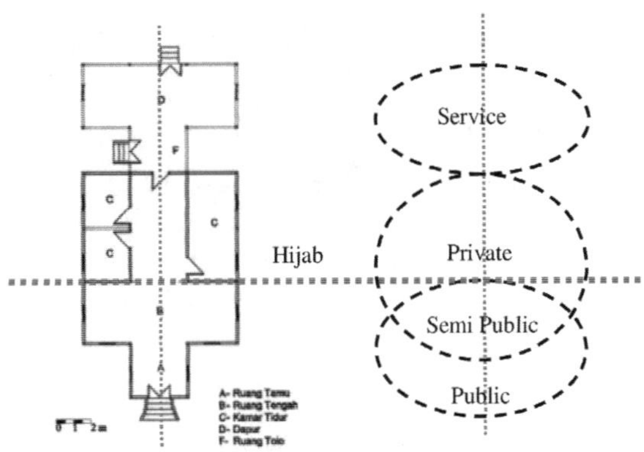

Figure 4: Space Zooning

The characteristic of Malay house is in accordance with Islamic guidance because the Malay community is known as the Islamic community. The additional spaces or additional parts (whatever their function and form are, must be carefully designed in order not to cover or not to be bigger than the main part of the house). This reflects the philosophical value that "the house has its owner, the village has its leader, the land has its king" or a saying said "the chicken has its mother, lemongrass has its foliage, the net has its clumps, the bound has its knot" (Effendi.T,[3]) This criteria should be considered to build a house, if the house is not built with the philosophy then the house can be said to be hostess (mad house). so, every time someone builds a house, it should be used as a reference, including every construction of a Malay house, it should be done symmetrically. This principle is central to reflection of balance, which contains cultural philosophy, customs and religion which prioritize the balance of the life in the world which relates to Malay Philosophy "right hand with left hand, left leg with right leg, right shoulder with left shoulder, and its location, shape (form) and measure are harmonious " (Effendi.T,[4]).

Malay culture is very appreciative of the development of Malay architectural design, so, every design is free to build as long as it does not violate the building identity principle mentioned above. Freedom is not only in the spatial arrangement and the addition of buildings that merge with the parent building,

IOP Conf. Series: Earth and Environmental Science **97** (2017) 012049 doi:10.1088/1755-1315/97/1/012049

but also to the symbol used as long as they are not opposed to culture, customs and religion. In building a good Malay house then the content and function of buildings in life need to manifest. The building should be built in accordance with cultural, custom and religious provisions so that the building can be declared as a house of Malay. Thus, the construction of Malay house is for the family. If the pattern is implemented then it will create" sakinah, mawadah, warahmah" family. Judging from the development process of modern home development today, the spatial distribution often does not follow the guidance of custom, religion and culture, instead only to build it in accordance with modern concept. The bathroom, for example, which is built in a bedroom. This is contrary to the value of Islam, which can be seen from prayer to enter the bathroom which asking God's protection from shaitan. It means the bathroom is a den of the shaitan. Other reason to separate bedroom from bathroom is related to the prayer when when husband and wife are having sexual intercourse.

بِسْمِ اللهِ اللَّهُمَّ جَنِّبْنِى الشَّيْطَانَ وَجَنِّبِ الشَّيْطَانَ مَا رَزَقْتَنَا

Meaning: *In the name of Allah, O God, keep the shaitan away from me, and keep him away from what You will give us (children, descendants). (Ayyub.A[2])*
Both of the above prayers have a very close relationship in manifesting a happy family according to the guidance of Islam, because prayer is clearly show our request to Allah SWT in order to be avoided from shaitan.
This fits perfectly with the characteristics of the spatial layout of the Malay house according to Islamic values.

4. Conclusions

The result proves how close the cultural and Malay values with Islam is, so that Malay people are identical with "Muslims". In addition, in order to keep the Islamic values applied in the activities and environment of our house, there are several things to note: Build a house close to the place of worship (mosque / mosque), Build it away from the place of sinners, Build at least three rooms, Keep the bathroom away from the main house and does not lead it or turn it to the Qiblah and Build delimiters to separate public space and private space.
Furthermore, this paper also offers how we can create a home that can bring happiness, comfort, peace and tranquility, simple healthy and having Islamic character for the family. It is important because the rapid development of architectural science has led people away from Islamic guidance. This is also important to remind us to build a house that can lead us to Allah's heaven.

5. Reference:

[1] As-Salam Al-Qur'an dan Terjemahannya Edisi 1000 Doa (2009) *Al-Mizan Publising House*
[2] Ayyub.A El-Faruqi.(2016) Panduan Lengkap Doa & Dzikir Rasulullah 24 Jam. Nabawi.
[3] Effendi,T. (2002). *Dasar-dasar bangunan tradisional Melayu Riau*
[4] Effendi,T. (2013). *Lambang dan Falsafah dalam seni bina melayu*
[5] Frizal.Y. (2015) *Mendefinisikan Kembali Arti Kampung Melayu*. Jurnal Arsitektur (Arsitektur Melayu dan Lingkungan) Vol.2 No.1 SBN 2503-3859
[6] Mudra, A. M. (2003). *Rumah Melayu*.
[7] Riau, Departemen Pendidikan dan Kebudayaan (1986) Arsitektur Tradisional Melayu Daerah Riau
[8] Samra,B. (2015) *Konsep Ruang Dalam Rumah Lama di Kawasan Senapelan Pekanbaru. Jurnal Arsitektur (Arsitektur Melayu dan Lingkungan)* Vol.2 No.1 SBN 2503-3859

RDBMS Applications as Online Based Data Archive: A Case of Harbour Medical Center in Pekanbaru

Bayu Febriadi[1], Ahmad Zamsuri[2]
Universitas Lancang Kuning, Pekanbaru, 28265, Indonesia
E-mail : bayufebriadi9@gmail.com, ahmadzamsuri@unilak.ac.id

Abstract : Kantor Kesehatan Pelabuhan Kelas II Pekanbaru is a government office that concerns about healthy, especially about environment health. There is a problem in case of saving electronic data, also in analyzing daily data both for internal and external data. The office has some computers and other tools that are useful in saving electronic data. In fact, the data are still saved in available cupboards and it is not efficient for an important data that is analyzed for more than one time. In other words, it is not good for a data is needed to be analyzed continuously. Rational Data Base Management System (RDBMS) application is an online based saving data and it uses System Development Life Cycle (SDLC) method. Hopefully, the application will be very useful for employees Kantor Kesehatan Pelabuhan Pekanbaru in managing their work.

Keywords: RDBMS, Data, Kantor Kesehatan Pelabuhan Kelas II Pekanbaru.

1. Introduction

Harbour Medical Center or *Kantor Kesehatan Pelabuhan* Pekanbaru is a government office that is concerning about environmental health. Nowadays, it manages registration of Umrah pilgrims and also notes pilgrims' vaccine records. All of data are written and they are saved in papers. As stated before, it is very harmful for daily routines in that office. This way of saving data is not efficient when it is just put on papers about pilgrims' registration. After registering, pilgrims will get vaccine and of course the previous data is needed again to do vaccine. If the data is still in written form, there will be problem in doing back up of the data. In fact, Harbour Medical Center Pekanbaru has computers and other supported tools in applying an online based application.

Web-based information system is also an efficient promotion tool and it is a source of information that is very useful for its users when they can find any important information needed in their life. Relational Data Base Management System (RDBMS) is a structured listing data and saved in a hard disk then its objective is to create accessible data quickly and easily (Kadir, 2008).

After doing preliminary observation at Harbour Medical Center Pekanbaru, the study found some problems to be solved, they are: (1) There is no electronic saving data system while computers and other supported tools are available, so daily data management is still saved in written form. It will be very not effective to analyze same data more than one time (continuously). (2) is no integrated system yet about saving archive Pekanbaru data that employees feel hard in working.

Then the writer line out the problem in carrying out this research as below:

1) How is applying online based application and data base as data archive media in helping analyze daily data at Harbour Medical Center?
2) How is implementing online based system at Harbour Medical Center Pekanbaru in order to be more efficient to serve the community?

2. Research Method

Research method explains some steps in solving problem to reach the goal, can be seen in figure 1 as follow:

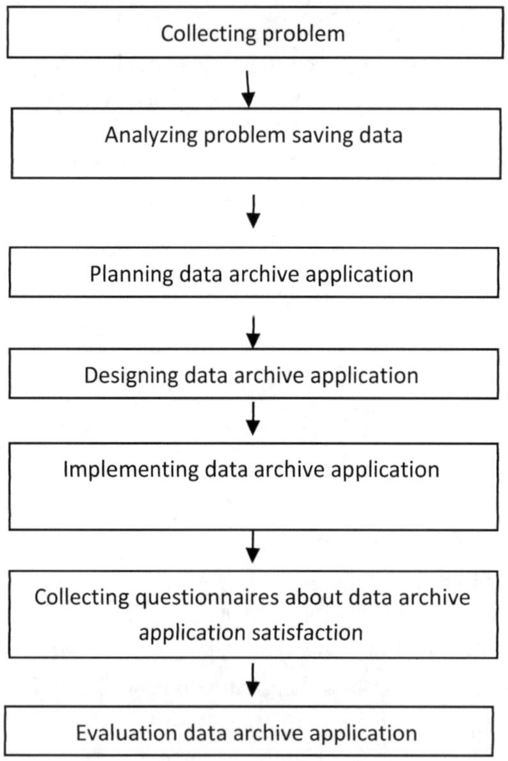

Figure 1. research method

Picture shows steps to be applied start from collecting and analyzing problem up to solving problem in archive data saving at Harbour Medical Center Pekanbaru.

3. Result and Discussion

Before starting, the writer had collected some questionnaires about data analysis by employees of Harbour Medical Center Pekanbaru and it was filled by international certificate vaccine (ICV) seekers, where the samples were taken one last week.

System planning steps used is unified modeling language (UML) method.

3.1 Usecase Diagram

Usecase diagram in planning the system Applications as Online Based Data Archive Media at Kantor Kesehatan Pelabuhan Pekanbaru can be seen in picture 1.2:

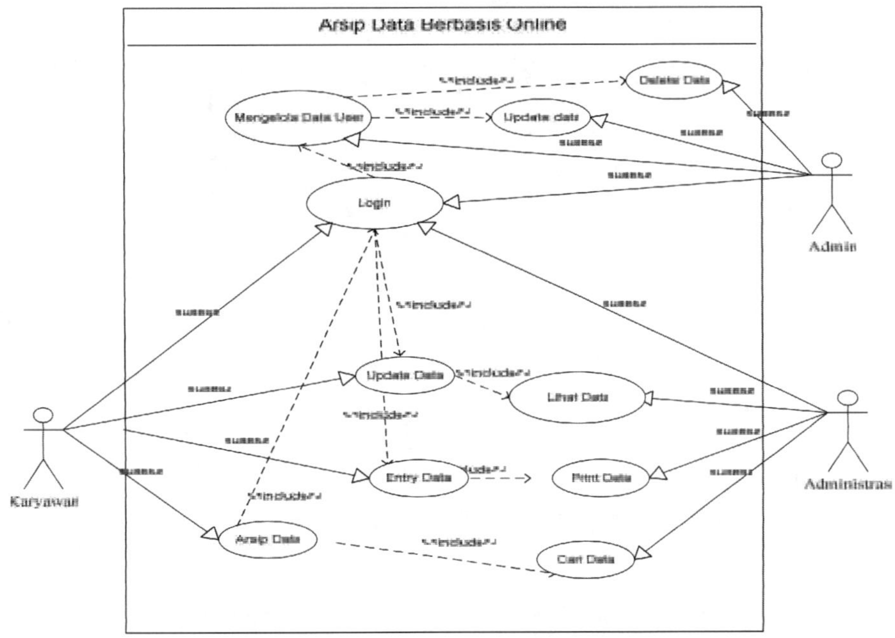

Figure 2. Use Case-Diagram

3.2 Activity Diagram

Activity diagram in planning the system can be seen in picture 1.3:

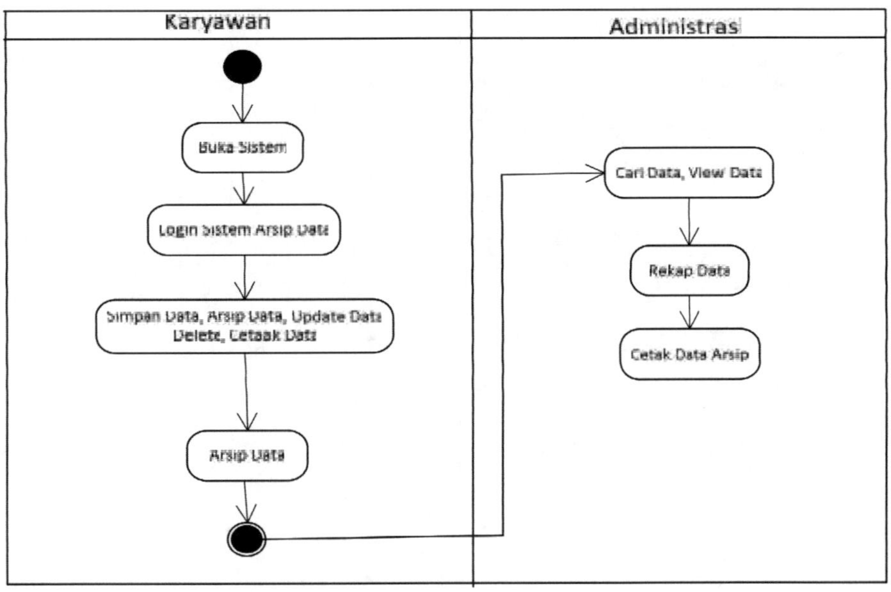

Figure 3. Activity Diagram

3.3 Class Diagram

Class diagram in planning the system can be seen in picture 1.4:

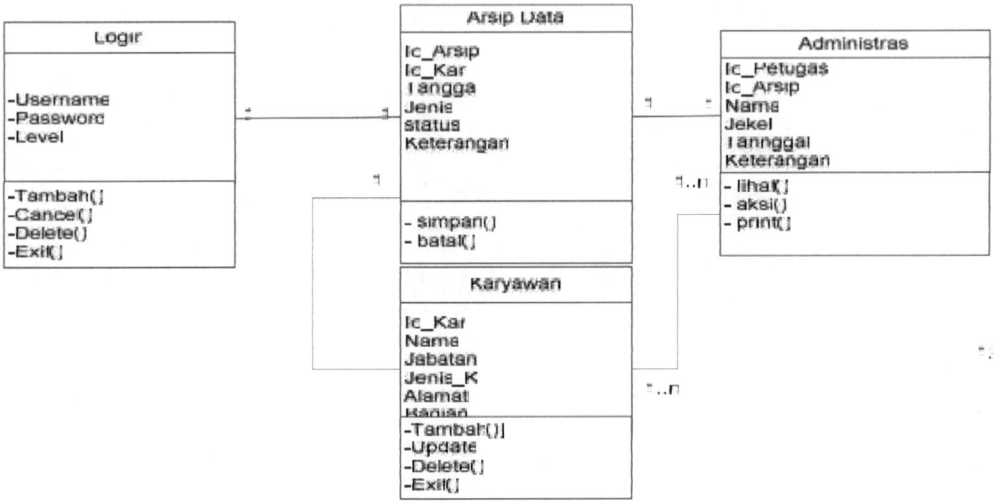

Figure4. Class diagram

4 Result

After implementing created application at Kantor kesehatan pelabuhan pekanbaru, then the writer give questionnaires about data archive application satisfaction to employees that use application.

Table 1

The questionnaire raw data

	1	2	3	4	5	6	7	8	9	10	Jlh
1	4	3	3	3	3	3	4	4	3	3	59
2	4	4	5	3	4	4	4	4	4	4	68
3	4	4	5	4	4	4	4	4	4	3	66
4	3	3	4	3	3	3	3	4	1	1	51
5	4	4	4	4	3	3	3	4	3	3	46
6	4	3	5	4	3	4	4	4	3	3	60
7	4	3	5	4	3	4	4	4	3	3	60
8	3	2	4	3	3	3	4	4	1	1	49
9	4	3	5	4	2	4	4	4	4	4	63
10	3	3	5	4	3	4	3	3	3	4	63
11	4	3	5	4	4	4	4	3	3	4	68
12	4	3	5	4	3	4	3	4	3	3	60
13	4	3	5	4	3	4	4	4	2	2	53
14	4	3	4	3	3	3	4	4	3	3	53
15	4	3	5	4	4	4	4	4	3	3	61
T	57	47	69	55	48	55	56	58	43	44	532
M	3.8	3.1	4.6	3.7	3.2	3.7	3.7	3.9	2.9	2.9	3.6

From the tabulation on table above, there is *mean* with score 3.6, it means that participants know and satisfied of the RDBMS application as electronic data analysis media. In compare to the preliminary observation, participants' knowledge about RDBMS is increase in using the application.

5 Conclusion

Result of evaluation the application at Kantor Kesehatan Pelabuhan pekanbaru concludes that:

1) Implementation of data archive application is appropriate to the objective of research to increase data analysis by applying RDBMS application in analyzing electronic data with score 3.6 point is bigger than before 2.1 point.
2) Planning by using Unified Modeling Language (UML) method is very helpful to employees in building data archive application.

6 Suggestions

Considering easiness of online based data archive application in analyzing electronic data, so writer gives some advices:

1) It is really need to give additional time to learn more about RDBMS in order to get better comprehension about the application.
2) Also, it is need to add more respondents in evaluating data archive application RDBMS to the system in order to evaluate the program well.

7. Reference

[1] Barthos, Basir. 2009 *Manajemen kersipan* Yogyakarta: Andi Offset

[2] Hanson. 2000. *Sejarah Pemograman berbasis Web.* Yogyakarta: Andi offset

[3] Hardjono. 2006. *Pengenalan Sistem Informasi Berbasis Web.* Jakarta: Erlangga

[4] Kadir, Abdul. 2005. *Pengenalan Sistem Informasi.* Jakarta, Yogyakarta: Andi

[5] Liang Gie.2000, *Administrasi perkantoran Modern.* Bandung: Informatika

[6] Nugroho, Andi.2004. Database Berbasis DBMS dalam Pengolahan Data Terintegrasi. Bandung: Informatika

[7] Oneto, Erima. 2008. Cascadading Management Style Sebagai media Program Berbasis Web. Bandung: Informatika

[8] Sutabri, Tata. 2005. *Sistem Informasi Managemen.* Jakarta: Andi Jogja

[9] Sutabri, Tata. 2005. *Aplikasi dasar pemograman PHP.* Yogyakarta: Andi

[10] The realities of using IT to transform the public sector. International Journal of Managin Service Quality, 13(6), 445–452.

Application of K-Mean Algorithm for Medicine Data Clustering in Puskesmas Rumbai

Taslim[1], Fajrizal[2], and Dafwen Toresa[3]
[1]E-mail: taslim.malano@gmail.com,
University of Lancang Kuning, Pekanbaru, Riau, Indonesia
[2]E-mail: Fajrizal@unilak.ac.id,
University of Lancang Kuning, Pekanbaru, Riau, Indonesia
[3]E-mail: dafwentoresa@gmail.com
University of Lancang Kuning, Pekanbaru, Riau, Indonesia

Abstract.
Through the government's health insurance program, efforts are made to ensure the health of the community through Puskesmas or community clinics. One of the most important components in health is the availability of medicines. The availability of medicines should be well managed to ensure that the medicines needed by the community are always available in sufficient quantities. Clustering on Data mining can be used to analyze the use of medicines during this time at a Puskesmas to be used as one of considerations for the Puskesmas to submit the demand of medicines in the period to come. The results of this study are expected to classify the level of medicines used in the pharmacy of Puskesmas in Rumbai Bukit Pekanbaru.

1. Introduction

Medicines is one of the important components in terms of good health to either prevent, reduce, eliminate or cure a disease or disease symptoms. That is why medicines need to be managed properly, effectively and efficiently. Planning on the needs of medicines is important to ensure the availability and distribution of medicines with the types and quantities sufficient so that medicines can be obtained quickly at the appropriate place and time at agencies related to health services, be it hospital, health centre, and so forth. The planning on the need of medicines will affect the procurement, distribution and usage of medicines in health care settings.Puskesmas Rumbai Bukit Pekanbaru is one of the public health service centres located in Pekanbaru which is government institution related to public health service in which one of its function isto provide medicine service for institution of health care settings in Pekanbaru. Effective and efficient analysis of the need of medicines is needed to ensure the availability of medicines in the pharmacy in Pekanbaru.

Clustering on the need of medicines is expected to be one of the considerations to ensure the availability of medicines in the health service of Pekanbaru. Data clustering is one method in data mining that can be used to get data mapping to classify into smaller groups based on the similarity of characteristics they have (Perim, Wandekokem, &Varejão, 2008). With these clustering results the distribution of medicines in health care agencies can be grouped according to the need based on the medicine distribution data in the previous year and it can be used as a reference for drug planning for the next year. It is hoped that the availability of medicines for the next year can be more secure and able to meet demand for medicines from health agencies.

One of the most well-known clustering methods among clustering algorithms is K-means (Patel & Mehta, 2011). The simplicity of this method makes the K-means algorithm applicable to various fields (K.Arai and A.R. Barakbah, 2007).

2. Phase stages in data mining

KDD is a nontrivial process of identifying the validity of data, potential, use, and ultimately yielding understandable data patterns. The stages in the stages of data mining are:

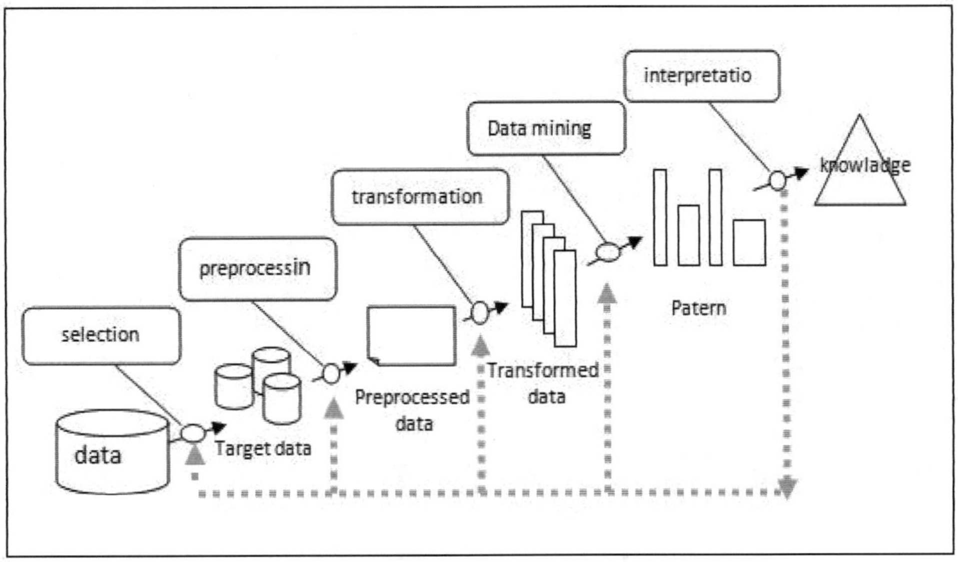

Figure 1: Stages of data mining

In this study, the data used are those taken from the Usage Report and Medicine Demand Sheet (LPLPO) Puskesmas Rumbai Bukit 2015. Then, the data are clustered to obtain patterns of medicine needs for the community of Rumbai in Pekanbaru.

2.1 Clustering

Clustering refers to grouping notes, observations, or cases into similar classes. A cluster is a collection of notes that resembles each other and differs from records in other clusters. Clustering is different from the classification which has no target variable for clustering. Instead, the algorithm clustering looks for the entire set of data segments into a relatively homogeneous subgroup or group, in which the similarity of records in the cluster is maximized, and the similarity of records out of this cluster is minimized.

Examples of grouping tasks in business and research include:

1) Determining product marketing targets
2) For accounting audit purposes, suspicious financial behavior is examined
3) It is used as a dimension reductional tool in which the data set has hundreds of Attributes

For grouping of gen expression, those in large number may exhibit the same behavior. In cluster analyzing, some factors need to note, for example such as determining

1) How to measure similarities
2) How to recode category variables
3) How to standardize or normalize numerical variable
4) How many clusters will be made
5) Interval-variable scale

Variable-scale interval is a continuous measurement which includes a linear scale. For instance the weight and height, latitude and longitude and weather temperature. The use of measurement units may affect cluster analysis [8]. For example, a changing measurement from meter to inch, or from kilogram to pound, may result in a very different cluster structure. For thatreason, it requires normalization of data, so that all data have the same weight. Steps for normalizing the data are:

- Calculating the mean value

$$s = \frac{1}{n}(|x - m| + |x - m| + \cdots . + |x - m|)$$

- Calculating the z-score

$$z = \frac{x - m}{s}$$

K-Means Algorithm
Clustering k -means algorithm [1] is a simple and effective algorithm for finding clusters in data with the following algorithms:

1) Determine the number of cluster
2) Determine the value of the location of the initial cluster.
3) Calculate the closest cluster center for each record
4) For each cluster k, calculate the centroid cluster and update Location of each cluster center
5) Repeat steps 3 through 5 until convergence or termination.

The k-means algorithm is known and widely used for the partitional method, which is to divide the set of data objects into a subset of non-overlapping clusters, so that each data object is exactly in one cluster

3. Results and Discussion
The source of data used in this study is data from LPLPO Puskesmas Rumbai Bukit 2014. The data used can be seen in table 1.

Table 1. Recapitulation of prescription data 2014

REKAPITULASI RESEP APOTEK

PUSKESMAS : RUMBAI BUKIT
BULAN : JANUARI
TAHUN : 2014

NO	NAMA OBAT	Satuan	Bulan											
			Jan	Feb	Maret	April	Mei	Juni	Juli	Agust	Sept	Okt	Nov	Des
1	Acyclovir krim	tube	4	5	4	21	23	15	19	19	24	24	91	75
2	Acyclovir 200 mg	tablet	20	110	20	95	110	132	23	0	0	0	0	0
3	Acyclovir 400 mg	tablet	10	0	0	488	0	0	210	100	200	195	220	115
4	Alat suntik 1 ml	set	0	0	0	0	0	0	0	0	100	0	0	0
5	Alat suntik 5 ml	set	0	0	0	0	0	0	0	0	0	0	100	0
6	Albendazol 400 mg tab	tablet	23	54	13	6	32	87	15	15	16	6	66	60
7	Alluporinol tab 100 mg	tablet	20	170	150	60	70	270	50	493	0	0	0	0
8	Ambroxol 30 mg	tablet	191	200	3028	225	127	72	120	20	149	340	1602	1035
9	Ambroxol sirup	botol	72	56	57	45	28	22	50	30	193	0	0	0
10	Aminofilin tab 200 mg	tablet	10	0	152	0	72	10	0	0	40	0	45	33

There are 133 types of medicines at pharmacies in Puskesmas of Rumbai Bukit. Some medicines and medicines spending each month can be seen in the following table.

Table 2. Monthly Recapitulation of Medicines

NO	Names of Medicines	units	Months											
			Jan	Feb	March	April	May	June	July	August	Sept	Oct	Nov	Dec
1	Acyclovir krim	tube	4	5	4	21	23	15	19	19	24	24	91	75
2	Acyclovir 200 mg	tablet	20	110	20	95	110	132	23	0	0	0	0	0
3	Acyclovir 400 mg	tablet	10	0	0	488	0	0	210	100	200	195	220	115
4	Alatsuntik 1 ml	set	0	0	0	0	0	0	0	0	100	0	0	0
5	Alatsuntik 5 ml	set	0	0	0	0	0	0	0	0	0	0	100	0
6	Albendazol 400 mg tab	tablet	23	54	13	6	32	87	15	15	16	6	66	60
7	Alluporinol tab 100 mg	tablet	20	170	150	60	70	270	50	493	0	0	0	0
8	Ambroxol 30 mg	tablet	191	200	3028	225	127	72	120	20	149	340	1602	1035
9	Ambroxolsirup	botol	72	56	57	45	28	22	50	30	193	0	0	0
10	Aminofilin tab 200 mg	tablet	10	0	152	0	72	10	0	0	40	0	45	33
11	Amoksisilin tab 250 mg	kapsul	1020	924	844	660	750	1456	1344	780	1142	1308	2002	1474
12	Amoksisilin cap 500 mg	kapsul	2245	2310	2370	1780	1503	1765	1795	1210	4340	2110	2754	2042
13	Amoksisilinsyrkering 125 mg	botol	110	110	77	83	96	166	162	91	145	161	193	150
14	Amlodipinbesilat 10 mg	tablet	150	130	180	550	755	200	20	40	371	90	400	110
15	Antasida tab doen	tablet	912	2147	1023	1090	1098	1145	990	268	529	693	2030	1724
16	Antasidadoensuspensi	botol	37	29	22	37	18	13	49	44	100	62	60	45
17	Anti bakteridoensalap	tube	14	21	20	15	10	19	24	7	12	17	35	25
......	....	....	....	....	....	....	....	....	....	....	....	....	....	....
....	....	....	....	....	....	....	....	....	....	....	....	....	....	....
133	Loperamid		94	122	128	20	0	0	0	0	0	0	0	0

From the above data (table 2), it leads to the process of data normalization so that the data has a smaller range of values that can accelerate the process of calculate of the normalization. This study uses z-score normalization. The next step is determine the number of clusters, that is 3 clusters in order to be able to determine whether a demand for the medicines are high", "high", or "low". For initial centroid values, it is randomly selected. The value of the centroid is taken from rows 50.60 and 70. In this step, updating of group cluster will be done. Next is clustering process d0 using Euclidean Distance, to get distance of matrix C1, C2 and C3. From clustering result after 33 times iteration, it is found out that data medicine is grouped into 3 groups, they are: very high, high, and low demands of medicines. The number of clustering results can be seen in the following table.

Table 3. Clustering 1

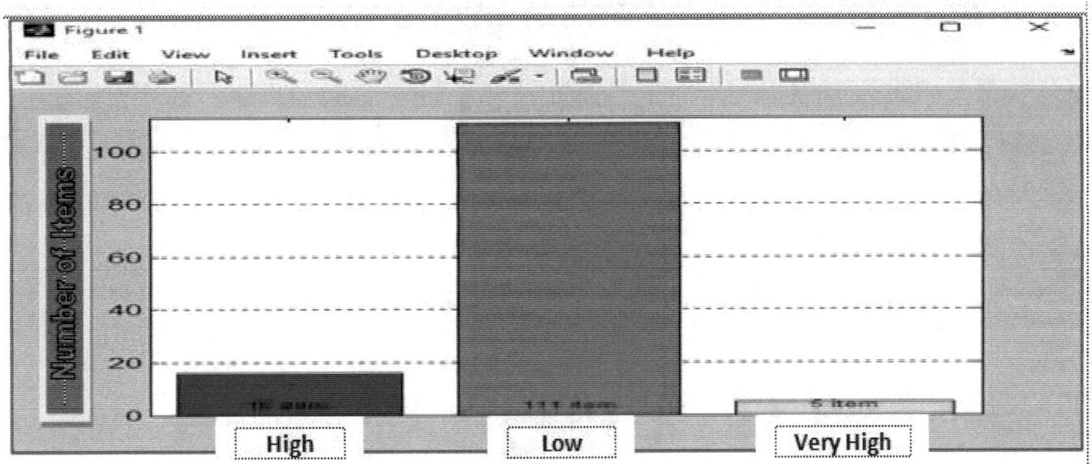

Some details of table on clustering medicine data can be seen in the following table.

Table 4. Clustering Results of medicine demand

1	Acyclovir krim	tube	4	5	4	21	23	15	19	19	24	24	91	75
2	Acyclovir 200 mg	tablet	20	110	20	95	110	132	23	0	0	0	0	0
3	Acyclovir 400 mg	tablet	10	0	0	488	0	0	210	100	200	195	220	115
4	Alatsuntik 1 ml	set	0	0	0	0	0	0	0	0	100	0	0	0
5	Alatsuntik 5 ml	set	0	0	0	0	0	0	0	0	0	0	100	0
6	Albendazol 400 mg tab	tablet	23	54	13	6	32	87	15	15	16	6	66	60
7	Alluporinol tab 100 mg	tablet	20	170	150	60	70	270	50	493	0	0	0	0
8	Ambroxol 30 mg	tablet	191	200	3028	225	127	72	120	20	149	340	1602	1035
9	Ambroxolsirup	botol	72	56	57	45	28	22	50	30	193	0	0	0
10	Aminofilin tab 200 mg	tablet	10	0	152	0	72	10	0	0	40	0	45	33
11	Amoksisilin tab 250 mg	kapsul	1020	924	844	660	750	1456	1344	780	1142	1308	2002	1474
12	Amoksisilin cap 500 mg	kapsul	2245	2310	2370	1780	1503	1765	1795	1210	4340	2110	2754	2042
13	Amoksisilinsyrkering 125 mg	botol	110	110	77	83	96	166	162	91	145	161	193	150
14	Amlodipinbesilat 10 mg	tablet	150	130	180	550	755	200	20	40	371	90	400	110
15	Antasida tab doen	tablet	912	2147	1023	1090	1098	1145	990	268	529	693	2030	1724
16	Antasidadoensuspensi	botol	37	29	22	37	18	13	49	44	100	62	60	45
17	Anti bakteridoensalap	tube	14	21	20	15	10	19	24	7	12	17	35	25
18	Anti fungidoensalep	pot	15	44	14	14	21	14	0	0	0	0	0	0
19	Anti hemoroidsuppkomb	supp	0	0	0	0	0	13	16	0	3	0	8	0
20	Asamaskorbat tab 50 mg	tablet	1419	3812	3783	3005	2784	3392	3890	2703	3955	3567	7090	6112
21	Asammefenamat tab 500 mg	kaplet	800	940	1040	1032	930	1000	570	420	820	715	1708	1510

4. Conclusion

From the results of data cluster of medicines, it can be concluded that medicines are grouped into low. The low demand of the medicines is caused by no demand at all for few months. For high demand, the average demand is 300 medicines. For the very high demand, the average demand is above 2000 medicines every month. This result of cluster analysis above still needs to be elaborated in order cluster data of medicines can be done more validly by setting the best centroid value.

5. References

[1] Arai. K and A.R. Barakbah "Hierarchical K-means : and algorithm for centroids initialization for K-means".(2007)

[2] Hans-Peter. K, Karsten, Peer.K, Alexey.P, Matthias.S, Arthur.Z, "Future trends in data mining" (2007)

[3] Ranjan, J, "Application of data mining techniques in pharmaceutical industry" (2007)

[4] Tipawan.S and Kulthida. T "Data mining and its application for knowledge management: a literature review from 2007 to 2012" (2012)

[5] Usama Fayyad, Gregory Piatetsky-Shapiro, and Padhraic Smyth "From Data Mining to Knowledge Discovery in Databases" (1996).

Effects of Urin Cow Dosage on Growth and Production of Sorgum Plant (Sorghum Bicolor L) on Peat Land

Sri Utami Lestari, Andi Andrian
Faculty of Agriculture, University of Lancang Kuning, Pekanbaru, Indonesia
Jl.Yos Sudarso Km.8 Rumbai Pekanbaru

Abstract
Sweet sorghum (*Sorghum bicolor* (L)), is a potential cultivated plant, especially in marginal and dry areas, sorghum has an important role as a source of carbohydrates, sorghum is expected as an alternative choice for peatland cultivation, with the use of peatlands is also expected Raising awareness of the environment by cultivating more environmentally friendly plants. The aim of this research is to know the influence and get the best dosage of cow urine on growth and production of Sorghum (Sorghum bicolor L) plant on peat soil. The experiment was conducted experimentally by using Completely Randomized Design (RAL), with one factor, namely: Cow urine administration, given in 5 treatments and 4 replications, resulting in 20 trials. Each experimental unit consists of 4 plants and 2 plants to be sampled. The factors studied were A0 = dose of cow urine 0 cc / l, A1 = dose of cow urine 25 cc / l, A2 = dose of cow urine 50 cc / l, A3 = dose of cow urine 75 cc / l, A4 = dose Cow urine 100 cc / l. Conclusion Giving of cow urine has significant effect on growth and production of sorghum plant which is seen on the parameters of plant height, leaf length, leaf width. While wet weight 100 seeds and dry weight of 100 seeds of sorghum plants have no significant effect. The best dose is given by A4 treatment with the best dose of 100 cc / l.

Keywords: sorghum, urine, peat

1. INTRODUCTION

1.1 Background

Sorghum is a plant that has great potential to develop, broad adaptability, drought-resistant, some varieties can diaksun, and all parts of the plant has economic value. Sorgum developed in Indonesia, generally the type of seed for human food. Not for animal feed, glue or for beverage industry.

Sweet sorghum (Sorghum bicolor (L)), is a potential cultivated plant, especially in marginal and dry areas, sorghum has an important role as a source of carbohydrates, sorghum is expected as an alternative choice for peatland cultivation, with the use of peatlands is also expected Raising awareness of the environment by cultivating more environmentally friendly plants.

The limitation of productive land causes agricultural extensification to lead to marginal lands. Peatland is one of the selected marginal land species, especially by large plantations, as it is relatively less populated and therefore the possibility of land use conflict is relatively small. Peat or peat soil or organic soil is a synonym that is the soil that has more than 50% organic material in the top 80 cm layer. The soil used for cultivation of sorghum is peat soil which is known to contain little nutrients and is acidic, for that Need improvement materials such as fertilizer.

Fertilizer is an organic material or an organic material derived from natural or man-made nutrients that plants need to grow. Fertilization is very effective on the soil less fertility, and to avoid competition of nutrient removal of plants with each other.
The use of inorganic fertilizers is quickly available to plants but if it continues to use it will damage the texture and structure of the soil, while the price is also expensive, whereas organic fertilizer requires decomposition process first but the price is not expensive and the amount is abundant. The specialty of organic fertilizers is to improve soil properties such as soil porosity, soil structure and water holding capacity, help neutralize soil pH, help neutralize toxins caused by heavy metals in the soil, one type of organic fertilizer is cow urine.
Cow urine has been allowed to waste as a waste is useful in agriculture as organic fertilizer, in addition to relatively cheap price, cow urine is also not difficult to find, and the process of making it very easy with the fermentation process. Based on these descriptions, the authors conducted a study entitled "The Effect of Cattle Urin Dose on Growth and Sorghum Plant Production (Sorghum bicolor L) On Peat Land".

1.2 Research Objectives
The aim of this research is to know the influence and get the best dosage of cow urine on growth and production of Sorghum (Sorghum bicolor L) plant on peat soil.

1.3 Hypothesis
Giving doses of cow urine had a significant effect on the growth and production of sorghum bicolor (Sorghum bicolor L) on peat soil.

2. MATERIALS AND METHODS

2.1. Materials and tools
The materials used are: sorghum seed varieties Numbu, lime dolomite, cow urine, NPK fertilizer (16: 16:16), Urea fertilizer, Decis 35 EC and polybag size 35 x 40.
he tools used are machete, hoe, bucket, meter, raffia strap, hammer, knife, jerry can, gembor, measuring cylinder, hansprayer, scales, camera and stationery.

2.2. Research methods
The experiment was conducted experimentally using Completely Randomized Design (RAL), with one factor, namely: Giving of cow urine, which will be sampled. It was given in 5 treatments and 4 replications, so it amounted to 20 experiments. Each experimental unit consists of 4 plants and 2 plants which the factors studied are:

A0 = dose of cow urine 0 cc / l
A1 = dose of cow urine 25 cc / l
A2 = dose of cow urine 50 cc / l
A3 = dose of cow urine 75 cc / l
A4 = dose of cow urine 100 cc / l
The complete random design mathematical model (RAL) used is: (Surtinah 2013).
$Yij = \mu + \tau j + \varepsilon ij$
Information :
Yij: The observation result on the therapeutic treatment and the k-j repeat
M: Middle value (Avg.
Tj: Effect of i th treatment
Eij: Effect of errors (residuals) due to the i th treatment and the j th repetition
Fingerprint data shows F. Count> F. The table is followed by DMRT test (Duncan's Multiple Range Test) at 5% level.

2.3. Implementation of Research

1) Land preparation
The land where the research was first cleaned from weeds and the remains of existing plants. Land area required for planting area of 9 x 7 m. Then done plot making with size 100 x 100 cm as much as 20 plot, with distance between polybag 50 x 50 cm.
2) Media Preparation
After the plot is completed, the ground is filled into polybags with a size of 35 x 40 polybag, weight of soil media 7 kg per polybag. The soil used is peat soil, on planting medium done liming using dolomite fertilizer. The dolomite dose used is 0.14 g / polybag
3) Labeling
Labeling is prepared and installed on each plot according to treatment. Installation of labels is done to facilitate the treatment in the implementation of research ..
4) Planting
Planting done a week after the labeling, making a planting hole with a depth of ± 3cm, then the seeds of sorghum are inserted as much as 3 grains per hole.
5) Provision of NPK 16: 16: 16 and Urea fertilizer
NPK Mutiara 16: 16: 16 NPK fertilizer was given 3.6 g and Urea was given 2.4 g of planting, given twice that first fertilization was given simultaneously with planting time, then second fertilization was given at age one month after planting. The dose of NPK Pearls 16:16:16 once administration is 1.8 g and urea with a dose of 1.2 g. Fertilization is done by means in the tugal as far as 7 cm from the plant
6) Treatment
Cow urine used is fermented cow urine for 30 days, cow urine treatment is given in accordance with the level of treatment. Treatment of cow urine begins at seedlings 2 weeks after planting and given 2-week intervals. Treatment was stopped 2 weeks before harvest.
7) Maintenance
Maintenance conducted in this research are:

- Watering done in the morning and afternoon by using gembor, the amount of water given in accordance with the conditions in the field with the same volume of each polybag and not done watering if it rains.
- Weeding done if there are weeds that grow around the plant or plot in a manual way by using the hand pull weeds that grow.
- Pest and disease control is done by spraying Decis 35 EC with concentration 2 cc / liter of water, spraying done a week after planting, with interval 2 weeks once using hand sprayer, spraying stopped a week before harvest.
- Thinning is done at the age of the plant 2 weeks after planting, treated plants are healthy plants to avoid pests.

8) Harvest
Harvesting is done at the age of 96 days, after the seeds are formed by looking at the visual characteristics such as the leaves are yellow and dried, the seeds are pithy and hard and the maximum flour content.
9) Observation
All observation parameters were performed at the end of the study, taking 2 plot samples. The parameters observed are as follows:
    - Plant Height (cm)
    The measurement of plant height starts from the base of the stem to the highest leaf tip perpendicular to the stem.
    - Leaf Length (cm)
    The measured leaves are the longest leaves and have opened perfectly, measurements made from the base of the leaf to the tip of the leaf.

- Leaf Width (cm)
  The measured leaf width is the widest part of the leaf starting from the right leaf edge of the left leaf is perpendicular to the mother's leaf bone.
- Wet Weight 100 Seed (g)
  Observation of 100 seed weight was done after harvesting with counted 100 seeds and then weighed with analytical scales.
- Dry Weight 100 Seed (g)
  Observation of weight of 100 seeds was done after all seeds were dried under sunlight until constant weight for 3 days and then counted 100 seeds and weighed with analytical scales.

3. RESULTS AND DISCUSSION

3.1. Results

1. Plant height (cm)

The result of variance analysis on plant height showed that cow urine administration had significant effect on the height of sorghum plant, the data are presented in appendix 5. The result of DMRT further test at 5% level on plant height is presented in Table 1.

Table 1. Average Height of Sorghum Plant Due to the Urine Cattle Urine Giving

Treatment	Plant Height (cm)
A_0	234,00 a
A_1	237,87 ab
A_2	240,37 ab
A_3	253,62 bc
A_4	259,37 c

Figures followed by the same small letters are not significant based on the DMRT advanced test at the 5% level.

Table 1, indicating that the higher the dose given the result is increased. Treatment of A4 with dose of 100 cc / l gave the best result at the height of sorghum plant that is 259,37 cm, although different is not real with treatment of A3. While the lowest plant height is A0 is 234 cm.

2. Leaf Length (cm)

The result of analysis of variance to leaf length showed that cow urine administration had significant effect on the length of leaves of sorghum plant, the data are presented in appendix 6. The result of further DMRT test at 5% level against leaf length is presented in Table 2.

Table 2. Average Length of Sorghum Plant Leaves Resulting from Urine Cattle Urine.

Treatment	Leaf Length (cm)
A_0	87,50 a
A_1	89,62 ab
A_2	91,50 abc
A_3	92,12 bc
A_4	95,75 c

Figures followed by the same small letters are not significant based on the DMRT advanced test at the 5% level.

Table 2, shows that the higher the doses given the results are increasing. Treatment of A4 doses of 100 cc / l gives the best results on the length of leaves of sorghum plants that is 95.75 cm. Although not significantly different in treatment A2 and A3, while the lowest leaf length is A0 that is 87.50 cm.

3. Leaf Width (cm)
Result of analysis of variance to width of leaf showing that giving of cow urine have significant effect to sorghum leaf width, data presented in appendix 7. DMRT further test result at 5% level to leaf width is presented in Table 3.

Table 3. Average Leaves of Sorghum Plant Leaves Resulting from Urine Cattle Urine.

Treatment	Leaf Width (cm)
A_0	9,73 a
A_1	10,75 ab
A_2	11,43 abc
A_3	11,81 bc
A_4	12,12 c

Figures followed by the same small letters are not significant based on the DMRT advanced test at the 5% level.

Table 3, indicating that the higher the dose given the result is increased. Treatment of A4 with dose of 100 cc / l gave the best results on the width of the sorghum leaf of 12.12 cm, although not significantly different from the treatment of A2 and A3. While the lowest leaf width is A0 that is 9.73 cm.
4. Wet weight 100 seeds (g)
The result of analysis of variance to wet weight of 100 seeds showed that cow urine administration had no significant effect on wet weight of 100 seeds of sorghum, the data are presented in appendix 8.

Table 4. Average Weight of Wet 100 Seeds of Sorghum Plant Resulting from Urine Cattle Urine.

Treatment	Wet Weight (g)
A_0	4,55
A_1	4,69
A_2	4,81
A_3	4,97
A_4	5,34

No further tests were performed

Table 4, shows that the average wet weight of 100 seeds is not significant. But from the data obtained, the higher the dosage of urine of cow to sorghum crop yields increased yield.

5. Dry weight 100 seeds (g)
The result of analysis of variance to dry weight of 100 seeds showed that cow urine administration had no significant effect on dry weight of 100 seeds of sorghum plant, the data are presented in appendix 9.

Table 5. Average Dry Weight of 100 Seeds of Sorghum Plant Resulting from Urine Cattle Urine

Treatment	Dry Weight (g)
A_0	4,11
A_1	4,11
A_2	4,18
A_3	4,47
A_4	4,50

No further tests were performed

IOP Conf. Series: Earth and Environmental Science **97** (2017) 012052 doi:10.1088/1755-1315/97/1/012052

Table 5, shows that the average dry weight of 100 unrefined seeds is not real. But from the data obtained, the higher the dosage of urine of cow to sorghum crop yields increased yield.

4. Discussion

The result of variance analysis showed that cow urine giving significant effect on the growth and production of sorghum plants seen in observation parameters are plant height, leaf length and leaf width, while wet weight 100 grain and 100 seed dry weight have no significant effect. Annexes 5,6,7,8, and 9.

In this study showed that in treatment A0 showed the lowest result on every parameter observed. The low rate of growth and production of sorghum in treatment A0 is thought to be due to the fact that the plants do not obtain nutrient suplay either from the fertilizer given as the treatment or from the growing medium. As we know the growing media used in this study is peat soil, where peat soil is poor soil nutrients. Thus, less nutrients can disrupt the growth so that the resulting production is low. This is in accordance with the opinion of Sutejo (2002) which states that incomplete macro and micro nutrients can lead to obstacles to the growth and development of plants. Lakitan (1993) also states that if the availability of essential nutrients is less than the required amount, then the plant will be disturbed metabolism because the plants have a positive correlation with the availability of nutrients so that in the cultivation of nutrient availability plants is a very decisive factor.

The best treatment in the study that is found in treatment A4. The increase in the treatment is thought to be due to the nutrients needed for sorghum plants available for growth and development. As we know cow urine has many advantages, including as a source of nutrients for plants that are easily absorbed by plants can also help water absorption is also one of the potential organic fertilizer as a source of nutrients for plants such as N, P and K. From the aspect of haranya, Cow urine fluid has a higher nutrient content than the solid dung. In addition to the nutrient content it has, in cow urine there is also Indole Asetat Asid (IAA) of 704.26 mg L-1 (Sutari, 2010).

Roshana and Seswita (2007) reported that auxin is very influential in root formation by increasing the number and length of roots. So that the maximum plant in the absorption of nutrients / plant nutrients. The results of the study also obtained a corresponding result that increased urine use can increase the components of plant growth such as improving the function of plants and the number of fruit weight. With the absorption of nutrients and nutrients are expected to maximize the maximum growth of vegetative plants sorghum.

The high growth of sorghum plant during vegetative period caused by liquid manure (cow urine) can accelerate and contain the hormone that can stimulate the growth of plants, and in liquid manure the content of N and K elements is quite big (Aisyah, 2011).

So with the composition of nutrients contained suspected to be able to stimulate the growth of sorghum plants as a whole. Syarief (1986) states that the Nitrogen element has an influence in the growth of plant height and gives green color to the leaves, enlarges the size of the fruit, the formation of proteins and other compounds. If the available nitrogen elements are balanced with other elements, more proteins can be produced and leaves can grow wider due to photosynthesis going well.

Further Nyakpa et al (1988) states that the optimal availability of N, P, and K nutrients for plants can increase the amount of chlorophyll, an increase in the amount of chlorophyll will increase the activity of photosynthesis which results in more assimilate. The results of photosynthesis are also used in the respiration process which will produce energy used for the formation of carbohydrates, proteins and fats in the formation of new cells.

As for the production parameters have no significant effect is suspected to be caused that the availability of nutrients on planting medium is not sufficient for the needs of crop production. Nutrient elements are a very important factor in the process of plant growth and development. According to Purnamayani et al. (2004), generally peatland is deficient in K. Furthermore, Coorley and Gray (1976) stated that potassium in plants acts as an activator of enzymes due to the presence of K + ions needed in enzyme activity. In addition Potassium also plays a role in assisting the transportation of assimilates from photosynthesis and contributes to the formation of carbohydrates and proteins and increases resistance to disease and fruit quality.

5. CONCLUSIONS AND SUGGESTIONS

5.1. Conclusion

Giving cow urine has a significant effect on the growth and production of sorghum plant which is seen on the parameters of plant height, leaf length, leaf width. While wet weight 100 seeds and dry weight of 100 seeds of sorghum plant have no significant effect. The best dose is given by A4 treatment with the best dose of 100 cc / l.

5.2. Suggestion

Based on the research that has been done, the researcher suggested that for further research by giving the addition of inorganic fertilizer higher than the dose given in this research (Urea: 2,4 g / plant, NPK: 3,6 g / plant) Better sorghum production.

6. Reference

[1] Aisha, S. N. Sunarlim and B. solfan. 2011. *Effect of fermented beef Urine with different dosage and interval of giving to plant growth of sawi* (Brassica juncea L.) *Journal of Agrotechnology*, State Islamic University of Sultan Syrif Kasim Riau.

[2] Coorley, RHV and BS Gray, 1976. *Growth and Morphology*, P.12-14. In RHV Corley, JJ Hardon and BJ Wood (eds) *Development in Crop Science* (1) Oil Palm Research. Elfisher Scientific Publishing Company. Amsterdam

[3] Assembled. B. 1993. *Plant Physiology Basics*. Grafindo. Jakarta.

[4] Purnamayani, Sabiham Hospital, Sudarsono, LK Darusman, 2004. *Zero Point Value (MTN) and Its Relation to Potassium Eruption of Jambi and Central Kalimantan Peatlands*. J. Environment Environment 6. 75-82.

[5] Surtinah, 2013. *Analysis of Agricultural Cultural Research Data*. Unilak Press. Pekanbaru.

[6] Sutari, W.S., 2010. *Test the quality of bio-urine fermented with microbes derived from plant material on growth and yield of green mustard plants* (Brassica juncea L.) Thesis Udayana University, Denpasar. Bali.

[7] Sutejo, and Kartasapoetra. 2001. *Fertilization and Fertilization Method.* Rineka Cipta Jakarta .

Learning Strategy in Class Management: A Reflection from Manado Case

Ismail Suardi Wekke[1], Alexander Yandra[2] and Budianto Hamuddin[3]
[1]Sekolah Tinggi Agama Islam Negeri (STAIN) Sorong, Indonesia
Email: iswekke@gmail.com
[2]Department of Public Administration, Universitas Lancang Kuning, (Unilak) Pekanbaru. Indonesia
Email: alexsy@unilak.ac.id
[3]Faculty of Education and Teacher Training, Universitas Lancang Kuning (Unilak), Pekanbaru. Indonesia
Email: budihamuddin@unilak.ac.id

Abstract: This article is a research conducted with qualitative approach with various case studies underlining a strategy that becomes the basis for classroom management. The article discusses and links to the learning revolution that becomes today's demands, including a discussion that analyses the condition of learners. The article based its data preliminary study conducted in Manado in the province of North Sulawesi in Indonesia. This region has its own characteristics with the encounter of Muslims and the Protestant community for century. Due to its uniqueness 3 Moslem schools and 3 Protestant schools in Manado were selected to study. Data collection was conducted for a year, from May 2016 to April 2017. The study employ four stages research steps: identification, data collection, data validity checking, and directed discussion. The stages include observation and in-depth interviews and conducting focus group discussions. Two important thought about the essence of learning strategy and learning revolution in class were shared briefly within this article.

Keyword: strategy, management, classroom, Manado, multi-religions.

1. Introduction

An ideal educational setting a class is not only seen as a group of people gaining lesson from a physical figure only, but they also possess psychic. In that part then the teacher mentors the existing psychic in a process called education. There are different desires and objectives to achieve. So they are not only gathering and lining up. A class needs to be managed well. The presence of teachers becomes the key of better classroom management so that teachers need to convey material with appropriate methods. Before arriving at the part of method, the strategy applied should also be considered in accordance with the method to be selected. The precision between the two will be an intermediary so that the delivered teaching materials will be understood by the students.

Further, learning failure becomes an educational disaster therefore this condition should be avoided to prevent the occurrence of humanitarian catastrophe. It is when the learning material that is run in the classroom is completely incomprehensible. This can be started from a teacher who does not see the context. So the students cannot accompany the program designed by the teacher. As if the real world with each classroom stands alone.

The main task of teacher is to connect the class with the reality that takes place in the reality of the world. If what the teacher says in the classroom stands alone, it certainly will not be useful for the learners. They actually live in a wider world in the real world; not only for limited classrooms, but the school environment is also limited. A report card or diploma is not the main goal. Certificates are merely educational administrative, not more than that. The phase which is needed to think about is how learners will live after school time.

IOP Conf. Series: Earth and Environmental Science **97** (2017) 012053 doi:10.1088/1755-1315/97/1/012053

Reflecting to the ideas above, this study tries to link with the reality where nowadays, the students who are sitting in school are part of generation that has a technological leap. They were born in a state of technological entirely different from the conditions of the previous years. The technology is growing day by day. Generations with information communication technology even every second receive information that overwhelms the old web. Not all information has been verified so that it is always open the possibility to be wrong. The world is in their hands. Nevertheless, when they are back at home, from the beginning, they have also been conditioned with the characteristics of communication technology media. Influence from TV, gadget or other electronic devices were something that need to put in mind regarding to educational settings. Yet, this conditions need to be considered when designing strategy. It is feared that the learning strategy will be useless because of incompatibility with the condition of the students with technological leap.

Another reality that needs to be stressed also the students who spend time not with their parents or without parents. Here, teachers need to present the activities that designed to nurture them in learning. Teachers need to clarify, then direct to the mastery of the ideal material. Where it then shows a valid and legit sources including the skills to obtain valid information. The strategy should lead them to learn independently and to select the right information in the real world.

To do so, this article describes how a strategy becomes the basis for classroom management in global then paralleled with Manado case. The discussion based its data from multicultural and dynamic ethnic where religions walk hand in hand and linked to the learning revolution that becomes today's demands, including a discussion that analyses the condition of learners to achieve the stated learning objectives.

2. Methodology

This article is a research conducted with qualitative approach with various case studies as mentioned earlier paralleled with one multicultural city in the region in North Sulawesi. This study selected North Sulawesi region due that this region has its own characteristics with the encounter of Muslims and the Protestant community. Thus, madrasah (school) is no longer simply teaching Arabic as part of the curriculum but also strengthening religious identity. This variable becomes a separate part in the preparation of this research design. The research was conducted in Manado City as a region that met the established research criteria.

To meet the ethical criteria of research, since the beginning the researchers have conveyed the identity of researchers, as well as declared the intention and purpose of research to madrasah managers (school manager). Research proposals are submitted to madrasah leaders and teachers to be an early consideration in providing answers to the posed research questions. The research proposal was also presented on two different occasions, the first was in a seminar on research proposals of Sekolah Tinggi Agama Islam Negeri (STAIN) Sorong, April 2016, and an educational seminar held by Institut Agama Islam Negeri (IAIN) Manado in May 2016. Both were to test the feasibility of the research proposal.

Data collection was conducted for a year, from May 2016 to April 2017. The research was conducted in four stages: identification, data collection, data validity checking, and directed discussion. For data collection, observation and in-depth interviews were conducted with the research subjects in accordance with the criteria of the research problem. Interviews were conducted on three groups: teachers, pupils, and madrasah management. All three were interviewed separately using the same question guides, while the observations were conducted six times with the same observation guidelines.

Furthermore, two-month extension of research was conducted in May and June 2017 to check again on two matters, they are, the condition of madrasah in Manado City area that has the same criteria, and checks on the same research subjects in the same madrasah. So it can be ascertained that the obtained data were valid data. The last part of the research carried out triangulation to check the validity of data through presentation. There were three activities to present papers to get input from peers and expert opinion, among others through ELITE conference at Universitas Islam Negeri Syarif Hidayatullah, Jakarta;

Once again, to ensure the data validity, it was carried out by focus group discussions through peer meetings in the same research conducted at Institut Agama Islam Negeri (IAIN) Manado in June 2017. All of them were conducted in order to ensure that research has explained the phenomenon of the research problem.

3. Findings: Preliminary

After a year of collecting data this study sees the learning strategy in many classes from the selected schools somehow becomes the basis for classroom management both in Moslem and Protestant schools in Manado. This study underlines some specific findings that need to be share due its importance in a multi ethnics or religions classroom management.

3.1 The Essence of Learning Strategy

Each strategy has its own uniqueness as well as strengthens and weaknesses. Each has its own advantages over the other strategies. So one strategy cannot be applied in all conditions or every classroom situation. At any given moment, there is a suitable strategy. However, on a specific situation, the strategy is completely unusable. There are different conditions from time to time, so that it is important to note, thus, an opportunity for the teacher to identify how the strategy can be effectively used in the face of a situation or similarity in the encountered conditions. By not generalizing any existing conditions, it is necessary to recognize which strategies are appropriate to the needs and conditions that occur.

Students get a chance in experiencing the process of learning new things. Experience after experience will emphasize the student learning cycle that will lead to a complete understanding of a phenomenon. For that, teachers have the task to manage the learning experience so that students will get the meaning of learning. From the learning process that is then replicated in real life to deal with the conditions encountered. Solving problems encountered, when the teacher was not on their side at the time. At that time, they know how to manage the real situation independently.

Teachers as facilitators need to look at the characteristics of learners. Every pupil have different tendencies and characteristics, therefore, it cannot be expected how the whole class members can have the same treatment or condition. The same condition is only in the part of desire to learn or knowing new things. This idea can be used to develop a learning program that can be used for the whole. In addition, teachers need to have a record of how learners learned before. As a class "leader" the teacher will direct the learning process toward the desired goal.

3.2 Learning Revolution in Class

The current state of the learning media can be managed with the help of the media. The availability of media is very abundant. Teacher's task is to choose the right media and use the existing media to be used together. Each media has different characteristics. Thus, it should be recognized how a media can help teachers to deliver material to learners. Nevertheless, the media do not instantly become a substitute for teachers but only as an additional tool to help the learning process.

Student interaction with existing technology does not necessarily eliminate the teacher's role. Indeed, the technology is only a tool to support. As a learning medium to help activity that cannot be done by a teacher alone, technology is not as a pillar of education. The teachers are part of education so that it cannot be eliminated with any technological progress. So, teachers familiar with the use of technology ~~so~~ to enrich learning experiences and class activities.

In the activity, learners need to see, hear, and experience. These three conditions should united in a learning material. For sixty minutes of study, it can be divided into three phases, so each has a portion although they are not exactly the same. Each activity will have an impact on learners' understanding. Each phase is integrated with another phase. Lectures can no longer be used as a reliable method, because students' attention cannot concentrate on listening activity only. The students cannot sit and remain silent, while the active ones are their teachers.

Similarly, the cultural background inhabited by each learner is different. With these different cultures, they will need to be incorporated into cultures that can cross each group to be part of the learning process. They need to be united in team unity so as to bridge the learning process (Evans & Vergnaud 1998). Cultural differences when they are managed productively will be an element of learning so that students can traverse themselves. They will then understand how others act and do not mind the differences.

Classes become the best area in providing leap for students. They will relate the learning experience to the real world. Easily, the learning process can be done if you stand on the existing reality. Students will go through the process of knowing, understanding, and analysing. The process can be done easily with the teacher's presence as a facilitator. Teachers are not the only source of information. Students easily get information from diverse sources. However, the teacher still has the function to verify and become a facilitator in sorting out existing information. Teachers are still required forever; the existence of a teacher cannot be replaced by any technological advancement.

Teacher has a human element that technology does not have. So technology cannot completely have the function of interaction and response mastered by the teacher. One of the teachers' duties is to develop a culture of literacy among students. As they become accustomed to a culture of literacy, it is easier to overcome all learning challenges. Even students can help themselves to get their own learning experience to use independently. The condition of the 21st century is the information literacy (Loo & Chung 2006).

Some institutions introduce natural schools. Classes are conducted unrestricted by a school wall that keeps the distance between students and the actual situation. This is simply to be an alternative where during this time, school even further away from nature. So the class needs to be presented in the real world. Students' best practice needs to be in natural conditions. Because they will eventually be in a world filled with problems and challenges, not just a matter and reading material, or just with the buttons only. So the class needs to be connected to the natural conditions to deliver to the understanding of learning.

Students should be guided to learn happily. Learning activities should not restrain the students. We witnessed the annual ritual, how the students rejoiced when they graduated. The simple question is "are they feeling sad when they are in the school?" So that when they graduate must be celebrated in such a way as to excite the happiness that exists after so long confinement. Joy with teachers should be presented in the classroom. They should be guided to love the class, so that the time with their class and school breakup is sad. They miss those moments back to be with the best teachers who have helped them learn

4. Conclusion

Exploring learning strategy in class management from around the globe then paralleled with Manado Case reflecting many perspective and ideas to strengthening the old though in educational settings. This study sees the implementation of the strategies in some schools in Manado and tries to underline some thinking that the learning strategies is something exists in reality and realised that when the learning experience is obtained, then that is the peak of the success in learning process. The students will put it in memory if it is done in a structured repetition. At that time, the educational function to maximize the potential of learners in facing the real situation has been achieved. The true education is to provide experience for real life. So when in school, students will be guided to face obstacles that will be faced at a time. School bench is not a fantasy that students will receive lessons that are not real, absurd, and useless at all. The real certificate is obtained from the society and the space of the vast devotion is in the community as well. So it is not enough to achieve a piece of paper and then stored neatly in the folder. After that, the students did not do anything. This is not what is expected, but the expectations of all parties are the best graduates, with the best skills, and the best abilities to apply in the real world.

Strategy is not a real thing that can be seen as it is. Learning materials mediate to see how the strategy works, strategies that will deliver the material to be used as teaching materials. In fact everyone has a different strategy though in the same material to be taught. The task of the teacher is to complete the strategy on a single subject so that students can arrive at the best understanding

5. References

[1] Banks, J. A., Cookson, P., Gay, G., Hawley, W. D., Irvine, J. J., Nieto, S., ... & Stephan, W. G. (2001). Diversity within unity: Essential principles for teaching and learning in a multicultural society. *Phi Delta Kappan*, *83*(3), 196-203

[2] Bondy, E., Ross, D. D., Gallingane, C., & Hambacher, E. (2007). Creating environments of success and resilience: Culturally responsive classroom management and more. *Urban Education*, 42(4), 326-348.

[3] Bos, C. S., & Vaughn, S. (2002). *Strategies for teaching students with learning and behavior problems*. Allyn & Bacon, A Pearson Education Company, 75 Arlington Street, Boston, MA 02116.

[4] Evans, J. St. B. T., Newstead, S. E. & Byrne, R. M. J. (1993).*Human reasoning: The psychology of deduction*. UK: Lawrence Erlbaum Associates Ltd.

[5] Garcia, T., & Pintrich, P. R. (1994). Regulating motivation and cognition in the classroom: The role of self-schemas and self-regulatory strategies. *Self-regulation of learning and performance: Issues and educational applications*, *127153*.Morrison, K. A., Robbins, H. H., & Rose, D. G. (2008). Operationalizing culturally relevant pedagogy: A synthesis of classroom-based research. *Equity & Excellence in Education*, 41(4), 433-452.

[6] Loo, A. and Chung, C. W. 2006. A model for information literacy course development: A liberal arts university perspective. *Library Review*, 55(4): 249–58.

[7] Sanders, J. A., & Wiseman, R. L. (1990). The effects of verbal and nonverbal teacher immediacy on perceived cognitive, affective, and behavioral learning in the multicultural classroom. *Communication Education*, *39*(4), 341-353.

[8] Susilo, C. (2010). Transforming the Public life in Manado: The Kampongs and Large Modern Commercial Project.

[9] Wekke, Ismail Suardi & Asmaun Sahlan. (2014). "Strategy in Creating School Enviroment: Lessons from High School Indonesia" in PROCEDIA: Social and Behavioral Sciences, Vol.143 [August], pp.112-116.

[10] Wekke, Ismail Suardi & Maimun Aqsha Lubis. (2008). "A Multicultural Approach in Arabic Language Teaching: Creating Equality at Indonesian Pesantren Classroom Life" in SOSIOHUMANIKA: Jurnal Pendidikan Sains Sosial dan Kemanusiaan, Vol.1, No.2, pp.295-310.

[11] Wekke, Ismail Suardi & Maimun Aqsha Lubis. (2011). "Educational Technology on Teaching and Learning of Integrated Islamic Education in Brunei Darussalam" in Ulumuna, Vol.XV, Iss.1 [June], pp.185-204.